The Chemistry *of* Coordination Complexes *and* Transition Metals

The Chemistry *of* Coordination Complexes *and* Transition Metals

Dr. P.L. Soni

Formerly Associate Professor in Chemistry,
ARSD College,
University of Delhi

Dr. Vandna Soni

Assistant Professor in Chemistry,
Maharaja Agrasen College,
University of Delhi

CRC Press
Taylor & Francis Group
Boca Raton London New York

CRC Press is an imprint of the
Taylor & Francis Group, an **informa** business

Manakin
PRESS

First published 2021
by CRC Press
2 Park Square, Milton Park, Abingdon, Oxon, OX14 4RN

and by CRC Press
6000 Broken Sound Parkway NW, Suite 300, Boca Raton, FL 33487-2742

CRC Press is an imprint of Informa UK Limited

British Library Cataloguing-in-Publication Data
A catalogue record for this book is available from the British Library

Library of Congress Cataloging-in-Publication Data
A catalog record has been requested

ISBN: 978-1-032-02461-5 (hbk)
ISBN: 978-1-003-18342-6 (ebk)

Manakin
PRESS

Brief Contents

Part–I: Coordination Chemistry

1. History of Complexes 3
2. Introduction 7
3. Correct Empirical Formula of Complexes 23
4. IUPAC System of Nomenclature 33
5. Isomerism and Stereochemistry 43
6. Bonding in Complexes 61
7. Preparation and Isolation of Complexes 91
8. Stability of Complex Ion 97
9. Kinetics or Inert and Labile Complexes 103
10. Substitution Reactions in Complexes 111
11. Coordination Complexes and Biological Systems 121
12. Uses and Applications of Coordination Complexes 143
13. Metal Carbonyls and Nitosyls 173
14. Organometallics 201

Part–II: Transition Metal Chemistry

1. General Characteristics of d-block Elements 229
2. Titanium Group 255

3. Vanadium Group 261

4. Chromium Group 269

5. The Cobalt Group 305

6. The Zinc Group 331

7. Inner Transition or f-block Elements: The Lanthanides 345

8. The Actinides **359**

Bibliophile **371**

Index **375**

Detailed Contents

Preface *xix*

Part–I: Coordination Chemistry

1. History of Complexes **3**
1.1 Complexes 3
1.1.1 History 3
1.2 Difference Between Double Salts and Complex Compounds 4
1.2.1 Addition Compounds 5

2. Introduction **7**
Coordination or Complex Compounds 7
2.1 Coordination Sphere 8
Ligands 8
2.2 Formation of a Complex 8
Coordination Number (C.N.) 11
Low Coordination Numbers 11
Four Coordination Complexes 12
Five Coordinate Geometry 12
Six Coordinate Complexes 13
Distortion 13
High Coordination Complexes 14
Eight Coordinate Complexes 14
2.3 Bridging Ligands 15
2.3.1 Non-Chelates 19
2.3.2 Chelates 19

3. Correct Empirical Formula of Complexes **23**
3.1 Precipitation Method 23
3.2 Molar Conductivity Method 24
3.3 Effective Atomic Number (EAN Rule) 25
3.4 Blomstrand-Jorgensen Chain Theory 26
3.5 Werner's Coordination Theory 26

4. IUPAC System of Nomenclature **33**
4.1 Nomenclature of Coordination Compounds 33
4.2 General Rules 33

4.3 Naming the Metal Ion 33
4.4 Naming the Ligand 34
Configuration Symbols 35
Naming the Inner Complexes 39

5. Isomerism and Stereochemistry 43
5.1 Isomerism 43
5.2 Structural Isomerism 43
Coordination Position Isomerism 44
Ionisation Isomerism 44
Compound 45
Hydrate Isomerism 45
Linkage Isomerism 45
Ligand Isomerism 46
Polymerisation Isomerism 46
Other Uncommon Isomerism 46
Spin Isomerism 47
Summation Isomerism 47
Valence Isomerism 47
Distortional Isomerism 47
5.3 Stereoisomerism 48

6. Bonding in Complexes 61
6.1 Nature of Metal-Ligand Bond 61
6.2 Lewis Acid-Base Theory 61
6.3 Valence Bond Theory 62
6-Coordinate Octahedral Complexes 62
Outer Orbital Complexes 63
Inner Orbital Complexes 63
Limitations of VBT (Valency Bond Theory) 69
6.4 Crystal Field Theory (CFT) 69
Magnitude of Δ_0 72
Effects of Crystal Field Splitting 74
Tetragonal Distortion or Jahn-Teller Distortion 75
Square Planar 78
Tetrahedral Complexes 79
6.5 Ligand Field Theory (LFT) 81
Complex Stability 82
6.6 Molecular Orbital Theory (MOT) 84
Bonding in $[Fe(CN)_6]^{4-}$ ion 87

7. Preparation and Isolation of Complexes **91**
7.1 Preparation 91
Trans Effect 91
Preparation of Metal Carbonyls 92
Preparation of Organometallic (Sandwich Compounds) 93
Preparation of Transition Metal σ Bonded Complexes 93
7.2 Isolation 94

8. Stability of Complex Ion **97**
8.1 Introduction 97
8.2 Factors Influencing Stability 98

9. Kinetics or Inert and Labile Complexes **103**
9.1 Introduction 103
Effective Collision 104
9.2 Inert and Labile Complexes 105
9.2.1 Classification of Inert and Labile Complexes on the Basis of Electrons in d Orbitals 106
9.2.2 Inert and Labile Complexes on the Basis of CFSE 106
9.2.3 Roie of a Cataiysi (Kinetics) 107

10. Substitution Reactions in Complexes **111**
10.1 Introduction 111
10.2 Mechanism of Substitution 112
10.3 Substitution in Non-Aqueous Solvents 113
10.4 Substitution When No Solvent Is Used 114
10.5 Mechanisms of Substitution 114

Select Topics

11. Coordination Complexes and Biological Systems **121**
11.1 Introduction 121
11.1.1 Essential Metals 121
11.1.2 Cyanide and CO Poisoning 123
11.2 Hemoglobin the Oxygen Carrier 124
11.2.1 Transport of Oxygen 124
11.2.2 Oxygen Carriers 125
11.2.3 Metal Ions Strong and Transportation in the Human System 125
11.3 Role of Some Metals in the Human Body 126
11.3.1 Selenium 126
11.3.2 Boron 127

11.3.3 Molybdenum 127
11.4 Sodium Potassium Equilibrium 128
11.4.1 Selectivity 129
11.5 Behavior of Calcium And Magnesium 129
11.5.1 Other Metalloenzymes 130
11.5.2 Vanadium in Biological Systems 130
11.5.3 Metallo enzymes as acid catalysts 131
11.6 Phosphate Storage and Muscles 131
11.7 Oxygen Binding and Oxygen Enzymes 132
11.7.1 Vitamis B_{12} or Cyanocobalamine 132
11.7.2 Food and Trace Elements 135
11.8 Trace Elements Absorption 135
11.8.1 Inhibitors in Absorption 135
11.8.2 Enhancing the Effect of Absorption of Metals (even trace elements) 137
11.8.3 Toxic Metal Ion 137
11.8.4 Enzymes Inhibition 139
11.8.5 Detoxification by Metal Ion and Chelation 139
11.9 Metal Complexes in the Cure of Cancer 142

12. Uses and Applications of Coordination Complexes 143
12.1 Introduction 143
12.2 Applications of Metal Complexes 143
12.2.1 Dye Industry 143
12.2.2 Biological Enzymes 144
12.3 Polymer-Metal Complexes as Catalysts 145
12.3.1 Catalytic Activities of Polymers-Metal Complex 146
12.4 Application of Polymer-Metal Complexes as Catalysts 147
12.4.1 Oxidation 147
12.4.2 Hydrogenation 147
12.4.3 Hydrolysis 148
12.4.4 Hydroformylation of Olefins 149
12.5 Initiation of Radical Polymerisation 149
12.6 Asymmetric Synthetic and Optical Resolution 150
12.7 The decomposition of H_2O_2 150
12.8 Polymer-metal complexes as Heterogeneous Biologically Active Catalysts 152
12.8.1 Porosity and Surface Area 152
12.8.2 Extent of Cross-Linking 153

12.8.3 Spacer Groups 153
12.8.4 Coordination Complex in Analytical Chemistry 153
12.8.5 Detecting Cations 154
12.8.6 Group I Cations 154
12.8.7 Confirmatory Test for Pb^{2+} Ions 155
12.8.8 Confirmatory Test for Ag^{+} Ions 155
12.8.9 Confirmatory Test for Mercury (1) ion 155
12.8.10 Group II Cations 155
12.8.11 Group reagent 156
12.9 Role of HCl in Group II Analysis 157
12.9.1 Group III Cations 157
12.9.2 Confirmatory Test for Fe^{3+}, Al^{3+}, Cr^{3+} 157
12.9.3 Group IV cations 158
12.9.4 Group Reagent 158
12.9.5 Confirmatory Tests 158
12.9.6 Group V cations 159
12.9.7 Group VI Cations 159
12.9.8 Zero Group Cations 159
12.9.9 Confirmatory Test for NH_4^+ ion 160
12.10 Detecting Anions 160
12.10.1 Group I Anions 160
12.10.2 Group of II Anions 160
12.11 Confirmatory Test for Cl^- Ion 161
12.11.1 Chromyl Chloride Test 161
12.11.2 Br^- and I^- 161
12.11.3 Test for NO^{3-} Ion 161
12.11.4 Test for $C_2O_4^{2-}$ Ion 161
12.11.5 Group III Anions 161
12.11.6 SO_4^{2-} Ion Test 161
12.11.7 PO_4^{3-} Ion Test 162
12.11.8 BO_3^{3-} Ion Test 162
12.12 Modern Techniques 162
12.12.1 Quantitative Analysis 162
12.12.2 Quantitative Vs. Qualitative 163
12.13 Extraction of Metal from Complex Ores 163
12.14 High Temperature Chemical Reduction 164
12.14.1 Photography 165
12.14.2 Artificial Silk 165

12.15 Metal Complexes in Medicine 165
12.15.1 Platinum Anticancer Agents 166
12.15.2 Hydrolysis 167
12.15.3 Attack on DNA 167
12.15.4 Monofunctional Adducts 167
12.15.5 DNA Base Recognition by Pt Complexes 168
12.15.6 Binding to Meta Thiamine 169
12.15.7 Metallo Thionein and Glutathione 169
12.15.8 Targeting Via Transferring Receptor: Ruthenium and Titanium 169
12.16 Gold Antipathetic Drugs 170
12.17 Bismuth Anticancer Drugs 171
12.18 Conclusion 172
12.18.1 Metal Complexes as Drugs and Chemotherapeutic Agents 172

13. Metal Carbonyls and Nitosyls **173**
13.1 Introduction 173
13.2 Preparation of Metal Carbonyls 174
13.3 Structure and Bonding in Carbonyls 177
Coordinate Mononuclear Metal Carbonyls 177
Applying EAN Rule to $[Ni(CO)_4]$ Complex 178
13.4 Iron Pentacarbonyl $[Fe(CO)_5]$ 179
Physical Characteristics 179
Structure and Bonding in $[M(CO)_5]$ 181
Physical Characteristics 183
13.5 Polynuclear Metal Carbonyls 184
Dimanganese decacarbonyl $[Mn_2(CO)_{10}]$ 184
Diironnonacarbonyl $[Fe_2(CO)_9]$ 185
13.6 Dicobalt Octacarbonyl $[Co_2(CO)_8]$ 186
Tetranuclear Metal Carbonyls $[M_4(CO)_{12}]$ 190

14. Organometallics **201**
14.1 Introduction 201
14.2 Definition 201
14.3 Classification 202
14.4 Nomenclature 202
14.5 Synthesis Of Organometallic Compounds 203
Synthesis of Sandwich Compounds 205
Synthesis of Alkene Complexes 205
Preparation of Zeise's Salt 206

Synthesis of Alkyne Complexes 206
Allyl Complexes 207
Metal-Benzenoid Complexes 207
Carbenes and Alkylidene Complexes 208
Carbene and Alkalidyne Complexes 208
Carbene Compounds 208
Physical Characteristics 209
Chemical Characteristics 209
Important Organometallic Compounds 210
physical Characteristics 210
Chemical Characteristics 210
Structure 211
Organometallic Compounds of Beryllium 211
Properties 211
Organometalliic Compounds of Mg 211
Group 3 or Group 13 Elements 213
14.6 Metal-Alkene Complexes of π-Bonded Organometallics 214
Preparation 214
Physical Characteristics 215
Bonding and Structure 215
Bonding in $[PtCl_3(C_2H_4)]^-$ 215
14.7 Sandwich Organometallics or Metallocenes 216
Preparation of Metallocenes 216
Physical Characteristics 217
Hydrogenation 217
14.8 Molecular Orbital Theory (MOT) 222

Part–II: Transition Metal Chemistry

1. General Characteristics of d-block Elements 229
1.1 Introduction 229
1.2 Placement in the Periodic Table 229
1.2.1 Electronic Configuration 232
1.2.2 Second Transition Series (4d-series) 233
1.2.3 Third Transition Series or 5d-series 233
1.2.4 Fourth Transition Series or 6d-series 234
1.2.5 Electronic Configuration of Transition Metal Ions 234
1.2.6 Atomic and Ionic Radii 235
1.2.7 Metallic Behaviour 236
1.2.8 Atomic Volume and Density 236

1.2.9 Melting and Boiling Points 237
1.2.10 Variable Oxidation States 238
1.2.11 Ionisation Energies 240
1.2.12 Reducing Action 240
1.2.13 Colour of Transition Metal Ions 242
1.2.14 Magnetic Properties 244
1.2.15 Diamagnetism 246
1.2.16 Ferromagnetism 246
1.2.17 Antiferromagnetism 246
1.2.18 Complex Formation Tendency 247
1.2.19 Stability of Complex Compounds 248
1.2.20 Catalytic Property 248
1.2.21 Interstitial Compound 249
1.2.22 Alloy Formation 250
1.2.23 Reactivity 250
1.3 Group Study of Physical and Chemical Characteristics of Transition Elements 250
1.3.1 Scandium Group 250
1.4 Physical Characteristics 251
1.4.1 Occurrence 251
1.4.2 Atomic Size 251
1.5 Chemical Characteristics 252
1.5.1 Electronic Configuration 252
1.5.2 Oxidation States 252
1.5.3 Halides 253
1.5.4 Complexes 253

2. Titanium Group 255
2.1 Introduction 255
2.2 Physical Characteristics 255
2.2.1 Electronic Configuration 255
2.2.2 Oxidation States 256
2.2.3 Extraction of Metals 256
2.2.4 Physical Constants 256
2.2.5 Atomic Size 257
2.3 Chemical Characteristics 257
2.3.1 Reactivity and Passivity 257
2.3.2 Behaviour of Ti Group Compounds in Different Oxidation States 258

2.3.3 Halides 259
2.3.4 Organometallic Compounds 260

3. Vanadium Group 261
3.1 Introduction 261
3.2 Physical Characteristics 261
3.2.1 Occurrence and Extraction 261
3.2.2 Electronic Configuration and Oxidation States 262
3.2.3 Atomic Size 262
3.2.4 Graded General Properties 262
3.3 Compounds of V, Nb and Ta 263
3.3.1 Halides 264
3.3.2 Oxides 265
3.3.3 Vanadates 266
3.3.4 Organometallic Compounds 267

4. Chromium Group 269
4.1 Introduction 269
4.2 Physical Characteristics 269
4.2.1 Abundance, Occurrence and Extraction 269
4.2.2 Electronic Configuration and Oxidation States 270
4.2.3 Oxidation States 270
4.3 General Chracteristics 271
4.3.1 Compounds in (+6) State 273
4.3.2 Chromium Trioxide or Chromic Anhydride 274
4.3.3 Halides 275
4.3.4 Zero, (–I) and (–II) State Compounds 277
4.3.5 Chromates, Molybdates and Tungstates 278
4.3.6 Tungsten Bronzes 279
4.4 The Manganese Group 279
4.4.1 Electronic Configuration and Oxidation States 279
4.4.2 Occurrence and Uses 280
4.4.3 Oxidation States 280
4.5 General Characteristics 281
4.5.1 Mn(III) Oxidation State Compounds 284
4.5.2 Mn(lll) Complexes 284
4.5.3 +IV Oxidation State Compounds 284
4.5.4 +V State Compounds 286
4.5.5 +Vl State Compounds 286
4.5.6 +VII State Compounds 287

4.5.7 Biological Use 288
4.6 The Iron Group 288
4.6.1 Electronic Structures and Oxidation States 288
4.6.2 Occurrence and Extraction 288
4.6.3 Steel Manufacture 290
4.6.4 Bessemer Process 290
4.6.5 Common Oxidation States 291
4.7 General Characteristics 292
4.7.1 Low Oxidation State Compounds 292
4.7.2 [$Fe(CO)_5$] 294
4.7.3 +11 Oxidation State Compounds 294
4.7.4 Oxides 294
4.7.5 Halides 295
4.7.6 Complexes 295
4.7.7 Test for NO^{3-} or NO^{2-} 296
4.7.8 Compounds with Oxidation State (III) 296
4.7.9 Oxides 297
4.7.10 Halides 297
4.7.11 Complexes 298
4.7.12 Cyclopentadienyl Compounds 301
4.7.13 Bioinorganic Chemistry of Iron 302

5. The Cobalt Group 305
5.1 Introduction 305
5.2 The Nickel Group 315
5.3 General Properties 316
5.4 The Copper Group of Coinage Metals 321
5.5 General Properties 323
5.5.1 Compounds in (+l) Oxidation State 324
5.5.2 Use of Ag Compounds in Photography 326
5.5.3 Compounds with (+II) Oxidation State 327
5.5.4 Compounds in (+III) Oxidation States 329
5.5.5 Compounds in (+V) State 329
5.5.6 Biological Use of Copper 330

6. The Zinc Group 331
6.1 Introduction 331
6.1.1 Occurrence 332
6.1.2 Extraction 332
6.1.3 Uses of Zinc, Cadmium and Mercury 333

6.1.4 Important and Common Oxidation States 333
6.1.5 Ionic Size and Some Physical Properties 333
6.2 General Characteristics 334
6.2.1 Oxides 335
6.2.2 Complexes 336
6.2.3 Mercury (+1) Compounds 337
6.2.4 Organometallic Compounds 338
6.2.5 Biological Role of Zn 338
6.2.6 Toxic Nature of Cd and Hg 339

7. Inner Transition or f-block Elements: The Lanthanides 345
7.1 Introduction 345
7.1.1 The Electronic Configuration 345
7.1.2 Abundance and Isotopes 346
7.1.3 Extraction 347
7.1.4 Lanthanide Contraction 350
7.1.5 Properties of (+III) Compounds 351
7.1.6 Compounds with (+IV) Oxidation State 352
7.1.7 Compounds of (+II) Oxidation State 353
7.1.8 Solubility 353
7.1.9 Colour and Spectra 354
7.1.10 Magnetic Moment 355
7.1.11 Coordination Complexes 357

8. The Actinides 359
8.1 Introduction 359
8.1.1 Electronic Configuration 360
8.1.2 Important Oxidation States 361
8.1.3 Occurrence 361
8.1.4 Preparation of Transuranium Elements 362
8.2 General Characteristics 363
8.2.1 Thorium, Protactinium and Uranuim 364
8.2.2 Chemical Characteristics 366
8.2.3 Neptunium, Plutonuim and Americium 367
8.3 Completing and Extending the Periodic Table and Its Series 369

Bibliophile 371

Index 375

Preface

Eversince, I joined the teaching profession way back in 1966 in SD College (Lahore) Ambala Cantt., I was obsessed with desire to write the book on the subject of chemistry. I joined ARSD college university of Delhi in 1968 to teach chemistry to B.Sc. (Gen.) course. In the year 1971 B. Sc. (Hons.) Chemistry got transferred to the colleges of Delhi University and some challenge, depth and seriousness flowed in the teaching of chemistry.

In the last 25 years in ARSD College I had the privilege to teach coordination chemistry to B. Sc. (Gen.) and B.Sc. (Hons.) classes, this gave me an opportunity to the have close contact with the subject and try to understand its intricacies and nuances. I feel certain amount of maturity and experience is a must before one ventures to enter the arena of book writing.

The first edition of the book was published in the year 2013, I felt satisfied and elated when all the teachers and students concerned gave positive and appreciating comments about the contents and the quality of the book. Here, I may add that the book was a runaway success in the international market specially Europe and the United States.

From the year 2016 I was getting a series of suggestions and requests from the readers to revise the book and keep it abreast of the recent development in the field of coordination complexes. Keeping this in view, the authors have not only updated the contents of the existing chapters but also added two new chapters to broaden the canvas of knowledge and make the book more comprehensive.

Recently the complexes have made inroads into many interdisciplinary avenues. Complexes have found awesome uses and applications in the biological system and incredible use in the industry. Keeping this in mind we have incorporated two new chapters namely (*i*) Complexes in the biological system.

(*ii*) Uses and application of complexes in many industries.

Who could believe that complexes will one day find use in Dyes and mordants, artificial silk and Zeigler Natta catalysts in the manufacture of polyethylene. These days the functioning of body chemistry, Haemoglobin, Vitamin B (cyanocobalamine) is all an interplay of complexes. The metals in complexes have provided a surface for the biological reaction. In the modern times the role of metals in recognized in all sorts of industrial productions and biological systems. We feel that the addition of these two chapters will equip all concerned for the better vision of the extent of the uses and applications of complexes in the various fields. More and more new fields will come under the purview of complexes with the passage of time.

Besides, writing the facts and theories in a systematic and lucid style it is imperative that author must put something of his own in the book. Only then life can be infused into the book, otherwise the book would be like any other book in the market. The author has tried to put his knowledge, understanding and experience into the book. I am sure readers will experience a new approach to the understanding of coordination chemistry in which all the interesting aspects of the metal complexes have been probed and highlighted.

Knowledge of complexes was given by Wisner in the year 1913. This book dwells extensively on all the aspects of metal complexes in a simple and lucid style. Metal complexes find applications in the field of Medicine, Pharmacy, Polymer Chemistry and Agriculture etc. The very existence of life in all possible forms is a gift of coordination chemistry. Polymers are considered the index symbol of progress of any country, coordination chemistry plays a key role in polymers. Blood that flows in the veins of animals and humans, chlorophyll in plants is a charisma of coordination chemists. Last of all, I feel the readers will enjoy, appreciate and draw benefit from my vast experience in the subject, the sequence of topics, theories covered, unfolding their intricacies systematically. Valuable suggestions are invited from the readers so as to enrich the contents of the book from time to time. My grateful thanks to my wife Kamini Soni,

Principal Navyug School, Sarojini Nagar, New Delhi who had been an inspiring invisible force, quietly encouraging me day to day in my endeavour to put my ideas and views before the vast community of readers.

My co-author Dr. Vandna Soni, Assistant Professor in Chemistry, Maharaja Agarsen College, Delhi University has given a useful contribution in systematising the facts and theories in the book in the present form.

The book has been published in two parts. Part one deals with nomenclature, stereo chemistry and the different theories regarding the bonding in complexes.

It includes metal carbonyls and nitrosyls, a select topic in complexes. The π-bonding carbonyls and nitrosyls show two way bonding (back bonding). Another select topic of organometallics also finds a place in this volume.

Part two deals with transition metal chemistry. A group-wise study of elements in 3*d*, 4*d* and 5*d* transition series has been discussed with emphasis on complexes and organometallics. Besides this inner tranisition elements chemistry has been taken up at length (Lanthanides and Actinides).

Lastly, I whole heartedly thank and praise the contribution of my Publisher Sh. Beenu Bhalla, Manakin Press for time to time suggestion and help in bringing this book in the present form.

Dr. P.L. Soni

PART–I

COORDINATION CHEMISTRY

Metal Complexes

Select Topics

Organometallics

1

History of Complexes

1.1 COMPLEXES

Coordination Complexes have been known from very early times, *e.g.*, Prussian blue (artist's blue ink), invisible ink, chlorophyll, haemoglobin etc. Earlier the formulation of these compounds was considered very difficult or complicated, therefore, the name complexes or metal complexes was coined. They are called metal complexes as metal atom or ion is one of their essential constituents.

For example:

(*a*) Prussian blue was formulated as: $KCN.Fe(CN)_2$, $Fe(CN)_3$

(*b*) Invisible ink as: $CoCl_2.6H_2O$

1.1.1 History

Complexes are formed as a result of Lewis acid-base concept in which metal atoms or ion acts as a Lewis acid and the ligand as a Lewis base *i..e.* it is simply a Lewis acid-base combination.

In the present context, they are no longer complex as every detail about them is well-known but the original name continues. These compounds contain coordinate bonds between their constituents hence they are known as coordination compounds. If anything is written on a piece of paper with $[Co(H_2O)_6]Cl_2$ having light pink colour it appears colourless. On slight warming of the paper it appears blue due to the reaction:

$$\underset{\text{Light pink}}{2[Co(H_2O)_6]Cl_2} \longrightarrow \underset{\text{Blue}}{Co[CoCl_4] + 12H_2O}$$

$$A^{\bullet} + B^{\bullet} \longrightarrow A\text{—}B \qquad \text{Covalent bond}$$

A: + B	$\longrightarrow$	$A \rightarrow B$	Coordinate bond
$AgCl + 2NH_3$	$\longrightarrow$	$[H_3N: \rightarrow Ag \leftarrow :NH_3]Cl$	

Other examples of the complexes that were known from the very early times are $CoCl_3.6NH_3$, $Cu(NH_3)_4SO_4$ and $K_2Pt\,Cl_6$.

These compounds find wide application in qualitative and quantitative chemical analysis.

1.2 DIFFERENCE BETWEEN DOUBLE SALTS AND COMPLEX COMPOUNDS

The complexes are quite different in properties compared to double salts. This is illustrated in Table 1.1.

Table 1.1

Double Salts	Complex Compounds
Double salts are molecular addition compounds which have a common anion but two different cations. They are formed by mixing two saturated solutions of salts in appropriate amounts which crystallise out on evaporation. *e.g.,* **Mohr's Salt:** $FeSO_4.(NH_4)_2SO_4.6H_2O$ **Potash Alum:** $K_2SO_4.Al_2(SO_4)_3.24H_2O$ On dissolution in water they split into constituent ions as hydration energy is greater than lattice energy *e.g.,* Mohr's salt solution will give test for Fe^{2+}, NH_4^+ and SO_4^{2-} ions.	These compounds are those molecular addition compounds in which there is partial splitting of ions on dissolution in water. *eg.,* $[Cu(NH_3)_4]SO_4 \longrightarrow [Cu(NH_3)_4]^{2+} + SO_4^{2-}$ This suggests two types of bonds: (*i*) Bond between Cu^{2+} and NH_3 remains intact even in solution form because in Cu^{2+} and NH_3 bond lattice energy is greater than hydration energy. (*ii*) $[Cu(NH_3)_4]^{2+}$ and SO_4^{2-} bond is distant and weak and hence SO_4^{2-} bond is set free in water. Here hydration energy is greater. In a complex compound there is a complex ion, $[Cu(NH_3)]^{2+}$. The Cu^{2+} ion cannot be tested from the complex ion but the presence of SO_4^{2-} ion can be tested by the $BaCl_2$ solution. This is because the charge is not on Cu ion but on the whole entity $[Cu(NH_3)_4]^{2+}$.

Besides the difference between double salts and complex compounds there is another category of similar looking compounds called **Alums**. Alums and double salts have very subtle difference.

Both alums and double salts are made of two salts showing in the form of crystals.

Double Salts	Alums
1. Mixing of two salt solutions to form the crystal representing both.	Two salt solutions forming a common crystal.
2. Both the salts have a common anion.	Two salts with a common anion.
3. The cations representing the two salts must be univalent and bivalent.	The cations of the two salts must be univalent and trivalent.
4. They contain six molecules of water of crystallisation	They contain twenty four molecules of water of crystalisation.
5. Crystalline state contains fine crystals.	Crystale are quite large.

6. Mohr's salt $FeSO_4.(NH_4)2SO_4.6H_2O$ $KCl.MgCl_2.6H_2O$	Potash alum $K_2SO_4.Al_2(SO_4)_2.24H_2O$ $(NH)_2SO_4Fe_2(SO_4)_3.24H_2O$
7. Tests for all the constituents ions can be performed in the solution.	$K_2SO_4.Cr_2(SO_4)_3.24H_2O$ Tests for the ions present can be performed in solution.

1.2.1 Addition Compounds

These are yet another category of compounds that are formed by the addition of metallic fluorides and hydrogen fluoride such as KF, HF, KF, 2HF, KF, 3HF etc. Here the metallic fluorides is to HF are in the ratio of 1:1, 1:2 and 1:3. However, ammonium fluoride and HF are in the ratio of 1:1.

Similar adduct is formed by

$$2FeCl_3 + POCl_3 \rightleftharpoons FeCl_4^- + [Cl_2Fe \leftarrow POCl_3]^+$$

$$\rightleftharpoons [FeCl_4]^- + [POCl_3]^+ + FeCl_3$$

or

$$FeCl_3 + POCl_3 \rightleftharpoons \underset{\text{Adduct}}{[FeCl_4]^- [POCl_3]^+}$$

Alkali metal chlorides and organic bases *e.g.* C_5H_5N form bases with $SeOCl_2$ while $SnCl_4$, $TiCl_4$, $AsCl_3$, $FeCl_3$, SO_2 etc form acids.

The adduct $[SnCl_6]^{2-} \cdot 2SeOCl^+$ behaves as an acid.

SUMMARY

Complexes have a history. Prussian blue and invisible ink were known from very early dates but their formulation was erratic. Double salts and alums have a marked distinction from complexes. Double salts or alums split into all the constituent ions on dissolution in a solvent (H_2O) but a complex would give only limited ions *e.g.,* $(Cu(NH_3)_4] SO_4$ will form $(Cu(NH_3)_4]^{2+}$ and SO_4^{2-} ions. There is no charge on Cu, therefore Cu^{2+} cannot be tested in $[Cu(NH_3)_4]^{2+}$.

■ ■ ■

2

Introduction

COORDINATION OR COMPLEX COMPOUNDS

When a central metal atom or ion is surrounded by a cluster of ligands a complex compound or an ion is formed. The charge on the complex ion is equal to the sum of the charges on the metal ion and the ligand. If the ligand happens to be a molecule or compound its charge is taken as zero. The formation of the complex is indicated by the square bracket '[ ]'. A solution of ferrous cyanide $Fe(CN)_2$ when mixed with a solution of potassium cyanide and evaporated crystals of $Fe(CN)_2 \cdot 4KCN$ separate out. Aqueous solution of these crystals do not give test for Fe^{2+} and CN^- ions but K^+ and $[Fe(CN)_6]^{4-}$ do give positive tests. It suggests ionisation is taking place in the following manner:

$$Fe(CN)_2 + 4KCN \longrightarrow Fe(CN)_2 \cdot 4KCN \rightleftharpoons 4K^+ + [Fe(CN)_6]^{4-}$$

Here, the molecule $Fe(CN)_2 \cdot 4KCN$ loses its identity yielding way to the formation of new compound known as coordination or complex compound.

The complex compound may contain a:

(*i*) Simple cation and a complex anion

(*ii*) Complex cation and a simple anion

(*iii*) Complex cation and a complex anion

e.g., $K_2[PtCl_6] \longrightarrow 2K^+ + [PtCl_6]^{4-}$

$Cu(NH_3)_4]SO_4 \longrightarrow [Cu(NH_3)_4]^{2+} + SO_4^{2-}$

$[Co(NH_3)_6]\,[Cr(C_2O_4)_3]$

The square bracket represent the complex ion [ ] according to Werner.

2.1 COORDINATION SPHERE

The central metal ion and the ligands which are closely attached to it inside the square bracket are called by Werner the first coordination sphere. The anions outside the square bracket form the second sphere of attraction.

The area in which complexation is taking place between the metal ion and the ligand is called the 'coordination sphere'. The complex may be a cation, anion or neutral, *e.g.*

Cation Complex $\quad \overset{2+\ \ +0\ =\ 2+}{[Cu(NH_3)_6]}$

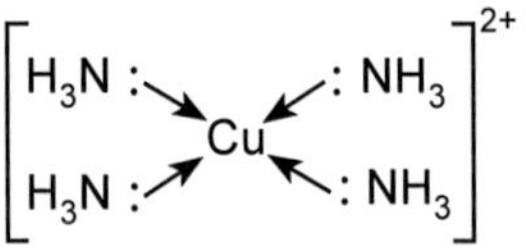

Cation Complex $\quad [Fe(CN)_6]^{-4}$

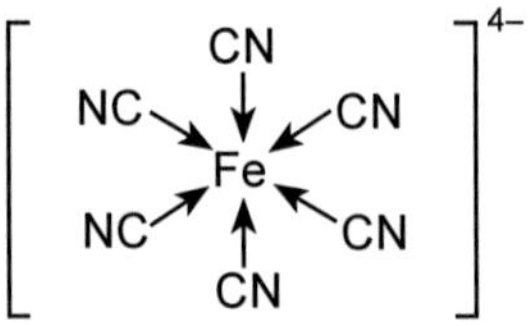

Neutral Complex $\quad \overset{2+\ +0\ \ +2-\ =\ 0}{[Pt(NH_3)_2Cl_2]}$

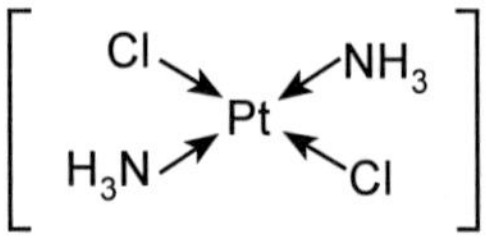

Ligands

The ligands are closer to the metal ion inside the square bracket. The arrangement between the metal ion and the ligand is fixed in space *i.e.*, linear, triangular, tetrahedral, square planer, square pyramid, trigonal bipyramidal, and octahedral.

Ligands: Ligands are anions or neutral molecules which can donate a pair or pairs of electrons. They are electron-rich molecules.

(*a*) **Anionic Ligands** $:\ddot{\underset{..}{Cl}}:^-$, $C \equiv N:^-$, SCN^-, CNO^-, F^-, Br^-, NO_2^-, SO_4^{2-} etc.

(*b*) **Neutral Molecules** $:NH_3$, $:C \leftrightarrows O$, H_2O, NO, CH_3NH_2 etc.

(*c*) **Cations** also act as ligands only in special conditions.

$\underset{..}{N}H_2.NH_3^+$ (Hydrazinium ion)

↓

(*d*) Ph $\underset{..}{N}H\ \overset{+}{N}H_3$ (Phenyl hydrazinium ion)

↓

2.2 FORMATION OF A COMPLEX

In the case of a complex formation, the central metal ion provides a large number of vacant orbitals of correct energy and orientation to accept the pair of electrons from the ligands. This implies that the energy difference between the vacant orbitals provided by the metal ions and the donating orbitals of the ligand must not be very large otherwise the coordinate bond formation will not occur.

For example: Ti^{4+} (Ar) $3d^{\circ}\ 4s^{\circ}\ 4p^{\circ}\ 4f^{\circ}$

In this, a total of 16 orbitals are vacant but $4f$ orbitals have a large energy difference from the donating orbitals of the ligand, therefore, they do not participate in bond formation.

Out of the remaining vacant orbitals only those are used which have a small energy difference and proper orientation with regard to donating orbitals of the ligands and thus result into a stable complex.

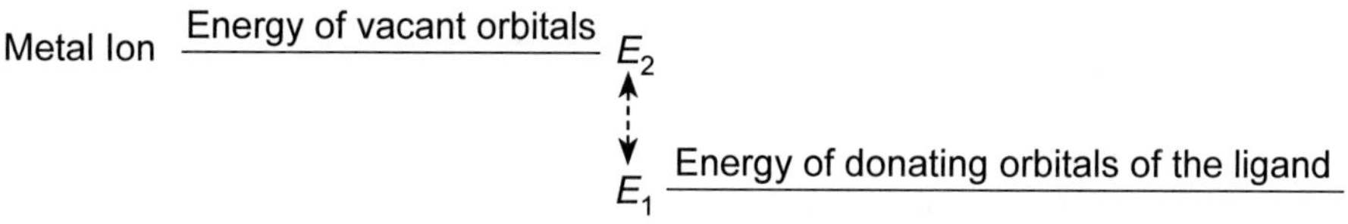

$\Delta E - E_2 - E_1$ (small)

Ligands: There are various types of ligands involved in the complex formation:

(*a*) **Monodentate Ligands:** These ligands provide only one donor site or they donate only one pair of electrons. For example, $:\ddot{Cl}:^-$, SCN^-, SO_4^{2-}, NO_2^-, CO, CN^-, H_2O, NH_3 etc.

(*b*) **Bidentate Ligands:** They provide two donor sites *i.e.,* two pairs of electrons are donated by the same ligand. For example:

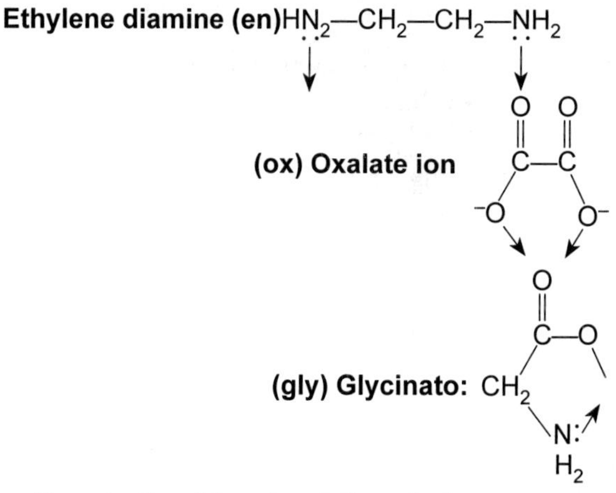

In these ligands the abbreviated form is important and even used in the complex formula.

(*c*) **Tridentate or Terdentate Ligands:** These ligands give three donor sites or three pairs of electrons are donated by the same ligand molecule.

For example: Diethylene triamine (dien)

$$H_2\ddot{N} - H_2C - CH_2 - \ddot{N}H - CH_2 - CH_2 - \ddot{N}H_2$$

(*d*) **Tetradentate Ligands:** They provide four donor sites or four pairs of electrons are donated by the ligand molecule.

For example: Triethylene tetramine (trien)

$$H_2\ddot{N} - H_2C - CH_2 - N\ddot{N} - CH_2 - CH_2 - \ddot{N}H - CH_2 - CH_2 - \ddot{N}H_2$$

(*e*) **Polydentate Ligands:** These ligands offer more than four donor sites although ligands with more than two donor sites may also be called polydentate ligands but with ligands having more than four donor sites, the term polydentate is a must.

For example: Ethylene diamine tetraacetate ion (edta)

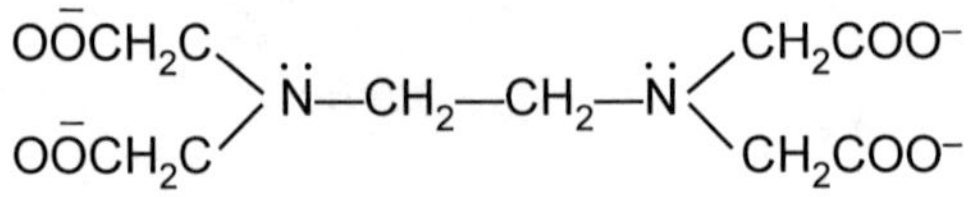

In reality, edta is hexadentate ligand. This ligand is known to form very stable complexes with metal ions such as Ca^{2+}, Mg^{2+} and Zn^{2+}. Edta is a very useful complexing agent and is used in complexometric titrations. Commercially, it is used to determine permanent hardness of water which is caused by chlorides and sulphates of Ca^{2+} and Mg^{2+} ions.

(*f*) **Ambidentate Ligands:** Ligands with more than one donor atom are called ambidentate ligands. A thiocyanate ion (SCN^-) can attach to the metal ion either by the 'N' atom or by the 'S' atom. These are called open-ended ligands.

M ⟵ :SCN thiocyanato complex

M ⟵ :NCS Isothiocyanato complex

More examples of ambidentate ligands include

M ⟵ NO_2 nitro

M ⟵ ONO nitrito

and SO_3^{2-} and PSO_2^{2-}; CN^- and NC^-, NCO^- and OCN^-, $SeCN^-$ and ^-NCSe ions.

(*g*) **Flexidentate Ligands:** These ligands have more than one donor atoms. They coordinate in such a way so that either one atom acts as a donor or two atoms act as donors e.g., in SO_4 ion ligand, either only one oxygen atom is bonded to the metal ion or two oxygen atoms are bonded

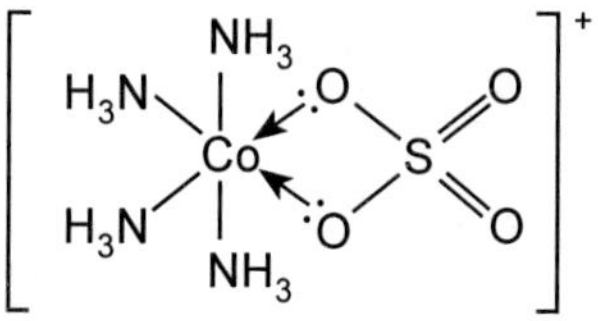

Structure of $[Co(NH_3)_4SO_4]$ (2 'O' atoms are bonded) to Co^{+3} ion)

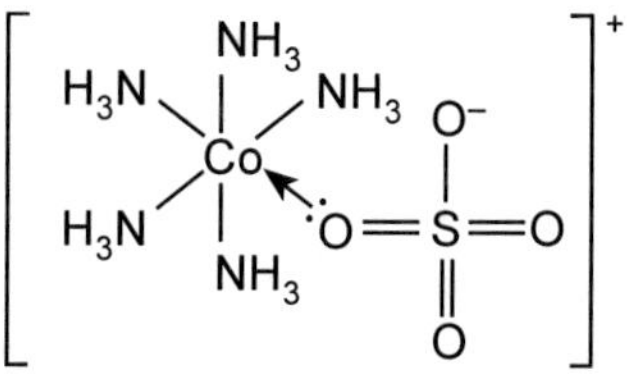

$[Co(NH_3)_5(SO_4)]$ one 'O' atom is bonded) to Co^{+3} ion)

Similar examples can be looked for CO_3^{2-} ion and N_2 molecule where CO_3^{2-} and N_2 can act as a monodentate as well as a bidentate ligand

Monodentate CO_3^{2-}	**Bidentate CO_3^{2-} ion**
$[Co(NH_3)_5(CO_3)]^+$	$[Co(NH_3)_3C1_2(CO_3)]$
Monodentate N_2	**Bidentate N_2**
$[Ru(NH_3)_5(N_2)]^{2+}$	$(_5(H_3N) \leftarrow N \equiv N \rightarrow Ru\ (NH_3)_5]^{4+}$

Coordination Number (C.N.)

It is the total number of atoms of the ligands which form a coordinate bond with the central metal ion. The numerical number of coordination bonds between the metal ion and the donor atoms of the ligand is called the **Coordination Number**, e.g.

$[Fe'''(CN)_6]^{3-}$ C.N. is six in the case of Fe^{3+}

$[Ni(en)_3]^{2+}$ C.N. of Ni^{2+} is six represent

Here ethylene diamine represents three ligands but each molecule has two donor atoms so that the total coordination bonds are six. Thus, this is the actual coordination number.

Low Coordination Numbers

Coordination complexes with coordination number one to two are called low coordination number complexes *e.g.* CN = 1 and CN = 2. They exist in the gas phase at high temperatures. The most commonly known complexes in the category are formed at ordinary conditions and are generally linear complexes e.g. Cu(I), Ag(I), Au(I) and Hg(II) as given below

Cu(I) : $[X \rightarrow Cu \leftarrow X]^+$

Ag(I) : $[H_3N \rightarrow Ag \leftarrow NH_3]^+$

Au(I) : $[R_3P—Au—PR_3]^+$

Here, Ligand may be R_3P, R_2S or X = halogen

Hg(II): $[H_3C—Hg—CH_3]$ or $[Cl—Hg—Cl]$

Ligands that are quite large in size are expected to accept linear geometry because of stereo chemical hindrance e.g.

$(N(SiMe_3)_2]^-$, = $[NPh\{B(MeS)_2\}]^-$

where $(Mes = 2, 3, 6—C_6H_2\, Me_3)$

Four Coordination Complexes

These complexes are formed in quite large numbers. They mostly opt for tetrahedral nd square planar geometries. These symmetrical structures are preferred over higher coordination complexes if the size of central atom is small or the ligands are large *e.g.* Cl^-, Br^-, I^-. Here, the steric effects override other factors. The preferred metal ions are of three types:

(*a*) Metal atoms or ions with noble gas configuration *e.g.* Be^{2+}

(*b*) Metal atoms or ions having pseudo-noble gas configurations $ns^2\ np^6\ nd^{10}$ (18 electrons)
e.g. Cu^{2+}, Zn^{2+}, Ga^{3+}.

(*c*) d-block metal ions which do not favour other geometries due to ligand field stabilisation energy (LFSE)
e.g. Mn^{2+}, Fe^{3+} and Co^{2+} etc,

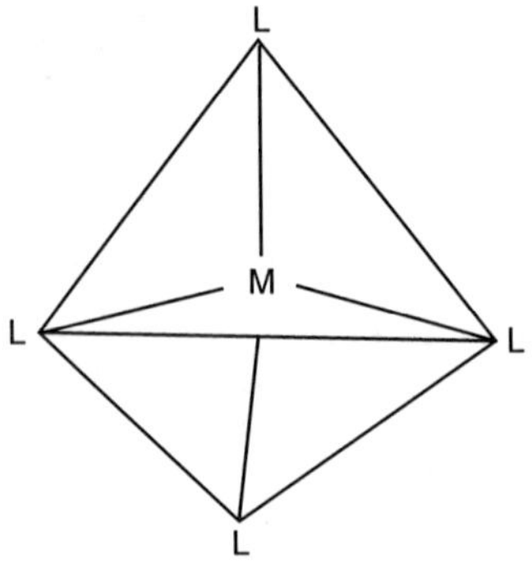

e.g. $VO_4^{3-}, CrO_4^{2-}, MnO_4^-, FeCl_4^-, CoCl_4]^{2-}, NiBr_4^{2-}, CuBr_4^{2-}$

Overall, tetrahedral is a spatial structure and is accepted when bulkier ligands are involved and square planar is accepted when smaller ligands are involved because square planar is more congested geometry.

Five Coordinate Geometry

Many complexes appear to have 5-coordinate stoichiometry e.g. $Ca[CoCl_5]$, $(NH_4)_3[ZnCl_5]$. MCl_5^{3-} in the crystals are made up of tetrahedral MCl_4^{2-} and Cl^- ions. $CuCl_5^{3-}$ is not stable in the solution form but has been isolated as salt $[Cr(NH_3)_6]^{3+}\ [C''uCl_5]$.

The geometry of the 5-coordinate complex can be either trigonal bipyramidal or square pyramid. A slight distortion may change one structure to another.

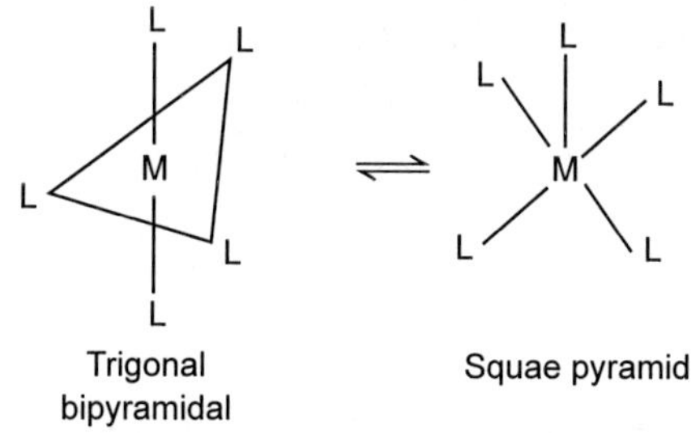

e.g. (*i*) $Fe(CO)_5$, PF_5, $Mn(CO)_5^-$, CrF_5, $[MoCl_5]$ $[CoCl_5]^{3-}$ and $[RuO_5]^{4-}$ are good examples of trigonal bipyramidal structures.

(*ii*) $[Xe\ OF_4]$, $[TiOCl_4]^{2-}$, $[Cr\ OCl_4]^{3-}$, $[MnCl_5]^{2-}$, $[Co(CN)_5]^{3-}$ and $[RuCl_2(PPh_3)_3]$ are examples which acquire pyramid geometries.

Six Coordinate Complexes

In the case of d-block metals in the configurations between d^1 to d^0 and d^{10} six coordinate structures are the commonest. Complexes with metal ions M^{2+} and M^{3+} acquire common octahedral or square bipyramid geometry. These metal ions have a large size. They also include +3 metal ions of *f*-block metals. The octahedral geometry is very symmetrical as all the bond lengths between metal ions and ligand are same and all bond angles of $\angle 90°$.

The common examples in this category are: $[Sc(H_2O)_6]^{3+}$ (d^0); $[Cr(NH_3)_6]^{3+}(d^3)$; $[Fe(CN)_6]^{3-}(d^5)$; $[Co(NH_3)_6]^{3+}(d^6)$; $[RhCl_6)]^{3-}(d^6)$; $[Ni(NH_3)_6]^{2+}(d^8)$.

f-block elements also display six coordinate complexes with the formula $[LnX_6]^{3-}$ where X = Cl, Br.

Distortion

Due to Jahn-Teller distortion or tetragonal distortion it may acquire the following structures

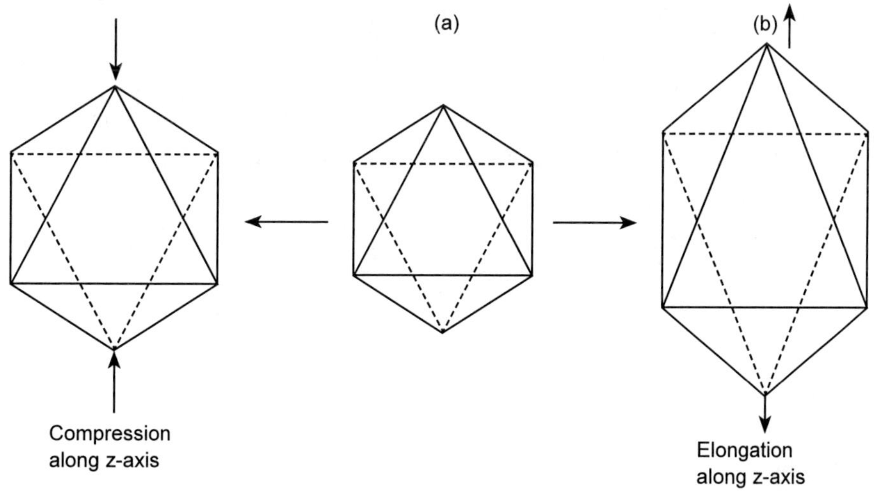

In the trigonal distortion the following change may occur.

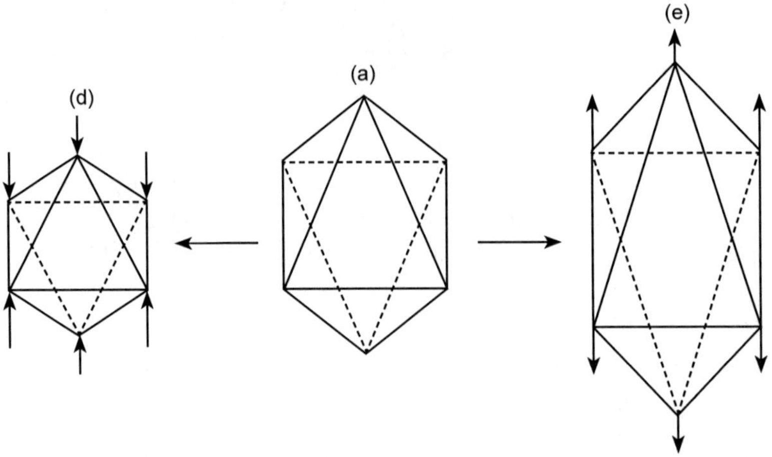

High Coordination Complexes

Higher than six coordination complexes are called high coordination complexes. They include 7, 8, 9, 10, 12 and 14 coordinate complexes. 7-coordinate complexes are observed with regard to 3d-metal complexes. 4d and 5*d* metal ions are larger metal ions and therefore can accommodate more than six legands. Seven coordinate species resemble five coordinate in terms of energy.

The most accepted geometries of the 7-coordinate complexes are pentagon bipyramid, monocapped octahedron, and monocapped trigonal prism shown below

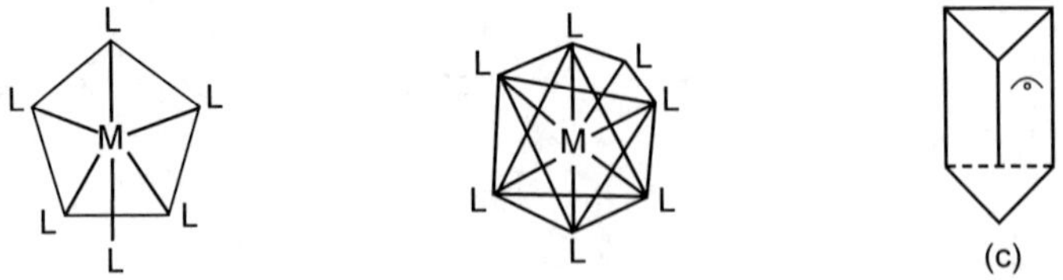

(a) Pentagon bipyramid (b) Monocapped octahedron (c) Monocapped trigonal prism

These are a number of intermediate structures between pentagon bipyramid to monocapped trigonal prism.

The interconversions are facile. *e.g.* $[ZrF_7]^{3-}$, $[TaCl_4(PR_3)_3]$, $[ReOCl_6]^{2-}$, $[UF_7]^{3-}$, $[NbF_7]^{2-}$ and $[UO_2(H_2O)_5]^{2+}$.

Eight Coordinate Complexes

They are observed most commonly in heavier metals ions such as Zr(IV), Ta(V), Mo(IV), Mo(V), Re(V, VI) representing d^0, d^1 and d^2 configurations. 8-coordinate complexes are stereo chemically non-rigid. They can be square antiprism in one crystal and do decahdedral in another *e.g.* $[Mo(CN)_8]^{x-}$ where $x = 3$ or 4.

9-coordinate complexes are observed in *f*-block elements as their cations are quite large and thus, can accommodate large number of ligands. They accept symmetrical tricapped trigonal prism geometry

e.g., $[MH_9]^{2-}$ where M = Ta, Re.

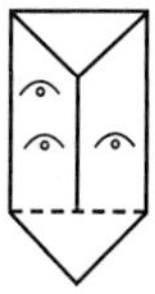

Tricapped trigonal prism geometry

Coordination 10 and 12 are encountered in *f*-block metal ions. In *d*-block metal ions they are quite rare, e.g. $(Th(C_2O_4)_4(H_2O)_2]^{4-}$. They are accomodated in bicapped square antiprism structure, having bidentate oxalate ligand. Coordination 12 is given by Ce^{4+} in $[Ce(NO_3)_6]^{2-}$ where NO_3^- acts s a bidentate ligand. Thorium has a similar example $[Th(NO_3)_6]^{2-}$. They have polymeric structure with a tricapped hexagonal antiprism geometry. Coordination number of 14 is represented by Th, Pa, U, Np and Pu.

Still higher coordination numbers are exposed to fragile geometries. With the slightest changes in energy one geometry gets converted to another similar geometry, *i.e.* the interconvessions of geometries are very quick. Therefore at any moment of time the geometry is not fixed. With the increasing number of sides of the base geometry the appearance approximates to a circle. Hence it does not represent any definite stable geometry.

2.3 BRIDGING LIGANDS

It includes those monodentate ligands which have more than one free electron pairs and these may get involved in bonding with two or more metal atoms through CT-bonds and thus act as a bridge between the metal atoms. The resulting complex is called a bridged or polynuclear complex having two or more metal centres. Some common examples of the bridging ligands are: OH^-, Cl^-, F^-, NH_2^-, SCN^-, CO, O^{2-}, SO_4^{2-}, O_2^{2-} etc. (overleaf).

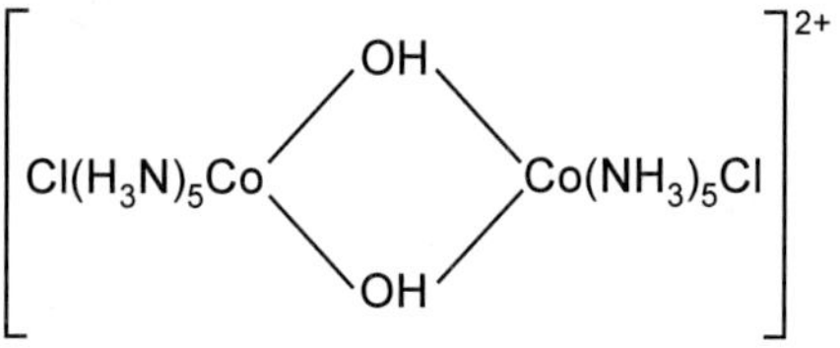

Coordination complexes *can be classified as* non-chelates *and* chelates.

Table 2.1: Some formulae and some abbreviations used for common ligands

Ligand Name	Formula	Abbreviation
Ethylene diamine	$H_2\ddot{N}-CH_2-CH_2-\ddot{N}H_2$	en
Propylenediamine	$\ddot{N}H_2-CH_2-CH(CH_3)-\ddot{N}H_2$	pn
Pyridine	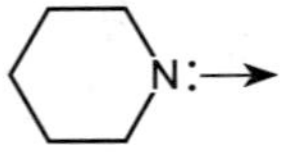	py
Diethylene triamine	$H_2\ddot{N}-CH_2-CH_2-\ddot{N}H-CH_2-CH_2-\ddot{N}H_2$	dien
Triethylene tetraamine	$H_2\ddot{N}-CH_2-CH_2-\ddot{N}H-CH_2-CH_2-\ddot{N}H-CH_2-CH_2-\ddot{N}H_2$	trien
Oxalato	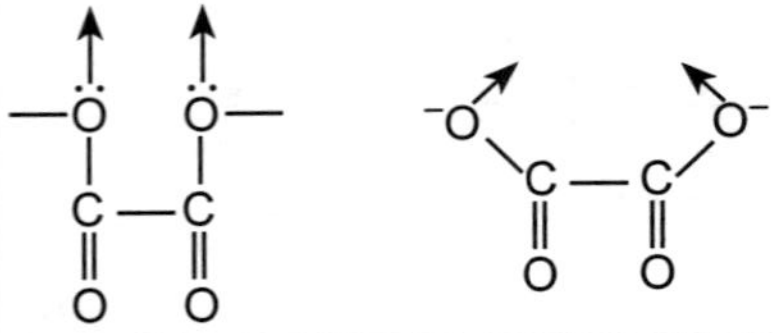	on

Ligand Name	Formula	Abbreviation
Acetylacetonato	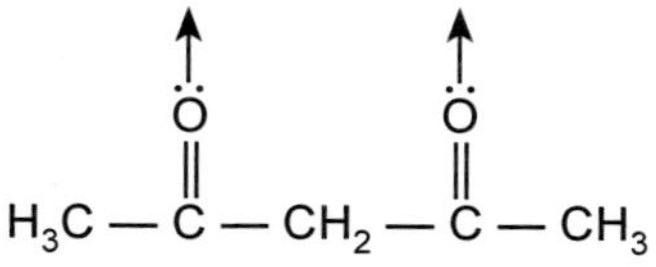	acac
Glycinato	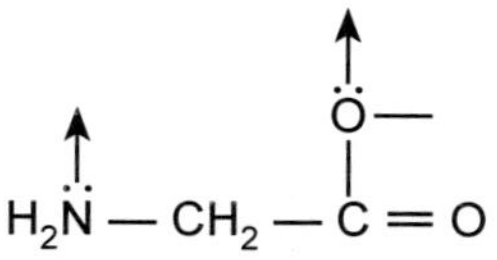	gly
Dimethyl glyoximato	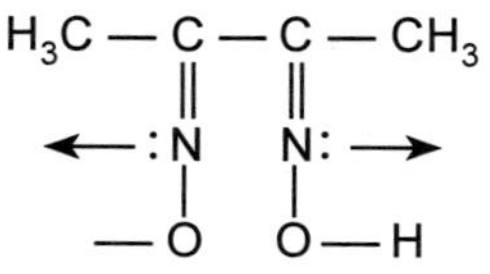	dmg
2,2′-dipyridine	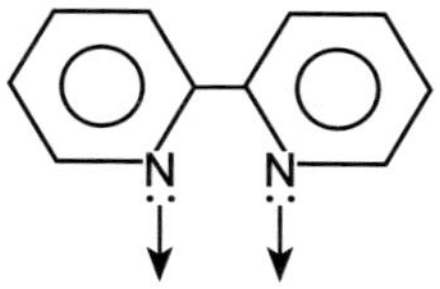	bipy
1,10-phenanthroline	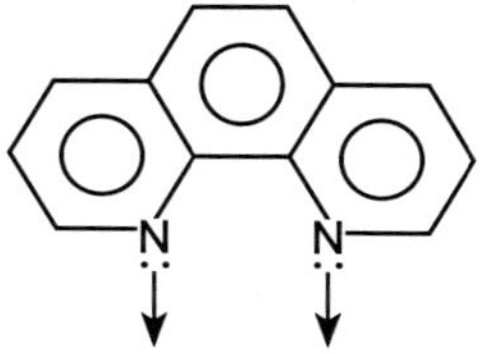	phen

Ligand Name	Formula	Abbreviation
8-hydroxyquinolinato	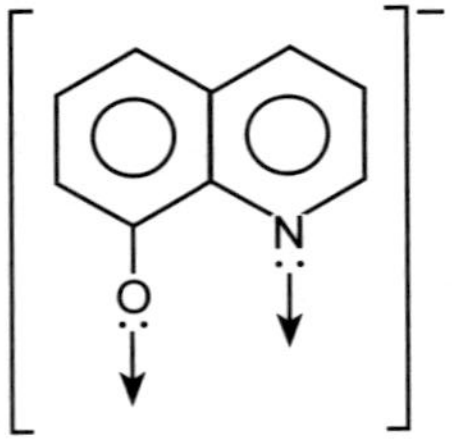	oxin
Bytylene diamine	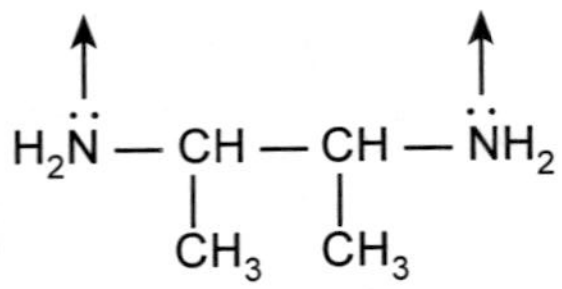	bn
Nitrilotriacetato	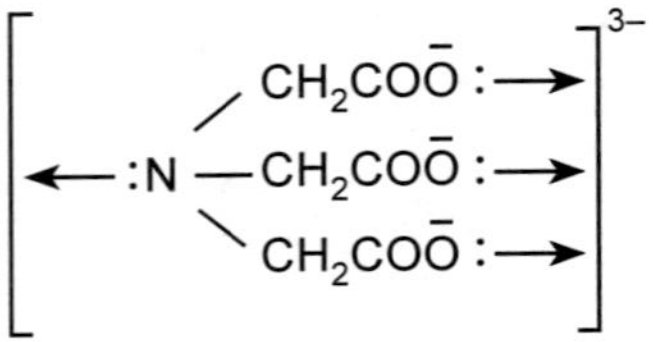	NTA
Ethylene diamine tetra acetato	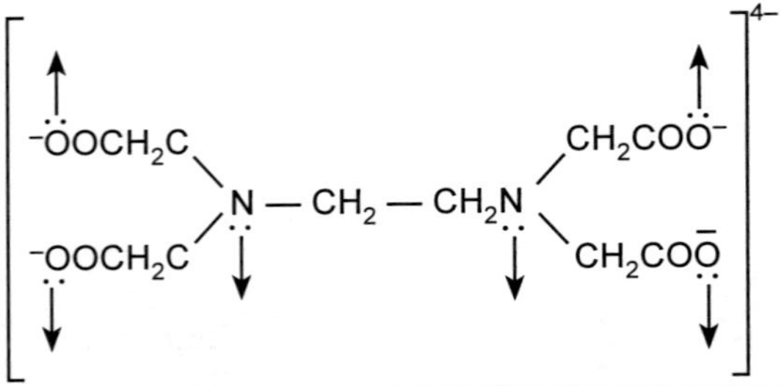	edta

2.3.1 Non-Chelates

All monodentate ligands form open chain complexes or non-cyclic complexes and are called non-chelated complexes. For example,

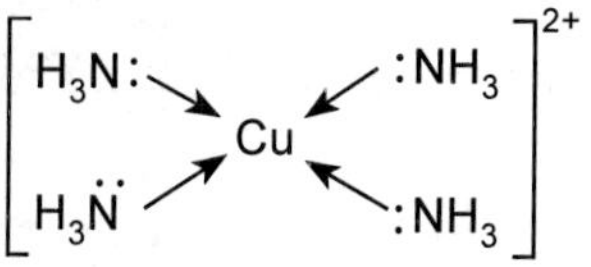

Tetraammine copper (II) ion

In these there is one attachment or bond between a metal atom and a ligand atom, molecule or ion, there is one donor atom per ligand.

2.3.2 Chelates

All bidentate and polydentate ligands are attached to the central metal atom in more than one direction forming a closed-chain, cyclic or ring structure called a chelate. The ligands involved are called chelating ligands. The coordinating atoms of the ligand are so arranged that they coordinate to the same metal atom.

In general five-membered ring chelates form a bonded complexes and six membered ring complexes form π-bonded complexes. These are more stable than the rest of chelates. For example,

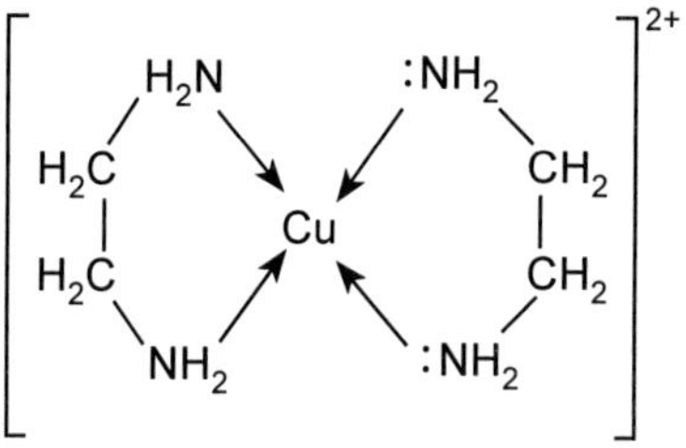

Bis(ethylenediamine copper (II) complex ion

These chelate rings are more stable due to the reduced strain.

Chelates are much more stable than the non-chelates. This statement is attributed to the following factors:

1. Just as inorganic compounds, ring compounds are more stable than the open chain compounds. Similarly, chelates are more stable than nonchelates (reduced strain).
2. In a chelate, a metal atom is bonded to the ligand in more than one direction resulting in a strong metal ligand bond which infuses greater stability.
3. In the formation of a chelate there is always increase in entropy value. Greater entropy leads to greater stability. This can be explained in a simple way as substitution given by the following reactions. The

easiest way to understand higher entropy is increase in the number of particles of the products, *e.g.,* aqueous solution of hydrated copper sulphate on addition of NH_3 and ethylene diamine as the substituting ligands gives the undermentioned reaction where there is always an increase of particles *vis-a-vis* entropy. This quantifies greater stability of the chelates.

On the L.H.S. there are (1 + 2) particles while on the R.H.S. there are (1+4) particles.

$$[Cu(NH_3)_4]^{2+} + 2en \longrightarrow [Cu(en)_2]^{2+} + 4NH_3$$

The term **strong field ligands** and weak field ligands needs to be elaborated.

Strong field ligands are those in which the charge of the lone pair of electrons for donation is focussed in only one direction. For example,

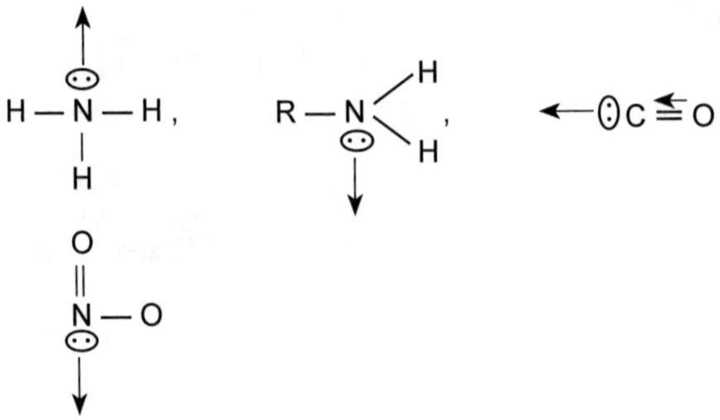

Weak field ligands are those in which the charge of the lone pair of electrons is diffused in more than one direction. For example,

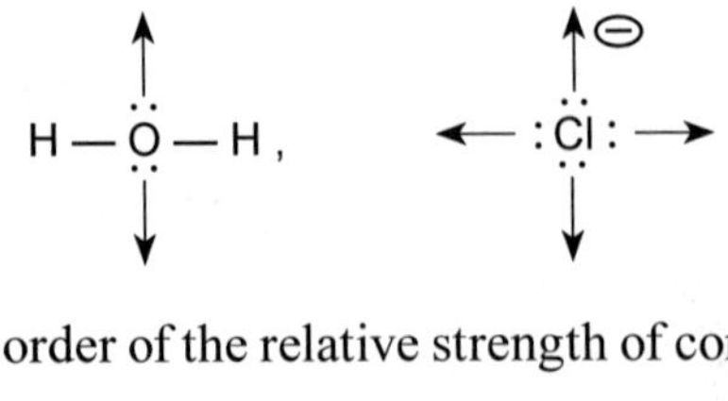

The approximate order of the relative strength of common ligands is given as:

$CO, CN^- > CH_3^- > H^- > phen > NO_2^- > NH_2OH > SO_3^{2-} \sim en > py \sim NH_3 > NH_2CH_2COO^- > CN^- > NCS^- > H_2O \sim C_2O_4^{2-} > OH^- > NH_2CONH_2 > F^- > N_3^- > SCN^- \sim Cl^- > Br^- > 1^-$

Aqueous solution of $CuS0_4$ forms a 4-coordinate aqua complex:

$$CuSO_4 + 4H_2O \longrightarrow [Cu(H_2O)_4]^{2+} + SO_4^{2-}$$

In a complex the strong field ligand replaces the ligand which is less strong and so on.

Ammonia is a strong field ligand in comparison to H_2O. Therefore, it substitutes H_2O molecules.

$$[Cu(H_2O)_4]^{2+} + 4NH_3 \longrightarrow [Cu(NH_3)_4]^{2+} + 4H_2O$$

$$1 + 4 = 5 \text{ particles} \longrightarrow 1+4 = 5 \text{ particles}$$

In this the number of particles on left-hand-side are equal to the number of particles on the right-hand-side hence there is no increase in the number of particles. Hence, Cu^{2+} and NH_3 complex is a non-chelated complex.

Now, when ethylene diamine is added to tetraamine copper (II) complex ion, ethylene diamine being a stronger ligand substitutes the ammonia ligand as given:

$$[Cu(NH_3)_4]^{2+} + 2en \longrightarrow [Cu(en)_2]^{2+} + 4NH_3$$

$$1 + 2 = 3 \text{ particles} \longrightarrow 1 + 4 = 5 \text{ particles}$$

In this the number of particles on the RHS shows an increase. Moreover, $[Cu(en)_2]^{2+}$ is a chelated complex. Likewise, in the formation of all chelated complexes there is always increase in the number of product particles and hence stability. In fact the reactions in complexes are simply substitution reactions and hence does not require balancing of equations.

2.3.2.1 Application of Chelates

Analytical Chemistry:

(*a*) Qualitatively ions such as Fe^{3+}, Ni^{2+}, As^{3+}, Co^{2+}, Cu^{2+}, Zn^{2+} etc. can be identified due to the formation of coloured complexes.

(*b*) Quantitatively Ca^{2+}, Mg^{2+} and Zn^{2+} form very stable complex with EDTA. The volumetric estimation of these ions is called Complexometric titrations, as complexes are formed at every stage of the reaction associated with colour change. Therefore, some of the chelating ligands are also used as **Indicators** in the titration of certain metql ions. $[Ni(dmg)_2]$ is used for estimation of Ni^{2+} gravimetrically.

(*c*) Chelating agents *e.g.*, edta are used to remove harmful **radioactive metals** from the body.

(*d*) *Water Softening:* By complexing the metal ions it i$ possible to suppress the precipitation reactions of Ca^{2+} and Mg^{2+} ions responsible for hardness of water.

(*e*) *Solvent Extraction:* In the presence of ligands capable of forming inner complexes, metals can be extracted into water-miscible organic phase.

(*f*) *Food preservation:* Fruit juices can be preserved with chelating agents.

Before proceeding further into the domain of metal complexes it would be pertinent to know and understand the correct formulation of complexes. The further foray into complexes is based on its correct formulation for which number of analytical processes are applied. It is important to know which of the anions form a part of the coordination sphere and which others serve as ionisable ions.

SUMMARY

In this chapter, an attempt has been made to understand the basic definition of a metal complex. Metal complex comprises of metal ion and a ligand. The charge on the complex ion is sum of the charges on the metal and the ligand. Types of ligands have been exemplified and formation and stability of chelates and non-chelates has been discussed with highlight on the concept of entropy.

■ ■ ■

3

Correct Empirical Formula of Complexes

Historical survey into the problem projects the names of the chemists who discovered them, i.e., relationship between name of chemists and formula

Name	Correct Formulation
Reinecke's salt	$NH_4[Cr(NH_3)_2(NCS)_4]$
Magnus green	$[Pt(NH_3)_4][PtCl_4]$
Erdmam's salt	$K[Co(NH_3)_2(N0_2)_4]$
Zeise salt	$K[Pt(C_2H_4)Cl_3]$

This was rejected as it had no scientific basis.

Later it was observed that Co(III) and Cr(III) complexes exhibit a wide spectrum of colours. Therefore, an attempt was made to involve **relationship between colour and formulation of complexes.**

Molecular Formula	Colour	Name	Formulation
$CoCl_3.6NH_3$	Yellow	Luteocobaltic chloride	$[Co(NH_3)_6]Cl_3$
$CoCl_3.5NH_3$	Purple	Purpureo cobaltic chloride	$[Co(NH_3)_5Cl]Cl_2$
$CoCl_3.4NH_3$	Green	Praseocobaltic chloride	*trans* $[Co(NH_3)_4Cl_2]Cl$
$CoCl_3.4NH_3$	Violet	Violeocobaltic chloride	*cis* $[Co(NH_3)_4Cl_2]Cl$
$CoCl_3.5NH_3.H_2O$	Red	Roseocobaltic chloride	$[Co(NH_3)_5(H_2O)]Cl$

This could not be sustained due to lack of scientific base.

3.1 PRECIPITATION METHOD

It was the first attempt to arrive at the correct formula of a complex on the scientific analysis. The ions outside the coordination sphere known as ionisable ions are set free, when the complex is dissolved in a solvent. By precipitating these ions with the help of a suitable reagent and weighing the precipitate gravimetrically we can calculate the possible ionisable ions which can lead to the correct formulation. For example,

Molecular formula	Number of Cl precipitated	Formulation
$CoCl_3.6NH3$	3	$[Co(NH_3)_6]Cl_3$
$CoCl_3.5NH_3$	2	$[Co(NH_3)_5Cl]Cl_2$
$CoCl_3.4NH_3$	1	$[Co(NH_3)_4Cl_2]Cl$
$CoCl_3.3NH_3$	0	$[Co(NH_3)_3Cl_3]$

This method has a limited scope as it is not always possible to find a suitable reagent for the precipitation of ions. Of course, it is successful in case of common ions which can be easily precipitated and analysed gravimetrically.

3.2 MOLAR CONDUCTIVITY METHOD

In certain type of complexes, it was found that molar conductivity value of the complex imparts a very useful information regarding the formulation of the complex. If the complex is more ionic it will show greater molar conductivity value compared to the one that is less ionic or more covalent. From the molar conductivity values we come to know of the total number of ions in the complex from which we can arrive at the correct formulation.

Molar Conductivity Values of Pt(V) Complexes

Molecular formula	Molar conductivity (ohm^{-1})	Number of ions indicated	Correct formulation
$PtCl_4.6NH_3$	523	5	$[Pt(NH_3)_6]^{4+}4Cl^-$
$PtCl_4.5NH_3$	**404**	**4**	$[Pt(NH_3)_5Cl]^{3+}3Cl^-$
$PtCl_4.4NH_3$	229	3	$[Pt(NH_3)_4Cl_2]^{2+}2Cl^-$
$PtCl_4.3NH_3$	97	2	$[Pt(NH_3)_3Cl_3]^+Cl^-$
$PtCl_4.2NH_3$	0	0	$[Pt(NH_3)_2Cl_4]$
$PtCl_4.NH_3.KCl$	109	2	$K^+[Pt(NH_3)Cl_5]^-$
$PtCl_4.2KCl$	256	3	$K_2[PtCl_6]$

The molar conductivity value is not fool proof for arriving at the correct formulation of a complex. They only indicate the total number of charges present in a complex. This information is further augmented by taking the cryoscopic measurements which indicate the number of particles. The number of particles formed may be different from the total number of charges. Molar conductivities and cryoscopic measurements both taken together help in arriving at the correct formulation of a complex. Cryoscopic measurements involve measuring how much the freezing point is lowered. The depression in freezing point depends upon the number of particles. More the number of particles in solution greater is the lowering of freezing point. Cryoscopic measurements can be used to find the dissociation of a molecule into the number of ions. If the molecule splits into two ions, the depression in freezing point will be twice that of a single particle and if three ions are formed the depression will be three times.

The use of the two measurements to arrive at the correct formulation of a complex is illustrated in Table below:

Molecular formula	Cryoscopic measurements	Molar conductivity	Correct formulation
$COCl_3.6NH_3$	4 particles	6 charges	$[Co(NH_3)_6]^{3+}$ $3Cl^-$
$CoCl_3.5NH_3$	3 particles	4 charges	$[Co(NH_3)_5Cl]^{2+}$ $2Cl^-$
$COCl_3.4NH_3$	2 particles	2 charges	$[Co(NH_3)_4Cl_2]^+$ Cl^-
$COCl_3.2NH_3$	1 particle	0 charge	$[Co(NH_3)_3Cl_3]$
$Co(NO_2)_3.KNO_2.2NH_3$	2 particles	2 charges	$K^+[Co(NH_3)_2(NO_2)_4]^-$
$Co(NO_2)_3.2KNO_2$ NH_3	3 particles	4 charges	$2K^+[Co(NH_3)(NO_2)_5]^{2-}$
$Co(NO_2)_3.3KNO_2$	4 particles	6 charges	$3K^+[Co(NO_2)_6]^{3-}$

3.3 EFFECTIVE ATOMIC NUMBER (EAN RULE)

In a metal complex, the metal ion generally belongs to the transition metal series and has large number of vacant orbitals. Each ligand donates an electron pair to this metal ion very readily. The electronic configuration of noble gases is very stable (Sidgwick). If the electron pairs from ligands added to the electrons of the central metal atom equal the number of electrons of the next noble gas then the complex is considered very stable and feasible.

For example, the effective atomic number of Fe^2+ in the complex ion $[Fe(CN)6]^{4-}$ equals that of Kr(36 electrons).

Metal Ion	Complex	Atomic number	Electron pairs gained	EAN	
Fe^{2+}	$[Fe(CN)_6]^{4-}$	24	12	36	(Kr)
Fe	$[Fe(CO)_5]$	36	10	36	(Kr)
Cu^+	$[Cu(CN)_4]^{3-}$	28	8	36	(Kr)
Co^{3+}	$[Co(NH_3)_6]^{3+}$	24	12	36	(Kr)
Pd^{4+}	$[Pd(NH_3)_6]^{4+}$	42	12	54	(Xe)

The EAN rule can correctly predict the number of ligands in many complexes. All carbonyls and nitrosyls invariably follow the EAN rule. There are significant exceptions to the rule.

(*i*) If the central metal ion has an odd number of electrons then addition of electron pairs from the ligand cannot result in a noble gas configuration (All noble gases are even numbered) but the tendency to acquire noble gas configuration is a significant factor but not a necessary condition for the formation of a complex. The complex must involve a geometrical structure *e.g.,* tetrahedral, square planar, octahedral irrespective of the electron pairs donated.

3.4 BLOMSTRAND-JORGENSEN CHAIN THEORY

Earlier it was felt that elements had only one type of valency. They suggested that there could be only three bonds to Co(III) in its complexes [Co(III) complexes were well-known]. Akin to carbon chain it was considered that 6 NH_3 molecule had NH_3—NH_3 chain. The 3Cf can be easily precipitated by $AgN0_3$. Therefore, they were believed to be some distance away from the Co(III) ion and are bonded through ammonia, *e.g.*, $CoC1_3.6NH_3$.

(i) 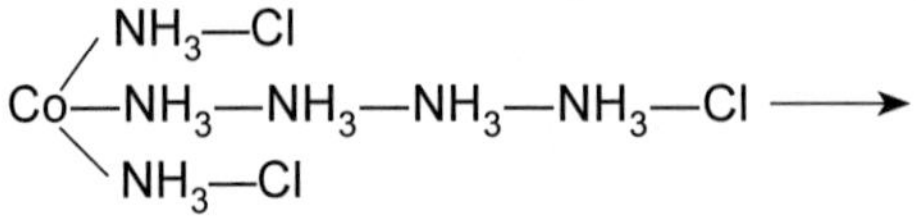

3Cl⁻ ions are precipitated by Ag^+ as the three Cl^+ ions are bonded to cobalt through ammonia = 3 AgCl

(*ii*) $[Co(NH_3)_5Cl]Cl_2$

In this one chloride that is inside the coordination sphere is directly attached to Co(III) ion and it does not ionise and precipitate as AgCI. The remaining 2 Cf ions are precipitated

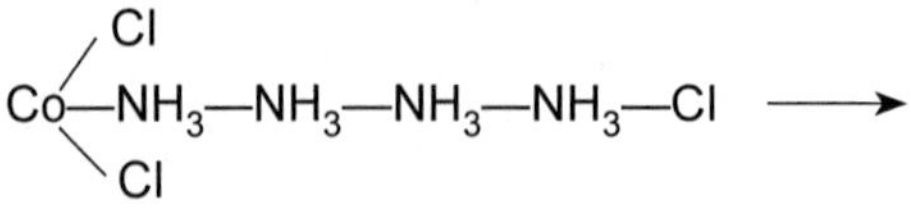

$2Cl^-$ ions are precipitated = 3 AgCl

Similarly other isomers are represented as given:

(*iii*) $[Co(NH_3)_4Cl_2]Cl$

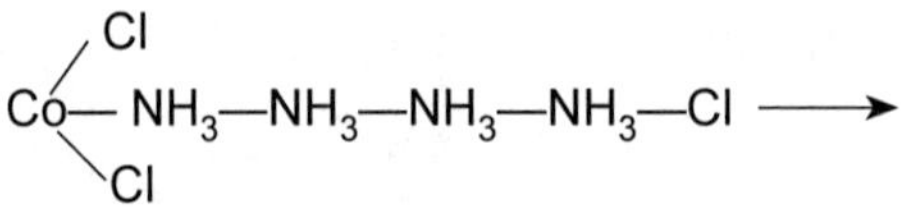

Cl^- ions is precipitated = 3 AgCl

(*iv*) $[Co(NH_3)_3Cl_3$

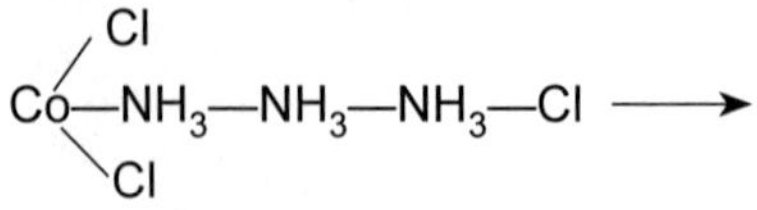

No Cl^- ion precipitate but the formula suggests that one Cl is precipitated = No AgCl

The chloride ions precipitated in both structures (*iii)* and (*iv*) are the same therefore, structure (*iv*) could not be properly represented. Hence, the theory of ammonia chain failed.

3.5 WERNER'S COORDINATION THEORY

Alfred Werner at the age of 26 prepared his famous 'Coordination Theory' which became a guiding force in inorganic chemistry. He is aptly called the father of coordination chemistry. He revealed the concept of valency in complexes for the first time. Three of his important postulates are:

(*i*) Most elements have two types of valencies in complexes:

(*a*) Primary valency corresponds to the ions and oxidation state of the metal ion or electrostatic charges. It is satisfied by the negative ions and its attachment with the central metal ion is shown by dotted lines as it is loosely bonded and hence ionisable.

(*b*) Secondary valency corresponds to the coordination number. It is non-ionisable (coordinate covalent bond). It is satisfied by negative ions or neutral molecules. The ligands which are within the coordination sphere are directly attached to the central metal ion and are shown by thick or permanent lines as they represent a strong bond.

(*ii*) Both primary and secondary valencies are satisfied in a complex. The negative ion may satisfy both primary and secondary valencies.

For example: $\mathbf{[Co(NH3)_6]^{3+}Cl_3}$

Primary valency $\longrightarrow$ 3^+ and 3^-

Secondary valency = 6

$\mathbf{K_4[Fe(CN)_6]}$

Primary valency = 4^+ or 4^-

Secondary valency = 6

(*iii*) The secondary valency is directed towards fixed position in space, *i.e.*, the complex ion represents a definite geometrical shape and is not haphazard. This opens up the field of stereochemistry or isomerism in complexes which in turn control the substitution reactions in complexes.

Further, Werner represented the formulation of complexes graphically taking into account the definite geometry acquired by the complex ion.

The common geometries are listed below:

Complex	No. of ligands	Most probable geometries
1. $[Cu(NH_3)_4]^{2+}$	4	Square planar or tetrahedral
2. $[Co(NH_3)_6]^{3+}$	6	Octahedral, hexagon or trigonal prism
3. $[Fe(Co)_5]$	5	Trigonal bipyramidal or square pyramid

The most common coordination number in transition rhetal complexes is ‘6’ and the most probable geometry is ‘octahedral’.

The next more important coordination number is ‘4’ and is represented by geometries such as ‘Tetrahedral’ or ‘Square planar’.

Werner gave a pictorial representation to the complexes and their isomers within the parameter of the most likely geometry.

e.g., $CoCl_3.6NH_3$

Werner observed that in the above complex $3Cl^-$ may act as primary valencies and six ammonia as secondary valencies. The stoichiometry of the complex may be deduced by precipitating the Cl ion as AgCl.

$$CoCl_3{\cdot}6NH_3 + AgNO_3 \xrightarrow[HNO_3]{dil} 3AgCl \text{ ——— } 3x \; \frac{\text{Weight of AgCl}}{}$$

Loss of one NH_3 should give $CoCl_3.5NH_3$ where one Cl ion may change from primary to secondary valency and the formula may be represented like $[Co(NH_3)_5Cl]Cl_2(NH_3)$ solvent molecule outside the coordination sphere is not considered. Therefore, it becomes $[Co(NH_3)_5Cl]Cl_2$ and ionises as:

$$[Co(NH_3)_5Cl]^{2+}2Cl^-$$

This is deduced by the $AgNO_3$ reaction *vis-a-vis*

$$CoCl_3{\cdot}5NH_3 + AgNO_3 \xrightarrow[HNO_3]{dil} 2AgCl \text{ ——— } 2x \; \frac{\text{Weight of AgCl}}{}$$

Similarly other isomers may be represented as $[Co(NH_3)_4Cl_2]Cl$ and $[Co(NH_3)_3Cl_3$ which may give AgCl precipitate in the following sequence.

$$CoCl_3{\cdot}4NH_3 + AgNO_3 \xrightarrow[HNO_3]{dil} AgCl \text{ ——— } 1x \; \frac{\text{Weight of AgCl}}{}$$

$$CoCl_3{\cdot}3NH_3 + AgNO_3 \xrightarrow[HNO_3]{dil} NoAgCl$$

Werner represented three isomers in the manner given below:

According to Werner dotted line (-----) represents primary valency and solid line (—) represents secondary valency. In these isomers, Cl^- ion precipitated as AgCl are in the ratio of $3x : 2x : x : 0$ by weight.

$[Co(NH_3)_6]^{3+}3Cl^-$

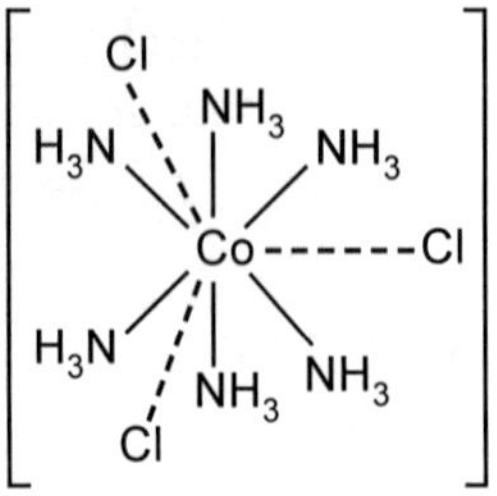

$[Co(NH_3)_5Cl]^{2+}2Cl^-$

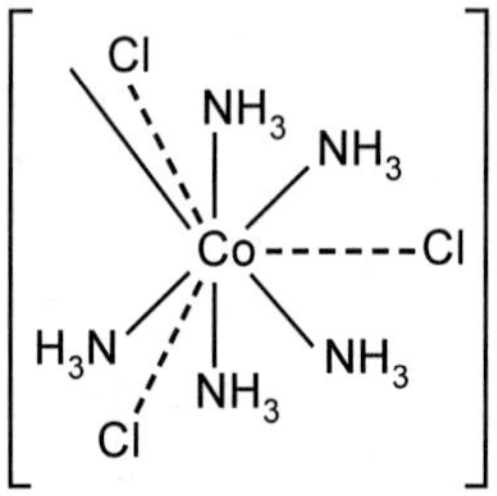

The Cl^- ion which changes from primary valency to secondary valency is represented by dashed-solid line.

$[Co(NH_3)_4Cl_2]^+Cl^-$

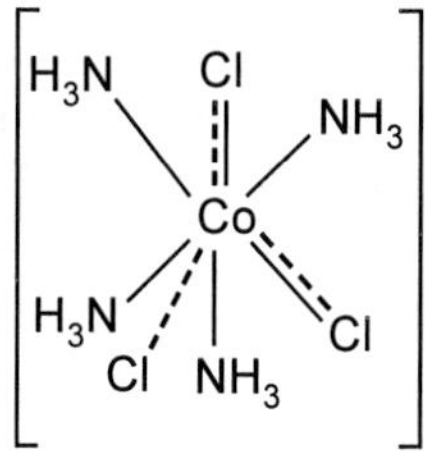

$[Co(NH_3)_3Cl_3]$

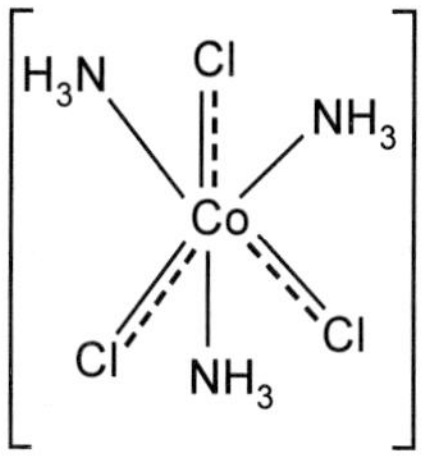

In all these isomers both primary and secondary valencies are firmly satisfied in the pictorial representation.

SUMMARY

Complexes were known from very early times but it was not possible to represent them by the correct formula which conveys all the important aspects. How to represent the strong bonds and the weak bonds in the same formula was the problem. Number of attempts were made but none was fool-proof. Jorgensen gave the famous chain theory but it failed. Werner's postulates paved the way for a satisfactory formulation. The strong bond between metal and ligand were put in square bracket [] and the weak bonds outside the square bracket *e.g.,* $[Co(NH_3)_6]C1_3Co \leftarrow (NH_3)_6$—strong bond, Co and Cl^- weak bond. Werner's postulates were found correct on the basis of (*a*) precipitation method (*b*) Molar conductance (c) Cryoscopic measurements. Werner in his graphic formula represented strong bonds by permanent lines and weak bond by dotted lines in a specific geometry, *e.g.,* $COCl_3 \cdot 6NH_3$.

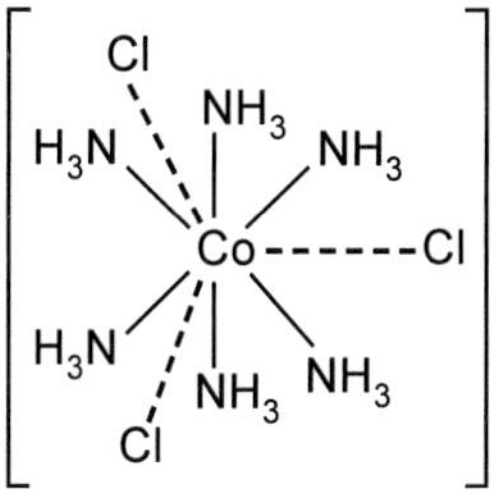

This could not be made possible for all the complexes therefore, the present formulation proposed by Sidgwick separating strong bonds and weak bonds by square brackets was involved. It was found satisfactory and was accepted. It covers all the aspects of Werner's theory except the orientation in space or geometry.

If the total electrons of metal and that donated by ligands equals noble gas configuration, the complex is possible and stable. All carbonyls obey EAN rule except $[V(CO_6)]$. If two different oxidation states of the metal form two complexes with the same ligand, then one of the two does not obey the rule. If ligands are different and the same metal forms two complexes with different coordination numbers *e.g.*, $[CoF_6]^{3-}$, $[CoCl_4]^{2-}$, then one of them does not obey EAN rule. Generally, complex ions whose central metal form is paramagnetic do not obey the rule but this theory cannot be stretched too far.

EXERCISE

1. What is a metal complex? Explain with example.

2. Explain the terms:
 (*i*) Coordination compound
 (*ii*) Coordination sphere
 (*iii*) Complexation
 (*iv*) Ligand

3. What is the correct formulation of:
 (*i*) Prussian blue
 (*ii*) Turnbull's blue
 (*iii*) Invisible ink

4. What are the three essentials of Werner's theory of complexes? Elucidate with examples.

5. How Blomstrand-Jorgensen represented the complexes with the help of chain theory?

6. How did Werner pictorially represent the different isomers of COCI3.6NH3?

7. In what way molar conductivity values helped to write the correct formulation of a complex? Give examples.

8. Illustrate primary and secondary valency in a complex $[Co(NH_3)_6]Cl_3$.

9. 'Both primary and secondary valencies must be satisfied in the complex'. Illustrate with an example.

10. Write the correct formula for the complex.
 (*i*) $PtCl_4.2NH_3.KCl$
 (*ii*) $PtCl_4.NH_3.2KCl$

11. 'The third postulate of Werner is directly related to the stereochemistry of the complexes'. Comment.

12. The light pink colour of $COCl_2.6H_2O$ becomes blue on warming. Explain with a suitable reaction.
13. What is the difference between a double salt, alum and a complex compound?
14. What is Fountain Pen Ink?
15. A complex retains its shape and properties in solution. Elaborate.

■ ■ ■

4

IUPAC System of Nomenclature

4.1 NOMENCLATURE OF COORDINATION COMPOUNDS

A comprehensive scheme of nomenclature of coordination compounds was not feasible prior to Werner's coordination theory. Like organic compounds it has two types of nomenclatures that are in use *vis-a-vis* (*i*) Common names and (*ii*) IUPAC names. Out of the two, IUPAC names are more scientific and systematic. The knowledge of complexes is about one century old. The last 30 years saw a big boom in the number of coordination compounds and their utility in industrial, medicinal and biological fields.

It is logical to mention that the common names of coordination compounds had not become very popular by the time they were substituted by IUPAC names.

4.2 GENERAL RULES

1. Coordination compounds are named like common compounds in which a positive ion is named first followed by negative ion *e.g.,* Na^+Cl^- sodium chloride.
2. In a complex ion there is a metal centre and the ligands. The ligands are named first and metal ion later.
3. The name of the complex should be taken as one word.

4.3 NAMING THE METAL ION

1. The name of the metal ion remains unchanged if the complex is neutral or a cation *e.g.,* copper, nickel, chromium, aluminium etc.

2. The name of the metal ion ends with the suffix 'ate' if the complex is an anion *e.g.,* chromate, borate, aluminate, mercurate, nickelate, cobaltate etc.
3. The oxidation state of the metal constituting the complex ion is written in Roman numerals in small brackets *e.g.,* (II), (III), (IV) etc.

4.4 NAMING THE LIGAND

1. If the ligand happens to be:

   (Water) H_2O aqua

   (Ammonia) NH_3 ammine

2. If the ligand attached to the metal bears a negative charge then name ends with letter 'o'. For example,

Ligand	Name
Cl^-	chloro
CN^-	cyano
SCN^-	thiocyanato
NO_3^-	nitrato
SO_4^{2-}	sulphato
NO_2^-	nitro

3. If the ligand attached bears a positive charge then it has a suffix '-ium'

   $NH_2NH_3^+$ Hydrazinium

   $PhNHNH_3^+$ Phenyl hydrazinium

4. If the ligand is twice, thrice or four times then the prefixes di—, tri- and tetra- are used provided the ligand molecule or ion is simpler and smaller (made of lesser number of atoms). The prefixes *bis-, tris-, tetrakis* etc. are used in case of bulky ligands (made of more number of atoms). In such cases the name of the ligand is given in small brackets ().

   e.g. (*a*) $[Co(NH_3)_4Cl_2]Cl$ Tetraamminedichlorocobalt (III) chloride

   (*b*) $[Co(en)_2Cl_2]_2S0_4$ Dichlorobis (ethylenediamine) cobalt (III) sulphate

5. If the metal ion is attached to two or more ligands then the names of the ligands are written in alphabetical order without considering the letter of the prefix *e.g.,* in tetraammine the letter taken is 'a' and not't' as tetra is a prefix and does not indicate the letter of the ligand.
6. Point of Attachment:

   Some ligands are such that they can be attached to the metal ion by two different electronegative atoms but one of the two atoms is attached at one time. These ligands are called 'Ambidentate' or open-ended ligands. They are named as given:

$M \longleftarrow SCN^-$		Thiocyanato or thiocyanato—S—
$M \longleftarrow NCS^-$		Isothiocyanato or thiocyanato—N—
$M \longleftarrow NO_2^-$		Nitro
$M \longleftarrow ONO^-$		Nitrito

7. Polynuclear or Polymetallic Complexes:

   If the complex contains two or more metal centres it is called a polynuclear complex. The bridging ligands which link the two or more metal atoms are indicated by the prefix 'p' followed by a small hyphen.

   If there are two or more bridging groups of the same kind, this is indicated by di-p-, tri-p- etc. If the bridging groups linking the two or more metal atoms are different then they are listed in alphabetical order and p is repeated two or more times (p-amido p-nitro). The alphabetical order of the bridging group with other groups is also maintained in consonance with the symmetry of the molecule.

   eg., 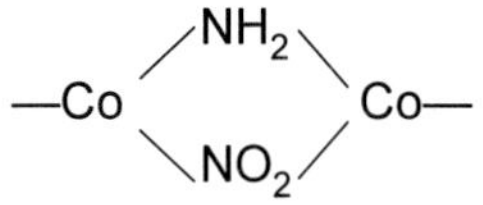 μ-Amido μ-nitro dicobalt.

   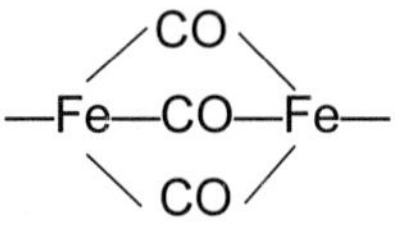 Tri-μ-carbonyl diiron

   The bridging ligand is given preference over the non-bridging ligands while naming as with di-p-chloro-tetrachloro.

8. Organic group implicitly considered to be anions are given their regular names without ending with 'o'.

eg., CH_3	Methyl
C_2H_5	Ethyl
C_6H_5	Phenyl

9. Optical isomers are designated by the symbols A or A and geometrical isomers by *cis* and *trans* in the prefixes.

Configuration Symbols

The configuration index comprises of a series of letters and digits which identify the positions of the ligating atoms in the coordinating polyhedron. The configuration index has the ability to distinguish between the geometrical isomers. The digits of the configuration index are chosen on the basis of the order of the ligating atoms. The ligating atom of the highest priority is given number 1 and next in priority is given number 2 and so on.

To incorporate the polyhedral symbols for the common coordination geometries for coordination number 2 to 9 are listed in the table below:

Coordination polyhedron	C.N.	Polyherdal symbol	Coordination polyhedron	C.N.	Polyhedral symbol
Linear	2	L-2	Cube	8	CU-8
Angular	2	A-2	Square		
Trigonal plane	3	TP-3	anti-prism	8	SAPR-8
Trigonal pyramid	3	TPY-3	Triangulare face bicappled		
Tetrahedron	4	T-4	trigonal prism	8	TPRT-8
Square plane	4	SP-4	Square face bicapped		
Trigonal bipyramid	5	TBPY-5	trigonal prism	8	TPRS-8
Square pyramid	5	SPY-5	Dodecahderon	8	DD-8
Octahedron	6	OC-6	Hexagonal bipyramid	8	HBP Y-8
Trigonal prism	6	TPR-6	Trans-bicapped		
Pentagonal bipyramid	7	PBPY-7	octahedron	8	OCT-8
Face monocapped					
octahedron	7	OCF-7	Square face tricapped trigonal prism	9	TPRS-9
Face monocapped square trigonal prism	7	TPRS-7	Heptagonal bipyramid	9	HBP Y-9

Some sample exercises in nomenclature of coordination compound have been solved below to acquaint the students with the systematic way of doing them:

e.g., $[Cu(NH_3)_4]SO_4$

(*a*) First of all balance the electrostatic charges or the primary valency

(*b*) Break up charges in the coordination sphere

(*c*) Apply the rules of the IUPAC nomenclature

(*d*) Consider the name as one word with only first letter in capital

$2^+ + 0 = 2 + 2-$

(*i*) $[Cu(NH_3)_4]\ SO_4$ (Cation complex)

Tetraammine copper (II) sulphate

$4 + 2^+ + 6^- = 4-$

(*ii*) $K_4[Fe(CN)_6]$ (Anion complex)

Potassium hexacyanoferrate (II)

(*iii*) $[Pt(NH_3)_2Cl_2]$ (Neutral complex)

Diamminedichloroplatinum (II)

(*iv*) All carbonyls and nitrosyls are essentially zero valent.

e.g. (*i*) $[Fe(CO)_5]$ Pentacarbonyliron (0)

(*ii*) $[Fe(CO)_2(NO)_2]$ Dicarbonylqinitrosyliron (0)

(*v*) Bridged or Polynuclear complexes

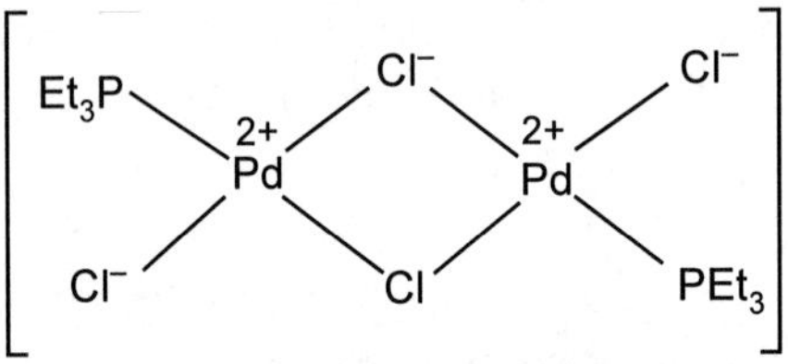

Trans-di-μ-chlorobis [chlorotriethylphosphinepalladium (II)]

Some important examples:

1.	$[Co(NH_3)_6]Cl_3$	Hexaamminecobah (III) chloride
2.	$[Co(NH_3)_5Cl]Cl_2$	Pentaamminechlorocobalt (III) chloride
3.	$K[B(NO_3)_4]$	Potassium tetranitfatoborate (III)
4.	$[CuCl_2\{O{=}C(NH_2)_2\}_2]$	Dichlorobis(urea)eopper (II)
5.	$Na_3[Ag(S_20_3)_2]$	Sodium dithiosulphatoargentate (I)
6.	$K_2[HgI_4]$	Potassium tetraiodomercurate (II)
7.	$Na[Au(CN)_2]$	Sodium dicyanoaurate (I)
8.	$K_2[Ni(CN)_4]$	Postassium tetracyanonickelate (II)
9.	$K_4[Ni(CN)_4]$	Postassium tetracyanonickelate (0)
10.	$K_2[OsCl_5N]$	Potassium pentachloronitridoosmate (VI)
11.	$Na[Pt\ Br\ Cl\ (NO_3)(NH_3)]$	Sodium amminebromochloronitro platinate (II)
12.	$[Fe(H_2O)_5Cl]Cl_2$	Pentaaquachloroiron (III) chloride
13.	$Li[AlH_4]$	Lithium tetrahydridoaluminate (III)
14.	$[Fe(CH_3NC)_6]Br_2$	Hexabis(methylisocyanide)iron (II) bromide
15.	$[Ru(HSO_3)_2(NH_3)_4]$	Tetraamminebis(hydrogen sulphito) ruthenium (II)
16.	$[Mg\ CH(CH_3)_2]Br$	Bromo isopropylmagnesium
17.	$[Pt\ Cl_2(C_2H_4)(NH_3)]$	Amminedichloro(ethene)platinum (II)
18.	$[Hg(C_5H_5)_2]$	Bis(cyclopentadienyl)mercury
19.	$[Hg(CH_3)_2]$	Dimethylmercury
20.	$[Pt\ Cl_3(C_6H_5)]$	Trichlorophenylplatinum (IV)
21.	$[Ru(NH_3)_5(N_2)]Cl_2$	Pentaamminedinitrogenruthenium (II) chloride
22.	$[Co(NH_3)_5(CO_3)]_2CO_3$	Pentaamminecarbonatocobalt (III) carbonate
23.	$K_3[Fe(CN)_5NO].2H_2O$	Potassiumpentacyanonitrosylferrate(II) dihydrate
24.	$Fe_2[Fe(CN)_6]$	Iron hexacyanoferrate (II)
25.	$Hg[Hg(SCN)_4]$	Mercury tetrathiocyanatomercurate (II)
26.	$NH_4[Cr(NH_3)_2(NCS)_4]$	Ammonium diamminetetraisothio-cyanatochromate (III)
27.	$Ph_4As[PtCl_2H(CH_3)]$	Tetraphenylarsonium dichlorohydrido methylplatinate (II)
28.	$Ba[BrF_4]_2$	Barium tetrafluorobromate (III)
29.	$Cs[ICl_3F]$	Cesium trichlorofiuroiodate (III)
30.	$[(H_3N)_5Cr\text{-}OH\text{-}Cr(NH_3)_5]Cl_5$	π-Hydroxobis[pentaamminechromium (III)] chloride

31. $[Pt(NH_3)_4][PtCl_4]$ — Tetraammineplatinum (II) tetrachloro platinate (II)

32. $[Pt(py)_4][PtCl_4]$ — Tetrakis(pyridine)platinum (II) tetrachloroplatinate (II)

33. $[Cr(NH_3)_6][CoF_6]$ — Hexamminechromium (III) hexafluoro cobaltate (III)

34. $[Cr(NH_3)_4(SCN)_2]$ $[Cr(SCN)_4(NH_3)_2]$ — Tetramminedithiocyanatochromium (III) diamminetetrathiocyanatochromate (III)

35. $K_2[PtCl_6]$ — Potassium hexachloroplatinate (IV)

36. 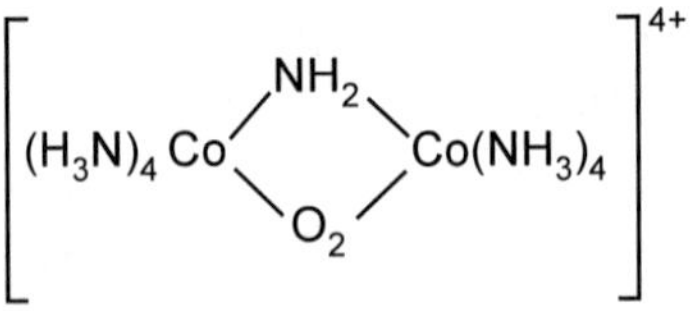
μ-Amido-p-superoxobis(tetraammine cobalt (III)) ion

37. $[HAuCl_4]$ — Tetrachloroauric acid (III)

38. 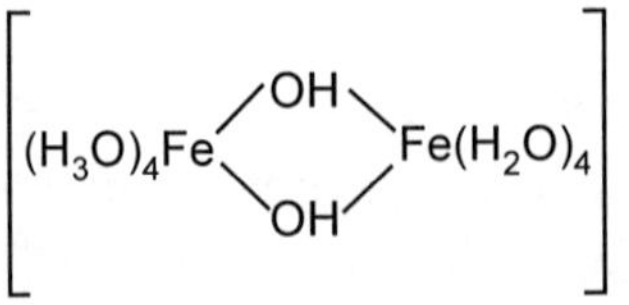
Di-μ-hydroxobis(tetraaquairon (III)) sulphate

39. 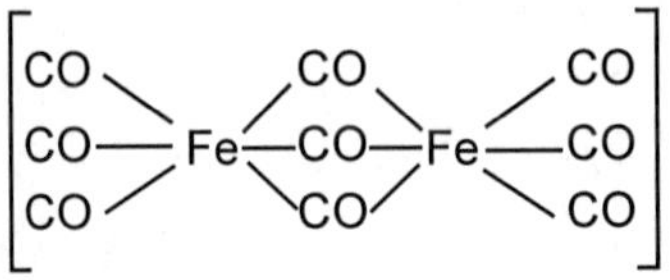
Tri-μ-carbonylbis(tricarbonyliron (0))

40. 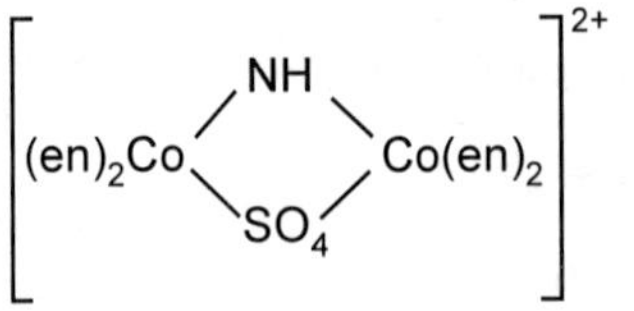
μ-Imid-μ-sulphatobis{bis(ethylenediaminecobalt (III)} ion

41. $[(CO)_5Mn\text{-}Mn(CO)_5]$ — Bis (pentacarbonylmanganese (0))

42. 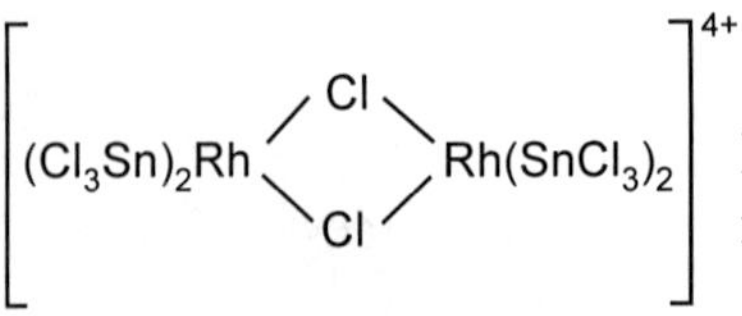
Di-p-chlorobis[bis(trichlorostannyl) rhodate (I)] ion

43. $[Be_4O(CH_3COO)_6]$ — Hexa-|Li-acetato(o,6)-p_4-oxotetraberrylium (II)

44. $K[CrF_4O]$ — Potassium tetrafluorooxochromate (v)

45. $[Be(CH_3COCHCOC_6H_5)_2]$ — Bis (benzoylacetonato)beryllium (II)

Examples of naming the cluster compounds.

$OS_3(CO)_{12}$	Dodecarbonyl-triangulo-tri osimium (3 Os—Os)
	or
	cyclotris (tricarbonylosmum (3 Os—Os)
$Cg_3[Re_3Cl_{12}]$	Caesium dodecachloro-triagalo-trirhenate $(3Re\text{-}Re)^{3-}$
	or
	Caesium cyclotris (tetsa chlororhenate) 3Re-Re (3–)
$[Nb_6Cl_{12}]^{3+}$	Dodeca-μ_3-chloro-octahedra hexaniobium (Z^+) ion
$[Mo_6Cl_6]^{4+}$	Octa-μ_3-chloro octahdra-hexamolybdenum 4^+ ion
$[Cu4\ I_4(PEt_3)_4]$	Tetra-μ_3-iodo-tetrakis (triethyl phosphine)-Tetrahedro-tetra copper
B_4Cl_6	Tetra-chloro-tetrahedro-tetraboron (6B-B)
	or
	Tetrahedro-tetrakis (choro boron)

Naming the Inner Complexes

Thi occurs in the case of ligands which possess salt forming groups and neutral donor groups through chelate ring structures satisfying both primary and secondary valencies in a complex giving an overall complex with no charge.

These inner complexes have definite colours. The non-ionic character give rise to high volatility, low water solubility and high solubility in organic solvents, *e.g.*

(*a*) 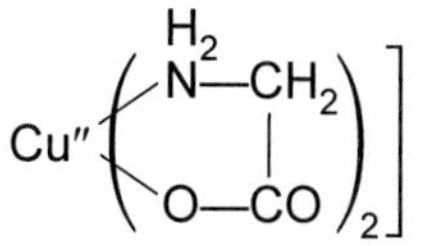

Bis (glycinato) copper (II)

(*b*) 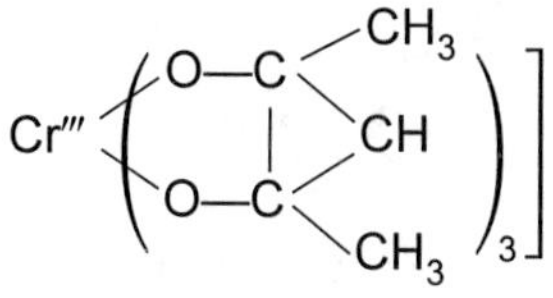

Tris (acetylacetanato) chromium (III)

(*c*) 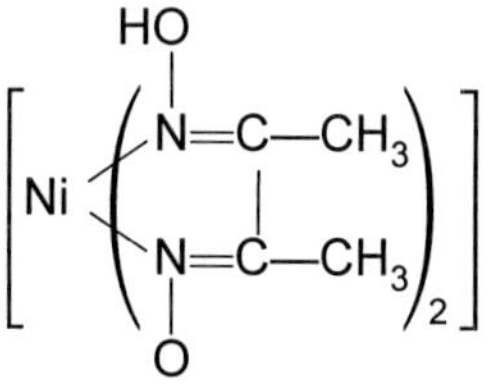

Bis (dimethyl glyoximato) nickel (II)

Some ligands form complexes which bear an overall charge include examples of Co(II) and B (III) complexes of acetyl acetanato anion where the number of acidic groups coordination is not equal to the primary valency of the central metal atom.

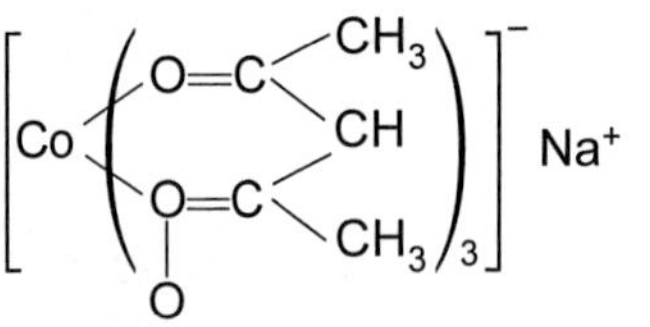

Sodium tris (acetyl acetonato) cobalt (II)

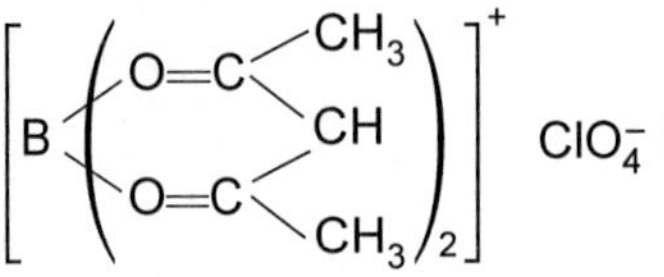

Bis (acetyl acetonato) boron (III) perchlorate

SUMMARY

Chapter begins with the rules of IUPAC names of the complexes. There are two set of rules but one in which Roman numerals are used is more prevalent. They are named like common compounds where positive ion is named first and negative later. Complex comprises of metal ion and a ligand. Ligand is named first and metal ion later. In all probabilities the alphabetical order of writing the names of the ligand is maintained in case of more than one ligand. Roman numerals indicate the oxidation states of the metals. In polydentate ligands numerical prefexes are used. The name of the bonding atom of the ligand is written in case of open-ended ligands like (← SCN) and (← NCS). In bridged or polynuclear complexes, symbol 'μ' is used and name of the bridging ligands is written first of all and the remaining complex later. The bridging groups may be one, two or three. If the complex unit on each side of the bridge is same then prefix *bis* is used. If the complex units may be different then they are named separately.

EXERCISE

1. What is the difference between oxidation state and oxidation number? Give examples.

2. Enumerate the oxidation number of the metal in each of these metal complexes:

(*i*) $[Co(NH_3)_6]Cl_3$ (*ii*) $[CoSO_4(NH_3)_4]NO_3$

(*iii*) $[Cd(SCN)_4]^{2-}$ (*iv*) $[Fe(CN)_6]^{4-}$

(*v*) $[OsCl_5N]^{2-}$ (*vii*) $[AlH_4]^-$

(*vii*) $[CuCl_2(CH_3NH_2)_2]$ (*viii*) $[Ag(S_2O_3)_2]^{3-}$

(*ix*) $[Mn_2(CO)_{10}]$ (*x*) $Na_2[Fe(CN)_5NO]$ ...... iron metal

3. What are chelated and non-chelated complexes? Why is the former more stable?

4. Write the correct IUPAC nomenclature:

(*i*) $[Pt(NH_3)_2Cl_2]$ (*ii*) $[Co(NH_3)_5NO_2](NO_3)_2$

(*iii*) $[Co(en)_2Cl_2]Cl$ (*iv*) $Na_2[Ni(edta)]$

(*v*) $K_4[Ni(CN)_4]$ (*vi*) $NH_4[Cr(NH_3)_2(NCS)_4]$

(*vii*) 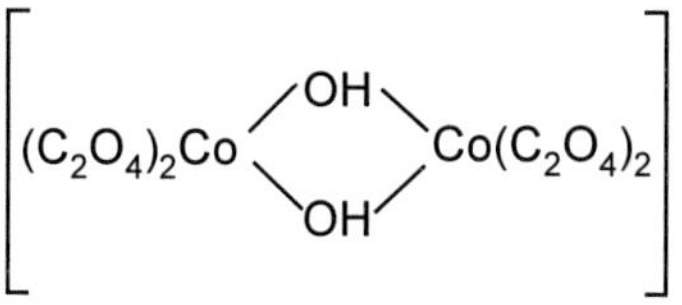

(*viii*) 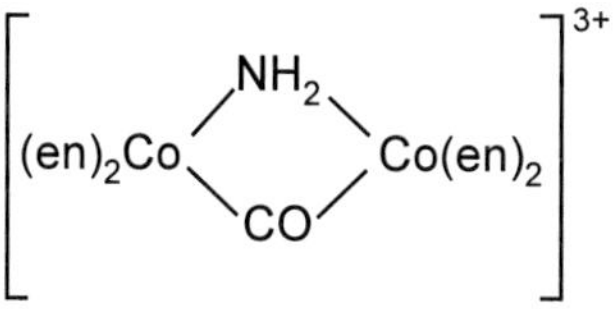

(*ix*) 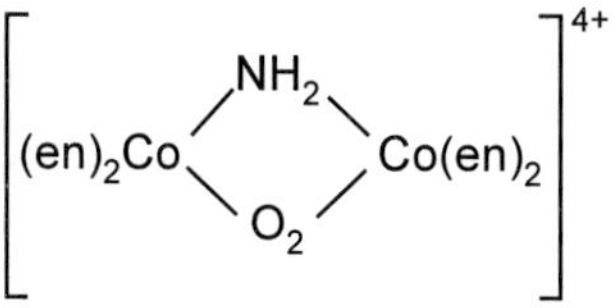

(*x*) $[Cr(gly)_3]$

5. Write correct formula for each of the following:

(*i*) Hexaamminecobalt (III) chloride

(*ii*) Potassium iron (III) hexacyanoferrate (II)

(*iii*) Sodium bis(thiosulphato)argentate (I)

(*iv*) Nickel hexacyanoferrate (IV)

(*v*) Tetraammineplatinum (II) amminetrichloroplatinate (II)

(*vi*) Bis(cyclopentadienyl)iron (II)

(*vii*) Potassium tetraoxomanganate (VII)

(*viii*) Potassium trioxalatoaluminate (III)

(*ix*) Tetrapyridineplatinum (II) tetrachloroplatinate (II)

(*x*) Diamminebis(ethylenediamine)cobalt (III) chloride

(*xi*) Tetra thiocyanato-N-zinc (II)

6. Give one example each of tridentate, tetradentate and poly dentate ligand.

■ ■ ■

5

Isomerism and Stereochemistry

5.1 ISOMERISM

Isomerism is a phenomenon in which a compound has the same molecular formula but their ligands are attached to the central metal atom in different ways.

Isomerism is of two types:

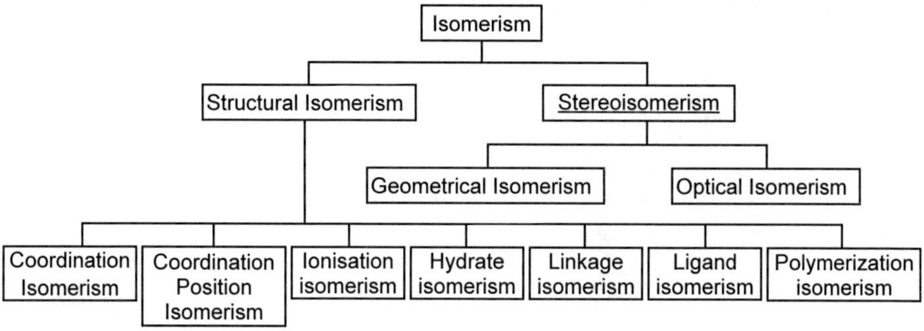

Structural Isomerism: It arises out of the difference in the structures of the complex.

Stereoisomerism: These isomers have the same position of atoms or groups but they differ in their spatial arrangements around the central metal atom in a particular geometry.

5.2 STRUCTURAL ISOMERISM

Coordination Isomerism

The ligands in the cationic and anionic pairs substitute either partially or completely. For example,

(*i*) $[Co(NH_3)_6][Cr(C_2O_4)_3]$ and $[Cr(NH_3)_6][Co(C_2O_4)_3]$
This occurs in case the two metal ions are of nearly the same size and ligands of comparable strength.

(*ii*) If one of the ligands is monodentate while the other is bidentate then the exchange of ligands with respect to metal ions may be complete as well as partial. For example,

$$[Rh(en)_3][IrCl_6];\ [Ir(en)_3][RhCl_6] \text{ and } [Rh(en)_2Cl_2][Ir(en)Cl_4]$$

In 6 coordinate complexes the partial exchange of ligands is in the ratio of 4:2 or 2:4, as all the ligands in the same plane are substituted simultaneously.

(*iii*) If the cationic and anionic species have the same metal ion then only partial exchange of ligand w.r.t. metal ion is possible.

$$[Cr(NH_3)_6][Cr(NCS)_6] \text{ and } [Cr(NH_3)_4(NCS)_2][Cr(NH_3)_2(NCS)_4]$$

(*iv*) The partial exchange of ligands is also cited in case the cationic and anionic species have the same metal but in different oxidation states. For example,

$$[Pt(II)(NH_3)_4][Pt(IV)Cl_6] \text{ and } [Pt(IV)(NH_3)_4Cl_2][Pt(II)Cl_4]$$

Coordination Position Isomerism

This is a special type of coordination isomerism which involves placement of ligands in different ways in a bridged complex.

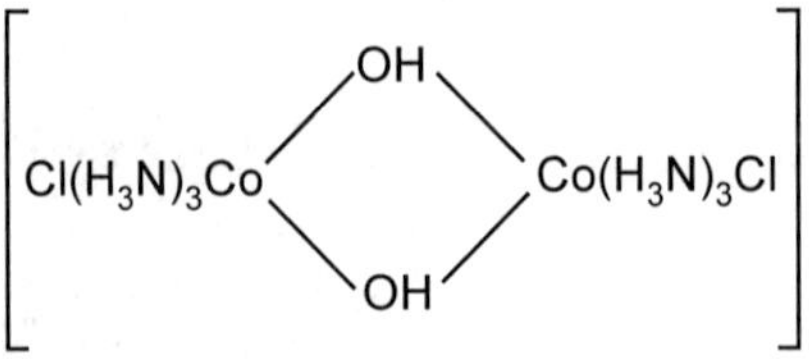

Symmetrical

and

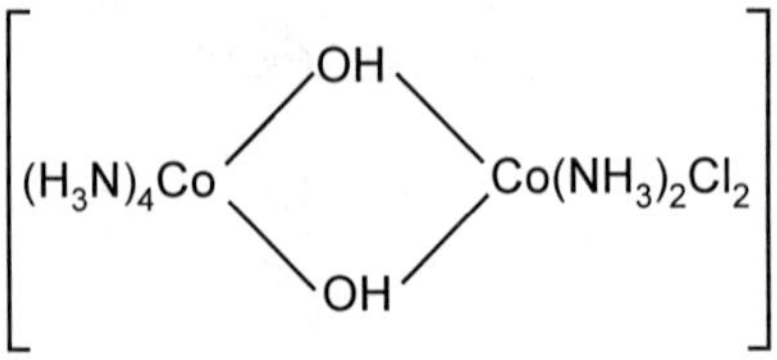

Unsymmetrical

Ionisation Isomerism

The complexes which have the same molecular formula but differ in their ionisation in solution represent ionisation isomerism. For example,

$$A\text{—}[Co(NH_3)_5Br]SO_4\text{—Purple} \longrightarrow [Co(NH_3)_5Br]^{2+} + SO_4^{2-}$$

$$B\text{—}[Co(NH_3)_5SO_4]Br\text{—Red} \longrightarrow [Co(NH_3)_5(SO_4)]^{+} + Br^{-}$$

Their existence and identification can be done by precipitating SO_4 ion as $BaSO_4$ on addition of $BaCl_2$ solution and likewise precipitating Br^- ion as AgBr (yellowish precipitate) with the help of $AgNO_3$ in presence of dilute HNO_3.

Other examples include:

(*a*) $[Pt(NH_3)_4Cl_2]Br_2$ and $[Pt(NH_3)_4Br_2]Cl_2$

(*b*) $[Co(en)_2\ NO_2Cl]SCN$, $[Co(en)_2NO_2SCN]Cl$ and $[Co(en)_2ClSCN]NO_2$

Compound

$$A \longrightarrow [Co(NH_3)_5Br]SO_4 + \underset{\text{Solution}}{BaCl_2} \longrightarrow \underset{\text{White}}{BaSO_4 i} + [Co(NH_3)_5Br]Cl_2$$

$$B \longrightarrow [Co(NH_3)_5(SO_4)]Br + \underset{\text{Solution}}{AgNO_3} \xrightarrow[HNO_3]{\text{dilute}} \underset{\text{Yellowish white precipitate}}{AgBr} + [Co(NH_3)5(SO_4)]NO_3$$

In these isomers, coordination number remains the same. Determination of conductance can also come handy in their identification.

Hydrate Isomerism

This is another kind of ionisation isomerism. In these, H_2O molecules act as ligands and can be easily substituted by the anions that are outside the coordination sphere. Accordingly, $CrCl_3.6H_2O$ exists in four different isomers:

Compound

The coordination number six is maintained, as expected, in all these isomers. The identification of these isomers is inferred from the weight of AgCl which is found to be in the ratio of 3 : 2 : 1 : 0 by moles.

Another example includes $[Co(en)_2(H_20)Cl]Cl_2$ and $[Co(en)_2Cl_2]Cl(H_20)$

$$A\text{—}\underset{\text{Violet}}{[Cr(H_2O)_6]Cl_3} + 3AgNO_s \longrightarrow \underset{\text{White}}{3AgCl i} + [Cr(H_2O)_6](NO_3)_3$$

$$B\text{—}\underset{\text{Bluish-green}}{[Cr(H_2O)_5Cl]Cl_2(H_2O)} + 2AgNO_3 \longrightarrow \underset{\text{White}}{2AgCl} + [Cr(H_2O)_5Cl](NO_3)_2$$

$$C\text{—}\underset{\text{Green}}{[Cr(H_2O)_4Cl_2]\ Cl(H_2O)_2} + AgNO_3 \longrightarrow \underset{\text{White}}{AgCl} + [Cr(H_2O)_4Cl_2]NO_3$$

$$D\text{—}[Cr(H_2O)_3Cl_3](H_2O)_3 + AgNO_3 \longrightarrow \text{No AgCl precipitate}$$

Linkage Isomerism

This type of isomerism is found in monodentate ambidentate ligands *i.e.,* the ligands in which two electronegative atoms have equal chance to act as donors to the central metal ion but at any time one of the two atoms acts as donor.

Linkage isomers

(*i*)	$M^+ \leftarrow \overline{S}CN$ and $M \leftarrow \overline{N}CS$	$[Pd(C_6H_5)_2(SCN)_2]$ and $[Pd(C_6H_5)_2(NCS)_2]$
(*ii*)	$M^+ \leftarrow \overline{N}O_2$ and $M^+ \leftarrow \overline{O}NO$	$[Co(NH_3)_5NO_2]^{2+}$ ion and $[Co(NH_3)_5(ONO)]^{2+}$
(*iii*)	$M^+ \leftarrow \overline{S}_2O_3$ and $M^+ \leftarrow \overline{O}SO_2S$	$[Co(NH_3)_5SSO_3]^+$ ion and

Ligand Isomerism

Some of the ligands especially organic molecules are themselves capable of existing as isomers. In the complexes formed by these ligands metal ion has no active role, it is only ligand molecules that decide the isomers:

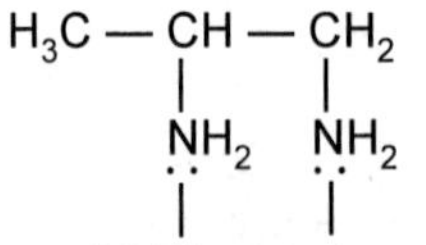

1,2 Diaminopropane
(pn)

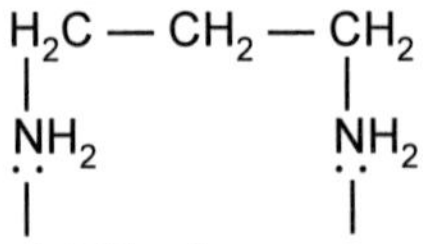

1,3 Diaminopropane
(tn)

They are neutral bidentate ligands.

Example: $[Co(pn)_2Cl_2]^+$ and $[Co(tn)_2CI_2]^+$ are ligand isomers.

Polymerisation Isomerism

Some complex compounds whose formulae appear to be a polymer of a simple complex compound. They appear to have the same ratio of different metal ions and ligands in them. They seem to be dimer and pentamer of (I) complex at given in the example:

(*i*) $[Co(NH_3)_3(NO_2)_3]$

(*ii*) $[Co(NH_3)_6][Co(NO_2)_6]$

Complex (*ii*) appears to be a dimer of (*i*) complex

The ratio of Co^{3+}:NH_3:NO_2 in both these is 1 : 3 : 3.

(*iii*) $[Co(NH_3)_4(NO_2)_2][Co(NH_3)_2(NO_2)_4]$

(*iv*) $[Co(NH_3)_5(NO_2)]_3[Co(NO_2)_6]_2$

Isomer (*iv*) has the metal to ligand ratio Co^{3+}: NH_3: NO_2 5 : 15 : 15 which also simplifies to 1 : 3 : 3.

Other Uncommon Isomerism

(i) Configurational Isomers

There are four coordination isomers which can occur in both square planar and tetrahedral geometries *e.g.,* nickel (II) complexes of the type Ni(II) $\{PPh_2(Et)\}_2Br_2$.

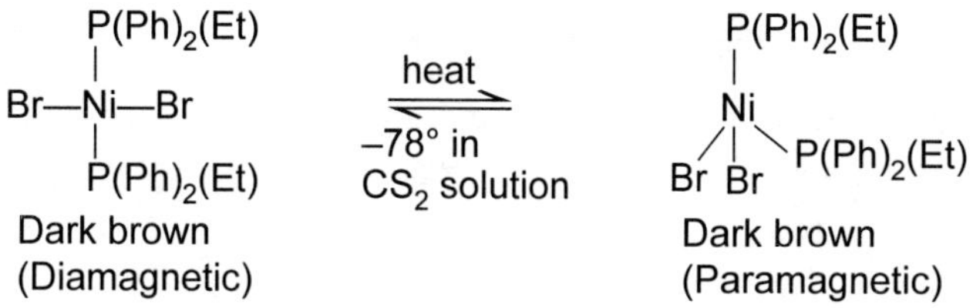

These isomers are called polytopal isomers or polytopism.

Spin Isomerism

In the cases where equilibrium exists between both high spin and low spin complexes the paramagnetic moment is shown to depend upon the temperature.

Summation Isomerism

Chan in 1967 showed the synthesis of the following pair:

$$\text{Cis–}[\text{Co en}_2(\text{H}_2\text{N—CH}_2\text{—CH}_2\text{—Br})\text{Cl}](\text{NO}_3)_2$$

$$\text{Cis–}[\text{Co (en)}_2(\text{H}_2\text{N—CH}_2\text{—CH}_2\text{—Cl})\text{Br}](\text{NO}_3)_2$$

In this the sum total formula of the isomer remains the same.

Valence Isomerism

In 1955 Jonassen gave the following example of the complex compounds as valence isomers.

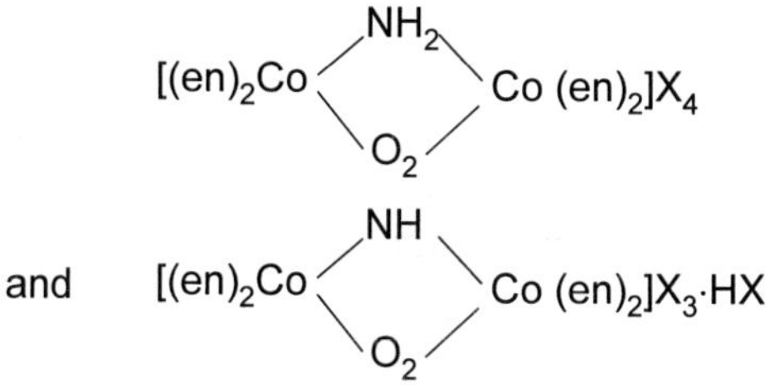

In these isomers the charge on the complex cation is different although both contain Co^{3+} with O_2^- as ligand.

Distortional Isomerism

The term distortional isomers was first proposed by Chatt in 1971 in relation to these two isomers:

$$[\text{MoOCl}_2(\text{PMe}_2(\text{Ph})_3]$$

Here, Me—O had different IR values. X-ray crystallography showed that the cis-mer configuration but differed in the orientation of phosphine ligands and Me—O bond length.

5.3 STEREOISOMERISM

Stereoisomers are those which have same ligands coordinated to the same metal ion **but have** different arrangement of ligands in space. They are also called space isomers.

Stereoisomerism is of two types:

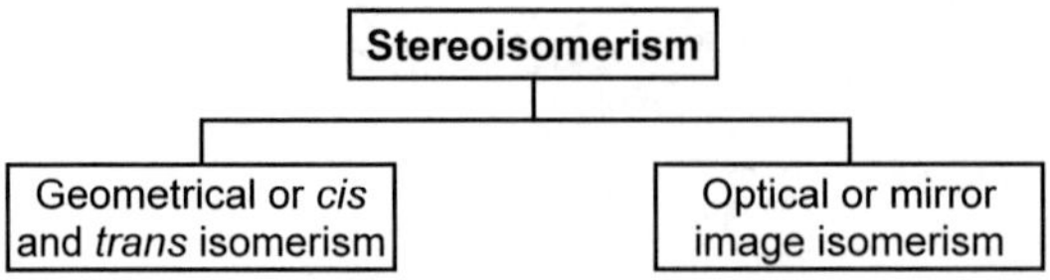

A large majority of the complex compounds known so far are 4-coordinate and 6-coordinate. Therefore, it is pertinent to discuss their stereochemistry in detail.

Moreover, 2- and 3-coordination complexes are rare and show no stereochemistry. The ligands in them are equidistant from each other.

4-Coordinate Complexes

These complexes can acquire two types of geometries (*i*) Square planar and (*ii*) Tetrahedral.

Square planar structure have ligands in the same plane while tetrahedral is a **spatial** structure.

Geometrical Isomerism

In geometrical isomerism, we look for the distance between two identical ligands. If the distance between the two identical ligands is less or they are adjacent it is a *cis*-isomer and it the distance between them is more or they are opposite to one another then it is a trans-isomer.

Optical Isomerism

In optically active isomers we plot a mirror image of the complex compound in a given geometry. If the mirror image is non-superimposable then the complex is optically active. The plane of polarised light can be rotated if the geometry is spatial.

Generalising the above facts we came across the idea that geometrical isomerism exists in a planar geometry and optical activity can exist in a spatial geometry. With the focus on 4-coordinate structures it can be concluded that square planar geometry exhibits geometrical isomerism and a tetrahedral will give only optical isomerism.

In a 6-coordinate complex the most probable geometry is octahedral. It has 4 ligands in a square plane and 2 ligands in the nodal plane. Therefore, it is both in part planar and spatial. Hence, geometrical and optical isomerism do exist in this structure.

Geometrical Isomerism in 4-Coordinate Square Planar Geometry having Monodentate Ligands

Complex

1. **|MA_2BC|:** These type of complexes show geometrical or *cis-trans* isomerism. Example:

   $[PdCl_2BrI]^{2-}$

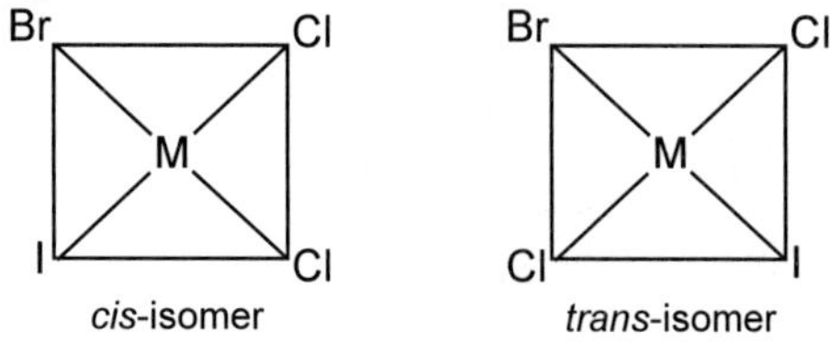

2. **|MA_2B_2|:** These complexes also show geometrical isomerism. In this *cis-trans* isomerism may be drawn w.r.t. ligand A or B the result is the same *i.e.,* if A is *cis* then B has to be *cis* and if A is *trans,* B has to be *trans* in a square planar geometry.

   e.g. $[Pt(NH_3)_2Cl_2]$

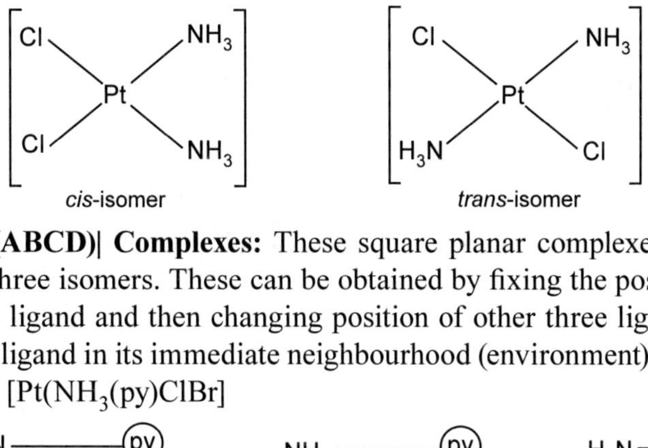

3. **|M(ABCD)| Complexes:** These square planar complexes may exist in three isomers. These can be obtained by fixing the position of any one ligand and then changing position of other three ligands so that the ligand in its immediate neighbourhood (environment) is different, *e.g.* $[Pt(NH_3(py)ClBr]$

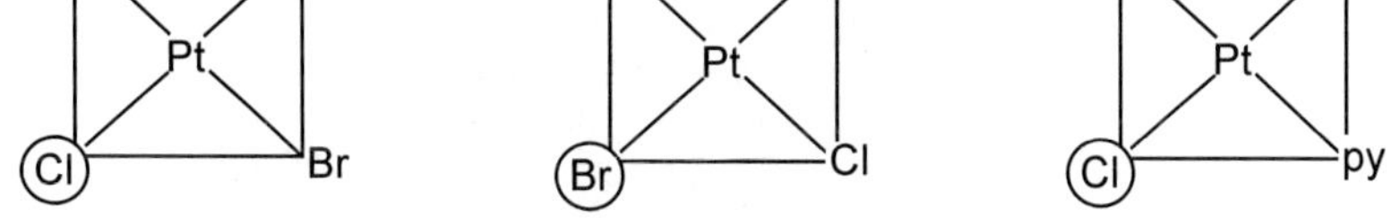

   Three isomers of the complex are possible. These isomers have very little difference in their properties and hence their separation becomes difficult. These isomers exist because of NH_3 ligand being flanked by environmentally different ligands in its neighbourhood.

4. **$[MA_4]$, $[MA_3B]$ and $[MAB_3]$ Complexes:** These square planar complexes do not show geometrical isomerism since the arrangement of four ligands round the central metal ion remains the same.

5. **$[M(AB)_2]$ Complexes:** In these complexes, M stands for the central metal ion and A and B are two different donor atoms of a coordination

bidentate ligand, *e.g.,* in $[Pt(gly)_2]$ gly represents glycinato ligand $H_2NCH_2COO^-$ which has 'N' and 'O' atoms acting as donors. This square planar complex exhibits geometrical (*cis-trans*) isomerism.

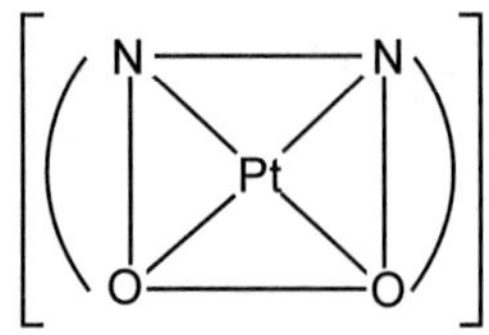

Cis-isomer

Two like 'N' or 'O' atoms have minimum distance between them.

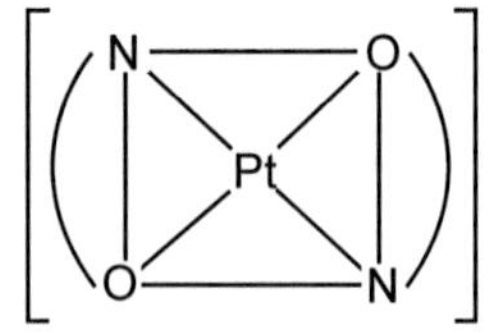

Trans-isomer

Two like 'N' or 'O' atoms have maximum distance between them.

4-Coordinate complexes having monodentate ligands in tetrahedral geometry

In this geometry, the distance between any two identical ligands does not alter the position of the ligands hence no geometrical isomerism is possible.

However, if the central metal ion is attached to unsymmetrical bidentate ligands they do show optical activity.

e.g., $[Be(C_6H_5COCHCOCH_3)_2]$

Non-superimposable Mirror Image

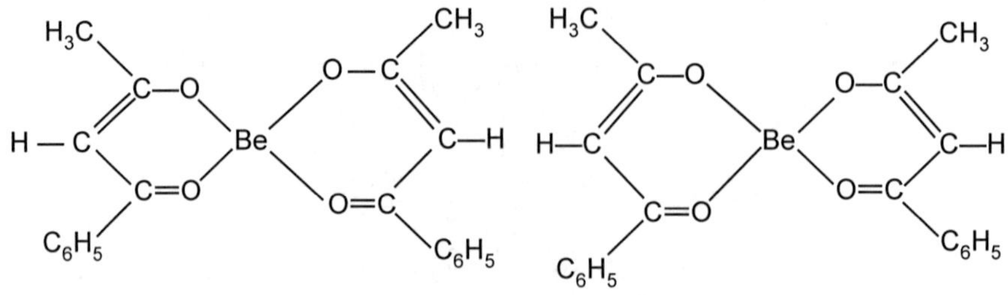

The first example of four different ligands attached to the central metal atom in tetrahedral structure was reported in 1963 but it could not be resolved in optically active isomers. In $[Be(Bzac)_2]$ and similar complexes of B(III) andZn(II), the bidentate ligands are unsymmetrical which give optically active isomers that can be easily resolved. Therefore, four different ligands around the central metal atom are not a must for optical activity. The unsymmetrical bidentate ligands go a long way to explain optical activity.

Optical Isomerism in 4-Coordinate Square Planar Complexes

Square planar complexes seldom show optical isomerism. Some rare cases in which it is exhibited are as follows:

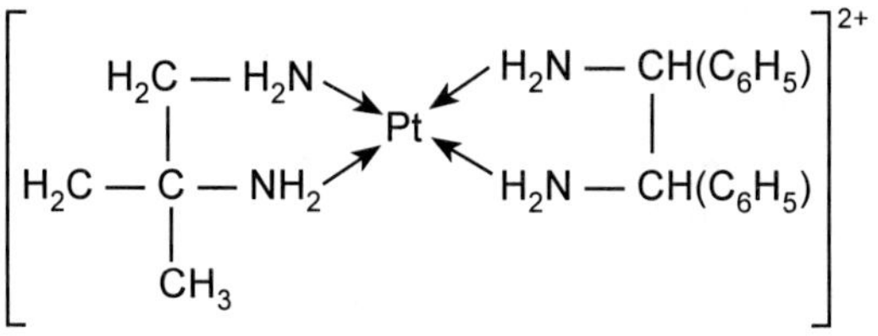

Isobutylenediaminemeso-stilbenediamineplatinum (II) ion.

Optical Isomer in 4-Coordinate Tetrahedral Complex having all the Four Different Ligands

For example, $[As(CH_3)(C_2H_5)(S)(C_6H_4COO)]^{2+}$ ion

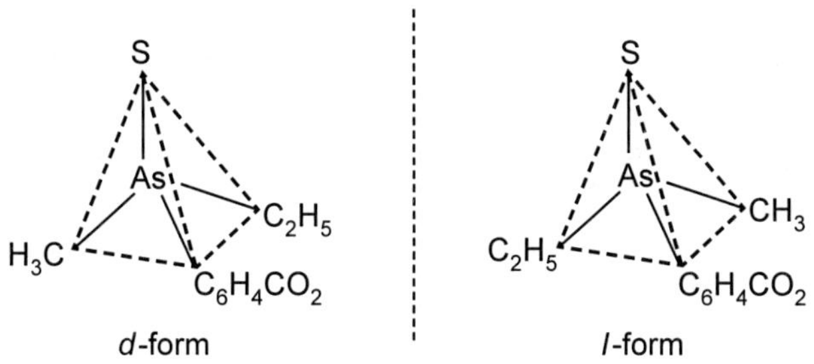

These examples are very few and rare.

6-Coordinate Complexes

It is by far the most common coordination number in complexes. The common and symmetrical structures imbibing 6-coordinate complexes are octahedral, square bipyramid, hexagon or trigonal prism.

Out of these geometries, 6-coordinate complexes almost exclusively acquire the most symmetrical octahedral geometry. The non-octahedral geometries are conventionally disposed off.

Octahedral geometry has a square base in one plane and two apeces in the plane perpendicular to the plane of the square. Thus, it can have ligands in the square base plane as well as in the apex plane. Hence, it can show both the geometrical and optical activity. This geometry exhibits 6 metal-ligand bonds and eight faces as in diagram:

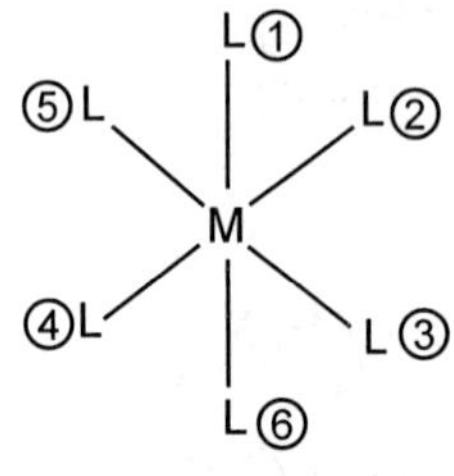

6 metal-ligand bond

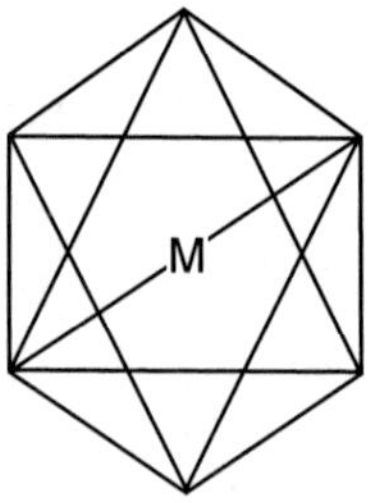

8 faces
four above the square
base and four below

It is a geometry of perfect symmetry with each bond angle of90° and same metal-ligand bond length. Due to unsymmetrical ligands it may suffer two forms of distortion of octahedral complexes. The first is tetragonal distortion due to elongated compression and the second possibility results in trigonal antiprism.

6-Coordinated Octahedral Complexes (Geometrical Isomerism)

[A]. Monodentate Ligands

1. **|MA_6] and [MA_5B]:** These complexes do not show any geometrical isomerism in octahedral complexes.
2. **[MA_4B_2] Complexes:** These exist in *cis*- and *trans*-isomers. Important example is [$Co(NH_3)_4Cl_2$]. The geometrical isomers are drawn as shown in which 2Cl ions occupy 1,2 or 1,6 positions.

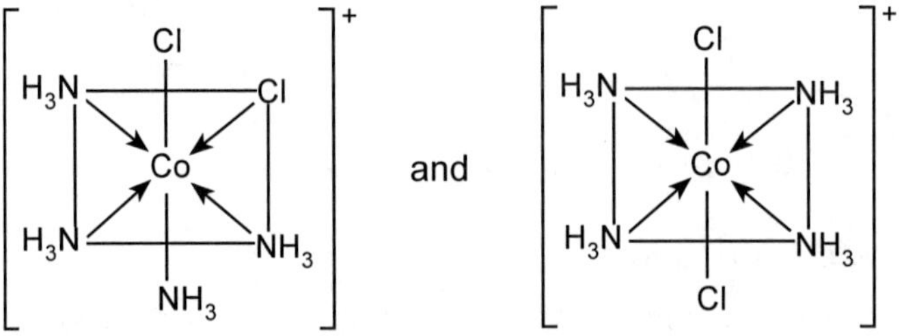

Cis-isomer *Trans*-isomer

Cis-isomer Trans- isomer

3. [MA_4BC] Complexes: Actual example of a complex is [$Co(NH_3)_4(H_2O)$ $Cl]^{2+}$. The *cis* and *trans* isomers are drawn as in the diagram.

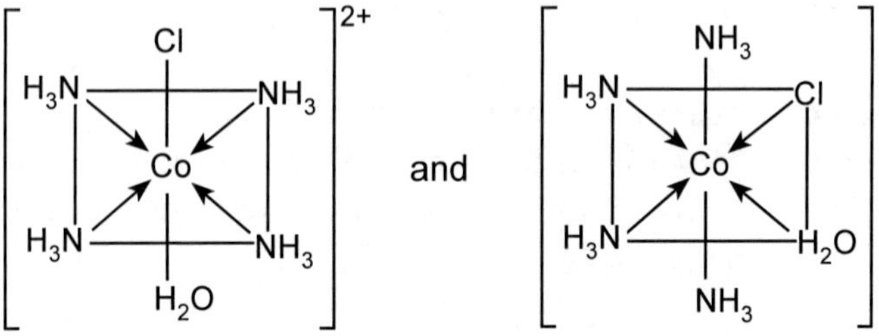

Cis-isomer *Trans*-isomer

The *cis* is qualified by less distance between any two NH_3 ligands afl *trans* by increase in distance between the identical NH_3 ligands.

4. **[MA_3B_3] Complexes:** [$Co(NH_3)_3Cl_3$], [$Cr(H_2O)_3F_3$], [$Cr(NH_3)_3Cl_3$ etc. Each of these complexes exists in *cis*- and trans-isomers. An exampel [$Co(NH_3)_3Cl_3$] has been illustrated.

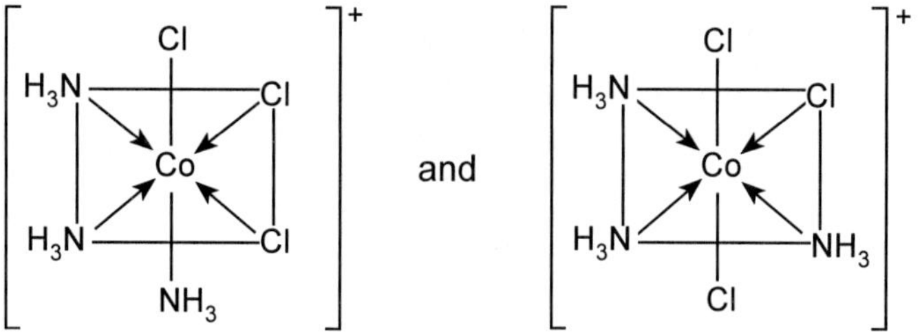

1,2,3-isomer
cis-isomer
(facial)

1,2,6-isomer
trans-isomer
(meridianal)

5. **[$MA_2B_2C_2$]:** [$Pt(NH_3)_2(py)_2Cl_2]^{2+}$ Ion: This ion may theoretically give five geometrical isomers but only three isomers have been isolated because of similar properties. The isomers are shown as:

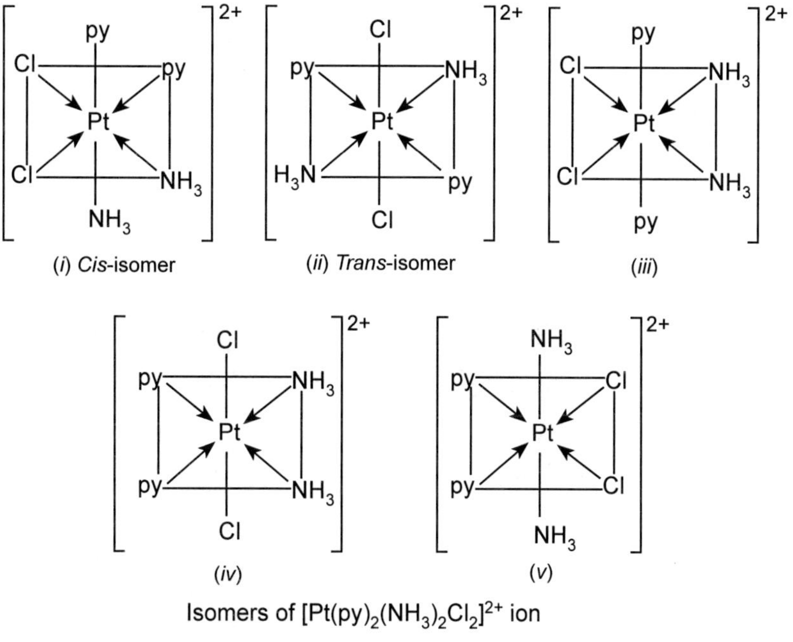

Isomers of [$Pt(py)_2(NH_3)_2Cl_2]^{2+}$ ion

6. [MABCDEF]: This octahedral complex having six different monodentate ligands has the only example [$Pt(py)(NH_3)(NO_2)(Cl)$ (Br)(I)].

Theoretically 15 geometrical isomers are possible but only three are isolated. They exist due to environmental differences.

[B]. Bidentate Ligands

There are two types of bidentate ligands;

(*i*) Bidentate ligands with similar donor atoms.

(*ii*) Bidentate ligands with dissimilar donor atoms.

1. **Identical Coordinating Atoms:** $[M(AA)_3]^{3+}$ A and A are identical donor atoms. The actual complex is $[Cr(en)_3]^{3+}$. This complex ion does not show geometrical isomerism but optical isomerism does exist, *e.g.*

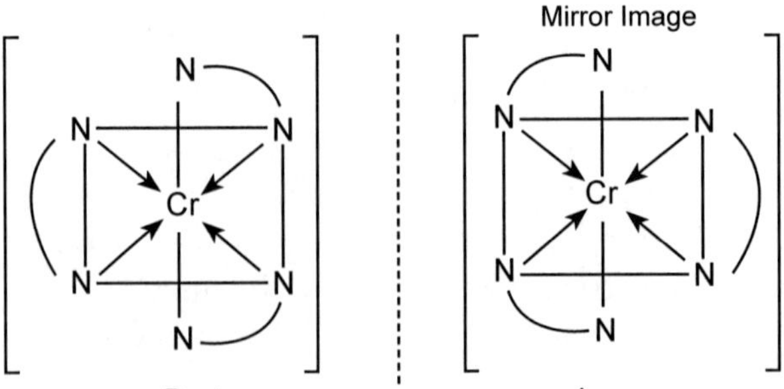

Dextro Laevo

2. **Bidentate Ligand with Non-identical Coordinating Atoms: The** real complex in existence is $[Cr(gly)_3]$ and the general formula representing it is $[M(A\text{-}B)_3]^{3+}$. This complex exists in *cis-* and *tram-* forms and both *cis-* and *tram-* are optically active.

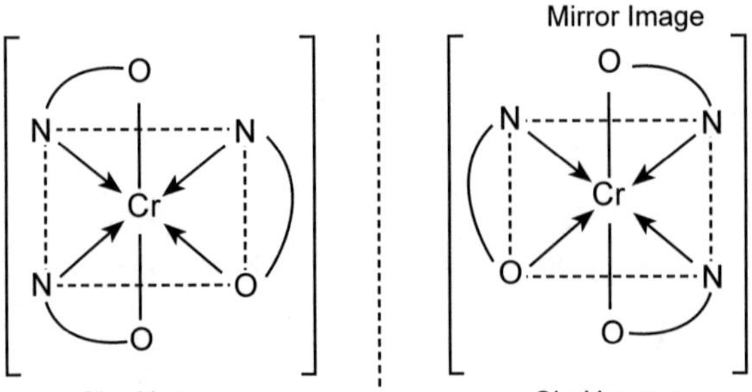

Cis-d-isomer *Cis-l*-isomer

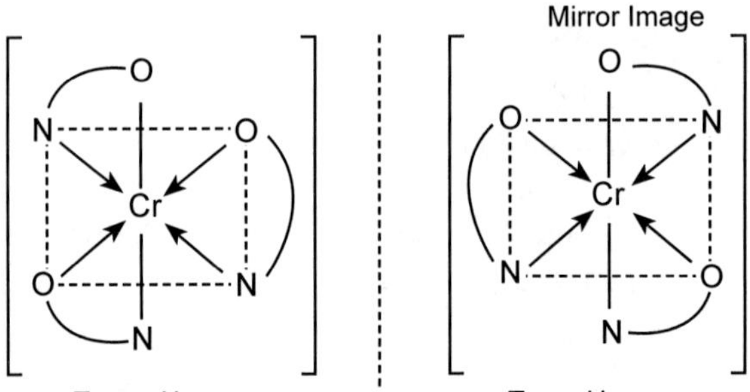

Trans-d-isomer *Trans-l*-isomer

[C]. 6-Coordinate Octahedral Complexes Containing Monodentate and Bidentate Ligands

1. **[M(A-A)$_2$B$_2$]:** The example considered here is $[Co(en)_2Cl_2]^+$ ion. The c/s-isomer is supposedly optically active as it does not show the centre of symmetry while *tram*-isomer is optically inactive due to the plane of symmetry.

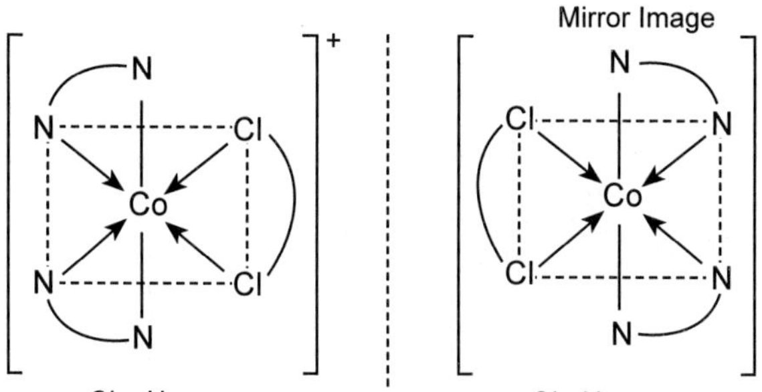

Cis-d-isomer *Cis-l*-isomer

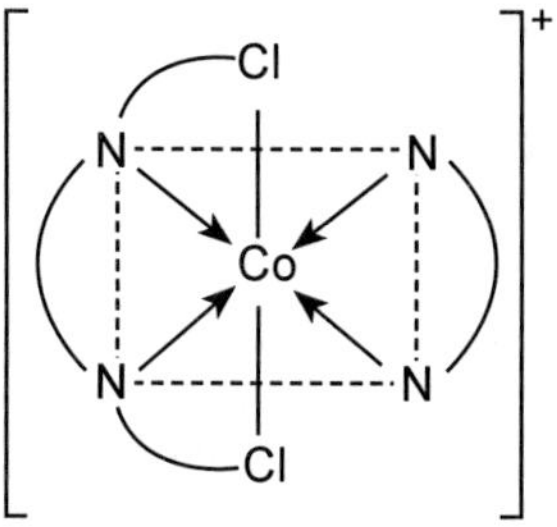

Trans-form
(gymmetrical) optically inactive

2. **[M(AA)$_2$BC] Complex:** The actual complex has the formula $[Co(en)_2(NH_3)Cl]^{2+}$ ion. It shows geometrical isomers. Due to absence of plane of symmetry c/s-isomer is optically active and trans-isomer is optically inactive due to symmetrical disposition.

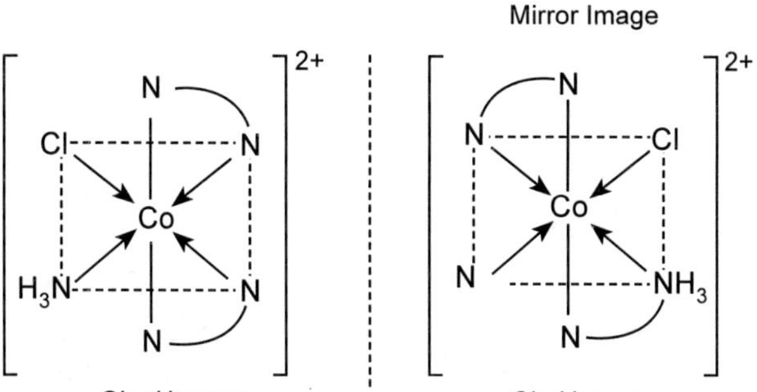

Cis-d-isomer *Cis-l*-isomer

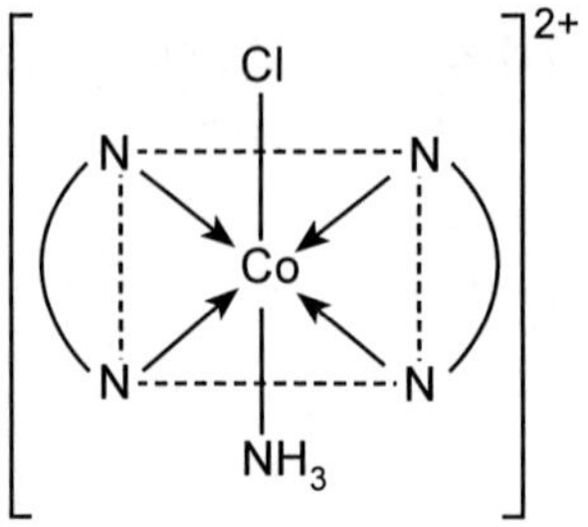

Trans-meso-isomer
symmetrical

3. **[M(AA)B_2C_2] Complex:** The actual ion is $[Co(en)(NH_3)2Cl_2]^+$. It exhibits geometrical isomerism in which *cis*-isomer is optically active and *trans*-isomer is optically inactive.

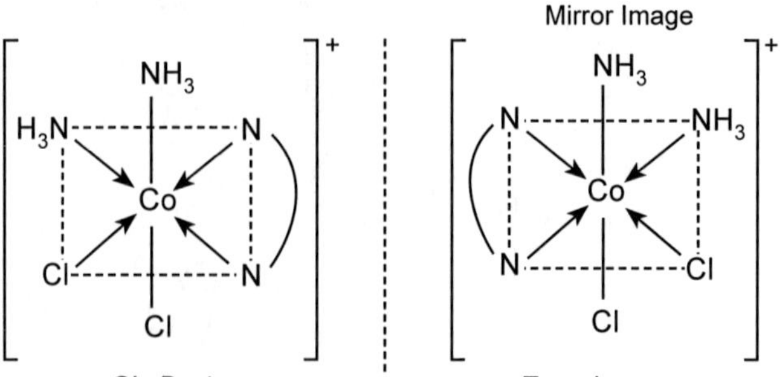

Cis-Dextro *Trans*-laevo

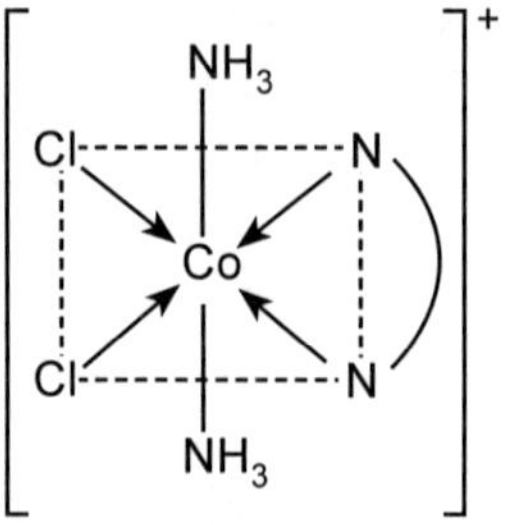

Trans-meso-isomer
symmetrical

[D]. Octahedral Complexes with Unsymmetrical Optically Active Bidentate Ligands.

The ligands in question are:

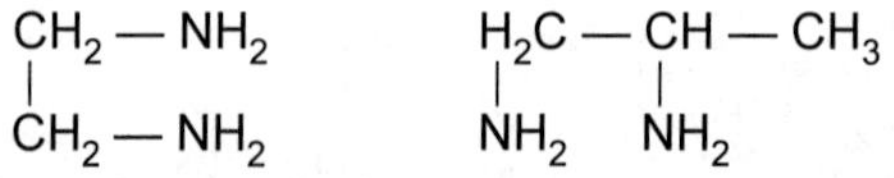

Ethylene diamine (en) Propylenediamine (pn)

The ligand (en) is symmetrical and optically inactive while the ligand (pn) is unsymmetrical and optically active. The complex $[Co(en)(pn)(NO_2)_2]^+$ ion exists in 4-geometrical isomers. The two c/s-isomers are unsymmetrical and hence optically active.

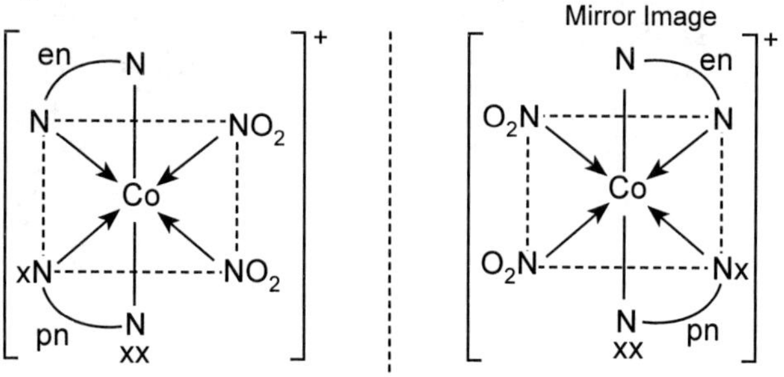

Cis-d-isomer *Cis-l*-isomer

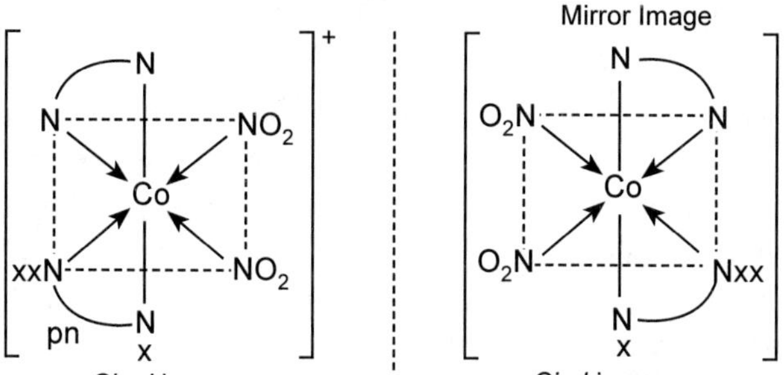

Cis-d-isomer *Cis-l*-isomer

d-and *l*-optically active isomers of other *cis*-isomer (Refer X and XX)

[E]. Octahedral Complexes Containing Polydentate Ligands

They include complexes containing edta hexadentate ligand. This ligand has 4-donor oxygen atoms and 2-donor nitrogen atoms. The formula of the ligand suggests 6-donor sites and four negative charges. The actual examples are [Co) edta)]~ ion. Since the ligand comprises of only one molecule or ion therefore geometrical isomerism is ruled out (no two identical ligands). However, it does show optical isomerism.

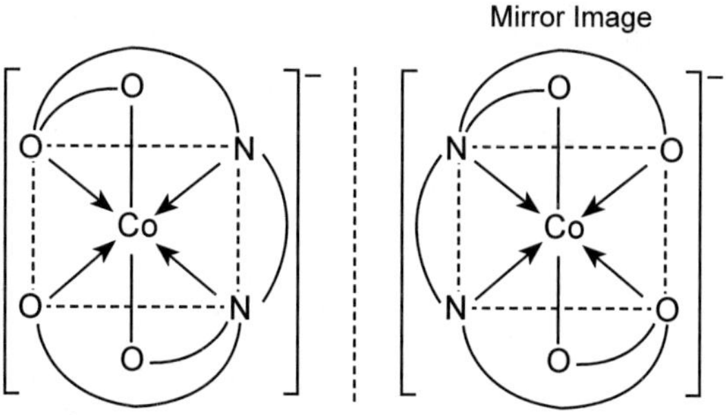

Dextro-form Laevo-form

[F]. Bridged Binuclear Octahedral Complex

The type of complex ion is formulated as $[Co(en)_4(NH_2)(NO_2)]^{4+}$. Here *cis*-form is optically active while *tram*-form is optically inactive.

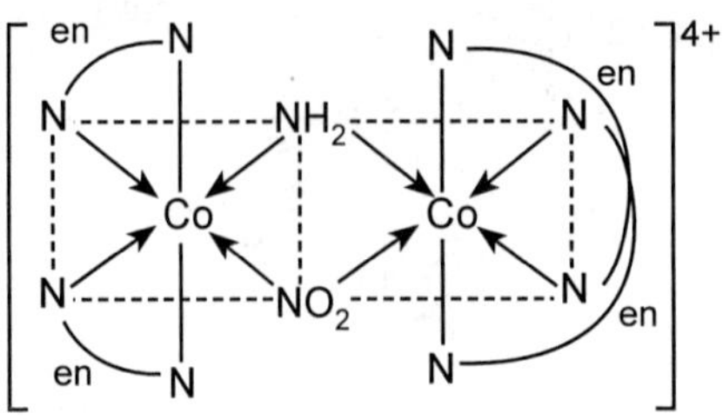

Cis-d-form

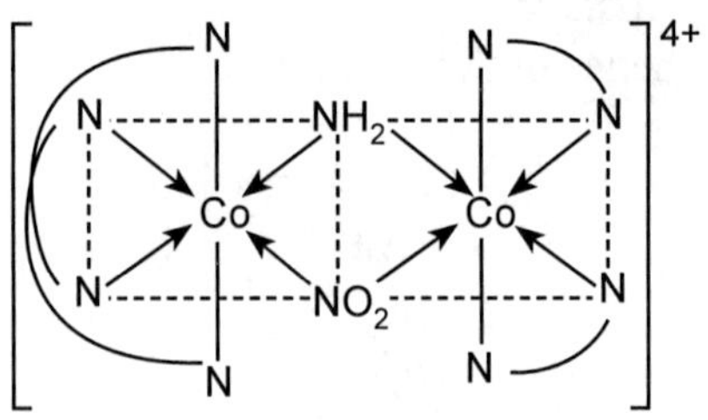

Cis-l-form

Trans-meso form is optically inactive.

SUMMARY

This chapter shows the isomeric form of two types of isomerism (*a*) structural and stereo isomerism. Structural isomerism is an outcome of metal-ligand bond which can be formed in various ways. Examples of coordination isomerism, coordination position isomerism, ionisation isomerism, linkage isomerism, ligand isomerism and polymerisation isomerism have been highlighted with suitable examples, stereo isomerism includes (*a*) Geometrical isomerism (*b*) Optical isomerism. Here, orientation of ligand in space with respect to the metal ions takes precedence.

Further, monodentate, bidentate and polydentate ligand orientation has been elaborated in 4-coordinate and 6-coordinate geometry. In geometrical isomerism *cis* and *trans* isomers have been explained when there are two identical ligands. Geometrical isomerism can exist in both monodentate and bidentate bigands. Optical isomerism *dextro (d)* and *laevo* (*l*) exists due to non-superimposable mirror images. Examples of six coordinate complexes having both mono and bidentate ligand have been discussed. Optical isomerism may exist over and above geometrical isomerism *i. e.*, one of the *cis* or *trans* isomer may show optical activity. In case of polydentate ligands the position of donor atoms oriented in the given geometry may help to identify *cis, tram* and optical isomers *e.g.,* [Co edta]T In bridged complexes *cis* and *tram* can be found from the complex. When two identical ligands are on the same side in the geometry it is *cis* otherwise *tram.* Tetrahedral complexes show optical activity but no geometrical isomerism and square planar shows no optical activity.

EXERCISE

1. Discuss all the isomers of an octahedral complex which has six monodentate ligands two A type and four B type.

2. Draw the possible isomers of a octahedral complex which has three monodentate ligands of type A and three monodentate ligands of type B.

3. Draw all the possible isomers of each of the following:

(*i*) $[Rh(en)_2Br_2]^+$

(*ii*) $[Be(Gly)_2]$
(*iii*) $[Ir(C_20_4)_2Cl_2]^{3-}$
(*iv*) $[Pt(en)Br_2Cl_2]$
(*v*) $[Cr(en)_3]^{3+}$
(*vi*) $[Credta]^-$
(*vii*) $[Pt_2P(C_2H_5)_3Cl_4]$

4. Indicate the type of isomerism in the following pairs:
(*i*) $[Cr(H_2O)_5Cl]Cl_2.H_2O$ and $[Cr(H_2O)_4Cl_2]Cl.H_2O$
(*ii*) $[Cr(NH_3)_6][Cr(SCN)_6]$ and $[Cr(NH_3)_4(SCN)_2]$ $[Cr(NH_3)_2(SCN)_4]$
(*iii*) $[Cr(NH_3)_5NO_2]Cl_2$ and $[Cr(NH_3)_5(ONO)]Cl_2$
(*iv*) $[Pt(NH_3)_4Cl_2]Br_2$ and $[Pt(NH_3)_4Br_2]Cl_2$
(*v*) $[Cr(NH_3)_6]tCo(NCS)_6]$ and $[Co(NH_3)_6][Cr(NCS)_5]$
(*vi*) $[Co(en)_3][Cr(CN)_6]$ and $[Co(CN)_6][Cr(en)_3]$

5. What test will you perform to distinguish $[Co(NH_3)_5Br]SO_4$ and $[Co(NH_3)_5(SO_4)]Br$?

6. Draw the various isomers of:
(*i*) $[Co(NH_3)_4Cl_2]^+$
(*ii*) $[Co(en)_2Cl_2]^+$
(*iii*) $[Co(en)_2NH_3Cl]^+$
(*iv*) $[Co(en)(NH_3)_2Cl_2]^+$

7. Draw the possible isomers of $CrCl_3.6H_2O$ and devise a method to distinguish between them.

8. Exhibit the type of isomerism in the pair:

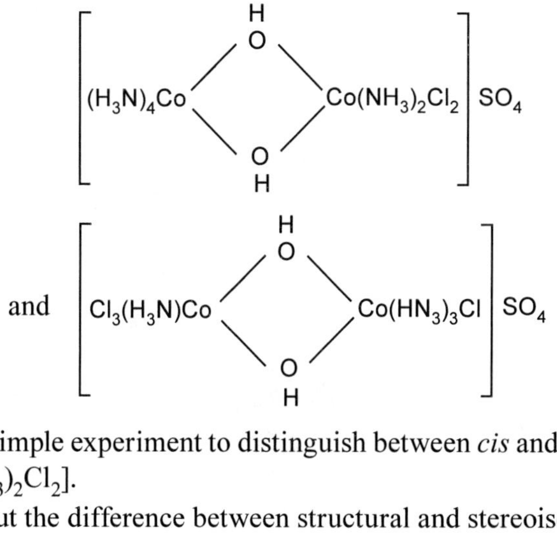

9. Give a simple experiment to distinguish between *cis* and *trans* isomers of $[Pt(NH_3)_2Cl_2]$.

10. Bring out the difference between structural and stereoisomerism.

6

Bonding in Complexes

6.1 NATURE OF METAL-LIGAND BOND

The elementary idea of a metal-ligand bond was put forward first of all by Werner which was advanced by the various theories from time-to-time. The actual bond between the central metal ion and the ligand was considered partly covalent and partly ionic. Certain theories support the covalent nature of bond between metal and ligand while there are other theories which emphasise on the model of ionic bonding. The final and most acceptable theory is of course the molecular orbital theory. The concepts and model of molecular orbital theory are a little hard to understand.

The theories suggesting covalent bonding model are:

1. Lewis Acid-base Theory
2. Valence Bond Theory

And the theories which are in support of ionic model of bonding are:

3. Crystal Field Theory (CFT)
4. Ligand Field Theory (LFT)

Molecular orbital theory covers all the possible aspects of bonding and has a final say.

6.2 LEWIS ACID-BASE THEORY

This theory states that in a metal-ligand bond all the metal ions act as a Lewis acid and all the ligands as Lewis bases.

Lewis Acid: Accepts a pair of electrons.

Lewis Base: Donates a pair of electrons.

In a complex ion formation *e.g.*

$$Ag^+ + 2NH_3 \longrightarrow [H_3\overset{0}{N}: \rightarrow \overset{+}{Ag} \leftarrow \overset{0}{N}H_3]$$

It has been already mentioned earlier that the metal ion is invariably from the transition series and hence has lot of vacant orbitals to accept a pair or pairs of electrons from the ligands.

The direction of the bond is from donor to acceptor and is in fact a coordinate bond (L $\rightarrow$ M). Hence, it is established that the nature of bond between ligand and metal is a coordinate bond.

This theory only tells about the nature of bond but nothing regarding the structure or geometry of the complex ion. The structure and geometry of the complex ion control most of its properties and reactions. Therefore, Lewis acid-base theory does not give a complete picture of metal-ligand covalent bond, *e.g.*

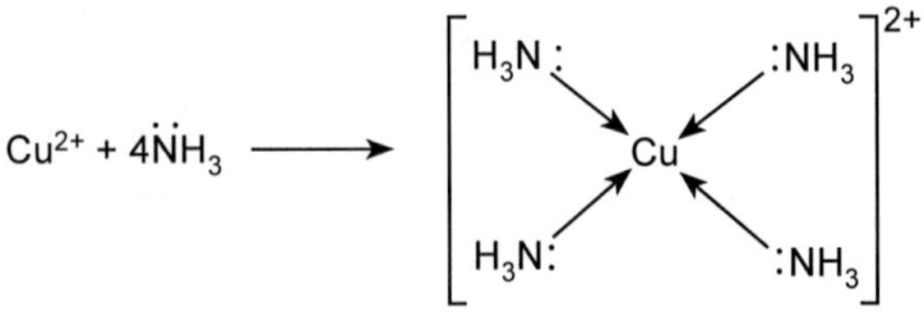

In the above example, there are four coordinate bonds between Cu^{2+} ion and 4 NH_3 molecules but the structure whether it is square planar or tetrahedral is not indicated.

6.3 VALENCE BOND THEORY

This theory gives a complete picture of the covalent bond between the metal ion and the ligand. It not only indicates the coordinate nature of bond but also indicates that with the help of the concept of hybridisation we can arrive at the correct geometry of the complex ion which is so essential,

The concept of hybridisation is supported by the experimental result which states all the metal-ligand bonds are of the same length and strength.

6-Coordinate Octahedral Complexes

There are two types of octahedral complexes according to this theory: the inner orbital complexes and the outer orbital complexes.

The valence shell configuration of the central metal ion is controlled by the ligands in complexes.

There are two types of ligands:

(*i*) weak field ligands

(*ii*) strong field ligands (discussed at *p.* 18)

Outer Orbital Complexes

The weak field ligands provide less energy to the central metal ion and are, therefore, unable to pair the unpaired electrons in the valency shell of the central metal ion and mostly result in paramagnetic and outer orbital complexes as outer (*d*-orbitals are used in hybridisation.

Inner Orbital Complexes

The strong field ligands provide enough energy to the central metal ion which is greater than the pairing energy of the electrons hence the unpaired electrons in the valency shell of metal ion get paired making available inner *d* orbitals of lower energy for hybridisation resulting in diamagnetic inner orbital complexes.

To explain the above facts in a more vivid manner the following examples are discussed:

$\overset{3+}{[Co}\overset{+6^-}{F_6}]^{3-}$ In this complex ion, the distribution of charge indicates that Co is in the +3 oxidation state. The valence shell configuration of Co^{3+} ion is mentioned below:

$Co^o \longrightarrow (Ar)\ 3d^7 4s^2$

$Co^{3+} \longrightarrow (Ar)\ 3d^6\ 4s^o\ 4p^o\ 4^o$

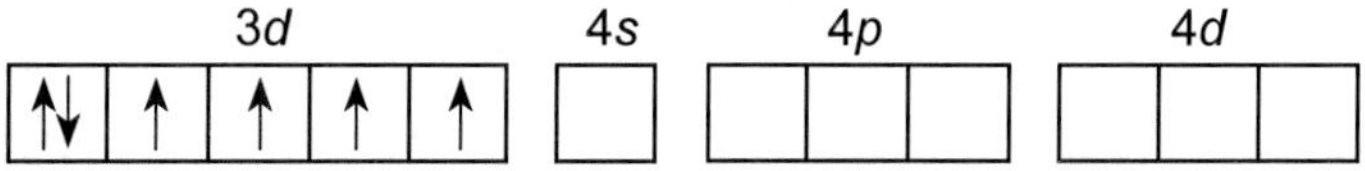

F $\longrightarrow 2s^2\ 2p^6$ *i.e.*, $:\ddot{\underset{..}{F}}:^-$ has 4 lone pairs of electrons in the valency shell. Being a monodentate ligand it can donate only one pair. F~ ion is a weak field ligand therefore it does not provide enough energy to the unpaired electrons of Co^{3+} ions in the *3d* shell to get paired. Hence, for bonding, the outer 4*d* vacant orbitals are used.

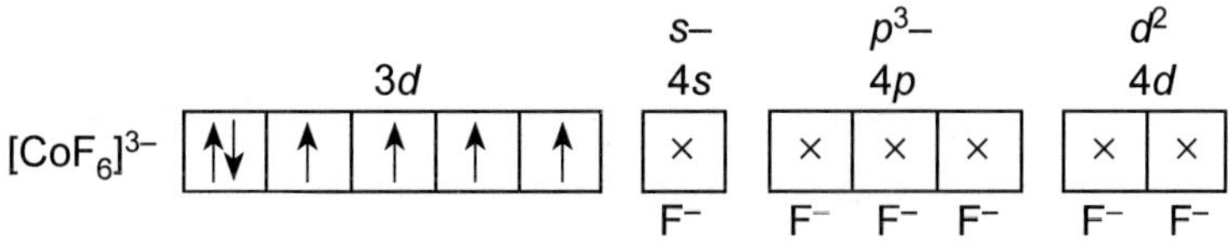

Since all the six Co—F bonds are of equal strength, therefore, the concept of hybridisation seems to be well-entrenched. The hybridisation involved is s-p^3-d^2 and the complex ion is paramagnetic and is an outer orbital complex. The hybridisation involved suggests the octahedral structure.

$[Co(NH_3)_6]^{3+}$ ion

Similarly in case of $\overset{3^+}{[Co}\overset{+0\ =\ 3^+}{(NH_3)_6]}$ ion Co holds the same configuration in the valency shell but NH_3 molecule being a strong field ligand alters the arrangement of electrons in its valency shell due to energy provided being more than the pairing energy of the electrons. Hence, maximum unpaired electrons get paired to make available inner low energy orbitals for bonding. The complex is thus known as Inner-orbital diamagnetic complex.

Co^{3+} ⟶ 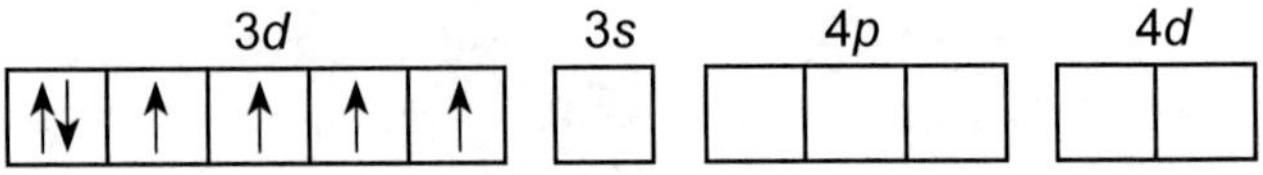

$6(NH_3)$, N is the donor N → $1s^2\ 2s^2 2p^3$

$[Co(NH_3)_6]^{3+}$ ⟶ 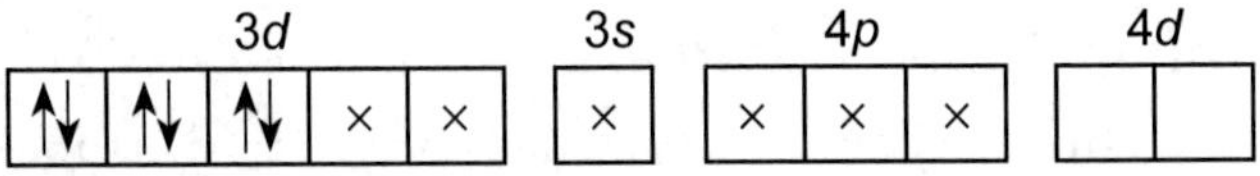

The hybridisation involved gives octahedral, diamagnetic ($\mu = 0$) and inner-orbital complex ion.

These two complex ions projected are very straight forward cases. Other examples of octahedral complexes need slight readjustments depending upon their characteristics.

In examples of the kind $[Ni(H_2O)_6]^{2+}$ and $[Ni(NH_3)_6]^{2+}$ we come across the deviations.

2+ +0 + 2+
$[Ni(H_2O)_6]$

Ni^{2+} 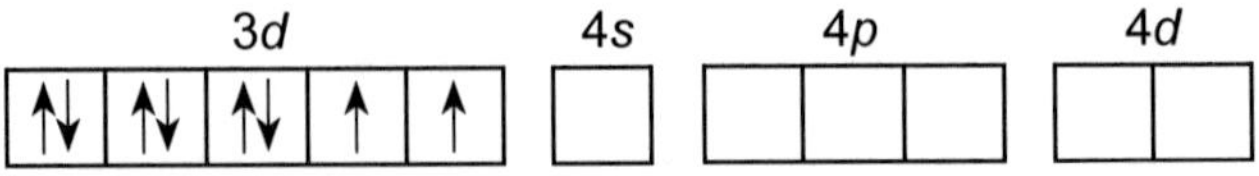

H—Ö—H is a weak field ligand.

$[Ni(H_2O)_6]^{2+}$ 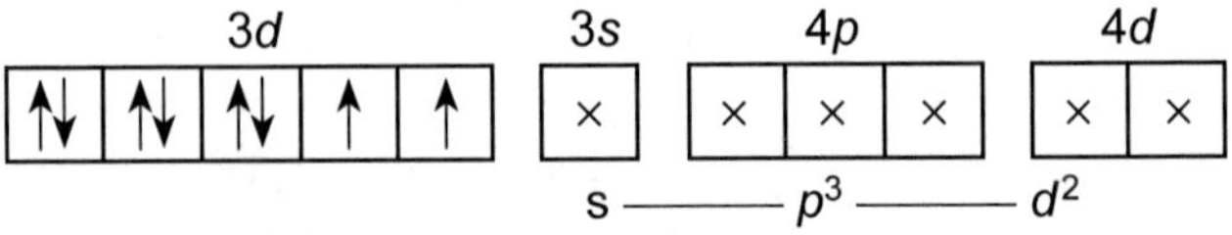

Therefore, it is an octahedral, paramagnetic and an outer orbital complex as expected.

But in

2+ +0 + 2+
$[Ni(NH_3)_6]$

Ni^{2+} 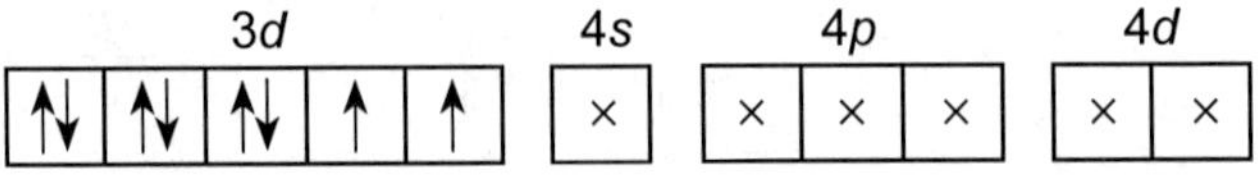

NH_3 is a strong field ligand. Therefore, on pairing the electrons in *3d* orbitals of Ni^{2+} ion we get only one vacant *3d* orbital. This prevents us from getting the correct hybridisation which can be either d^2sp^3 or sp^3d^2. Therefore, in $[Ni(NH_3)_6]^{2+}$ ion we have no option but to get octahedral, paramagnetic and outer orbital complex. Therefore, generalising this trend for Ni^{2+} ion we can say that all Ni^{2+} octahedral complexes are paramagnetic and outer orbital complexes irrespective of the strength of the ligand.

In example such as $(Ti(H_2O)_6]^{3"}$ we came across another aspect in which out of the vacant *3d* and *4d* orbitals, low energy *3d* orbitals are preferred for hybridisation *i.e.,* d^2-s-p^3 and paramagnetic.

In example like $[Fe(CN)_6]^{3-}$ inspite of strong field (CN^-) ligand the complex ion projects d^2sp^3 hybridisation with paramagnetic octahedral complex ion. Rest of the examples in 6-coordinate octahedral complexes, fit into the above mentioned deviations.

4-Coordination Complexes

Either fit into sp^3 tetrahedral or dsp^2 square planar structures depending upon the metal ion involved and the ligand.

The examples to be discussed here:

$2^+ + 0 = 2^+$

1. $[Cu(NH_3)_4]$

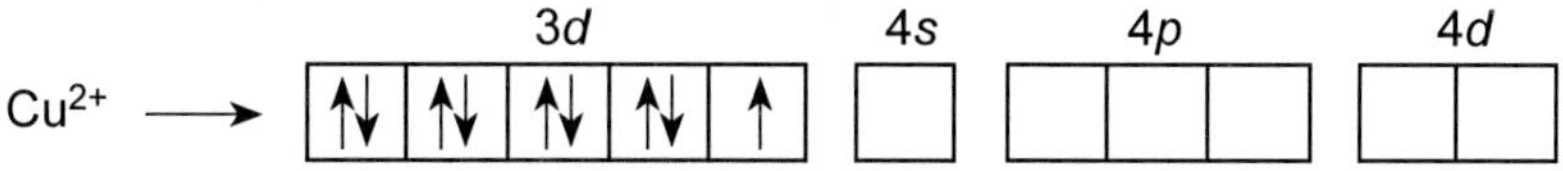

$4NH_3 \longrightarrow$ NH_3 is a strong field ligand. $4NH_3$ molecules require 4 vacant orbitals to form four coordinate bonds between $Cu^{2'}$ and NH_3 ligand. Ammonia gives enough energy to $Cu^{2''}$ to excite its unpaired electron to the vacant 4 orbitals. This is attributed to the fact that energy absorbed is quantised.

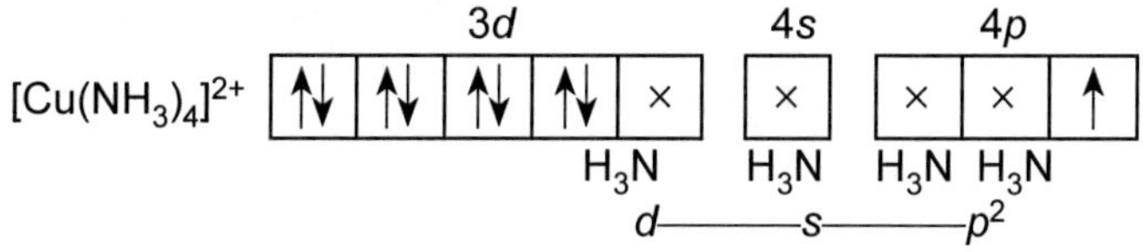

This complex ion gives dsp^2 hybridisation has a square planar geometry and is paramagnetic.

2. $[Ni(CO)_4]$

Nickel carbonyl has a zero valent nickel. CO group is a very strong field ligand. Therefore, maximum electrons in it are paired, complex and invariably diamagnetic.

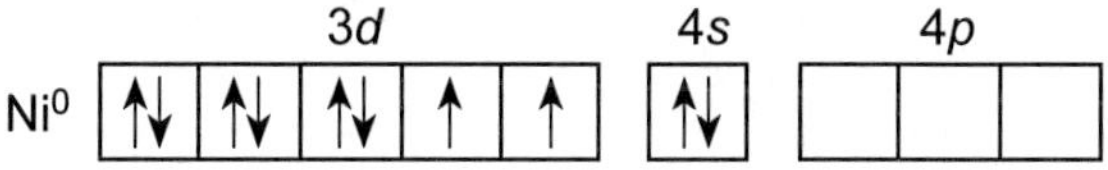

4 : CO

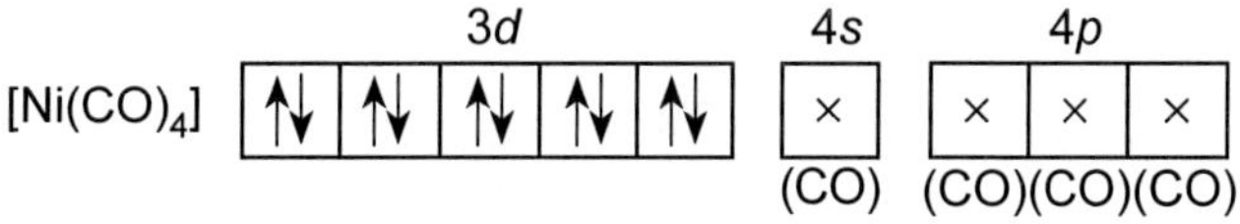

It involves sp^3 hybridisation, tetrahedral geometry and is diamagnetic.

Tetrahedral

$[Ni(CO)_4]$

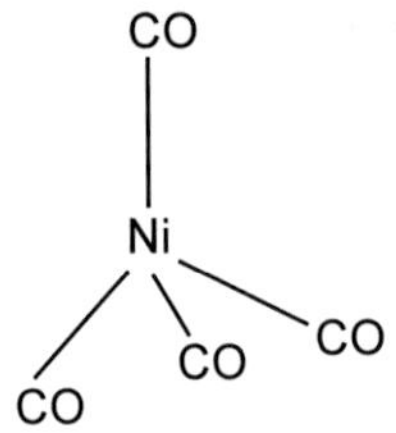

$2^{+} + 4^{-} = 2^{-}$

$[Ni(CN)_4]$

Here again the complex ion has Ni metal in the +2 oxidation state. CN^- a strong field ligand and pairs the unpaired 3*d* electrons in Ni^{2-} resulting into dsp^2 hybridisation yielding diamagnetic and square planar complex.

Ni^{2+} (Ar) $3d^8\, 4s^0\, 4p0$

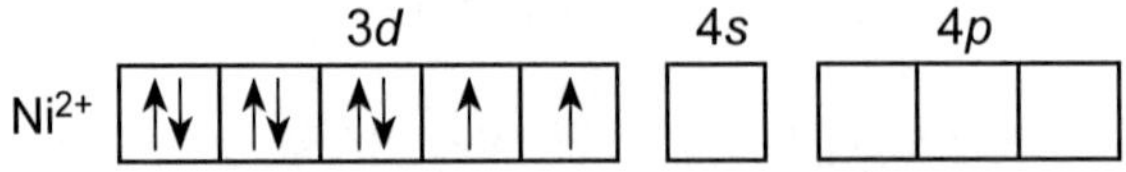

CN^- strong field ligand

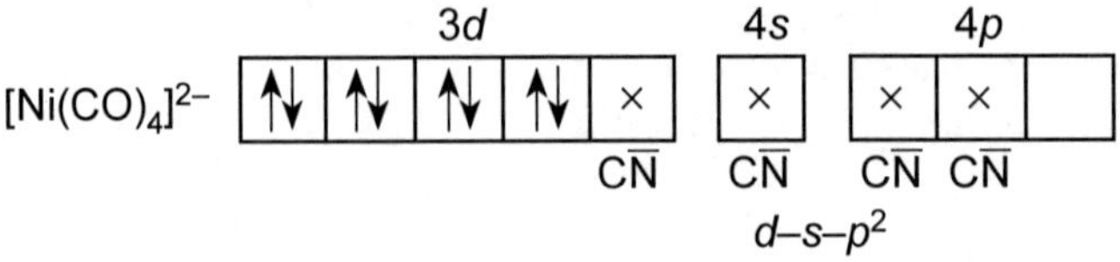

Square Planar

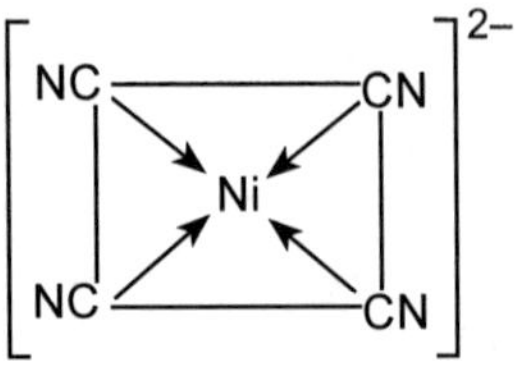

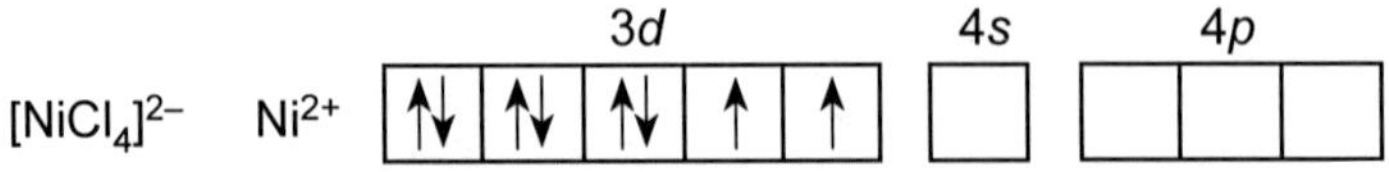

$:\ddot{\underset{..}{Cl}}:^-$ is a weak field ligand

Hence, $[NiCl_4]^{2-}$

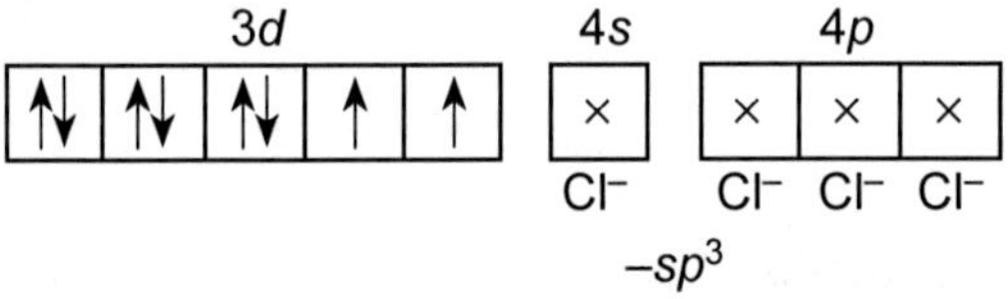

For 4-coordinate complexes having a bulky and weak field Cl^- ligand, it may be generalised that such complexes sterically are most likely to opt for the more spacious tetrahedral geometry.

The rest of the 4-coordinate complexes fall into the categories of hybridisations and structures as explained in the examples discussed.

5-Coordinate Complexes

There are fewer number of 5-coordinate complexes for the reason that the most likely geometry, *i.e.,* trigonal bipyramidal is less symmetrical. This is due to the fact that the triangular base angle is 120° while the apex angle is 90°.

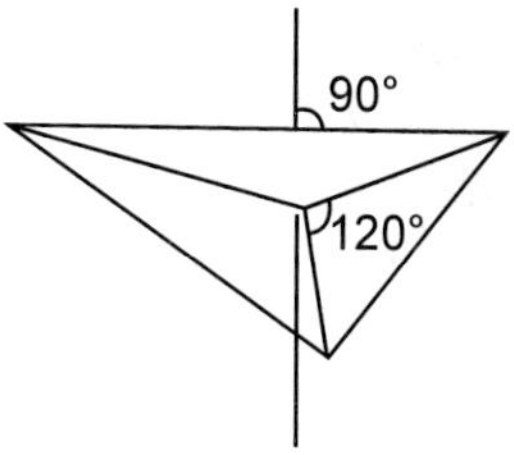

The less probable square pyramid geometry is even less symmetrical than trigonal bipyramidal.

The common examples are:

Fe(CO)$_5$]

In metal carbonyls, metal is always zero valent.

$Fe^{\circ} \rightarrow (Ar)3d^6\, 4s^2\, 4p^{\circ}\, 4d^{\circ}$

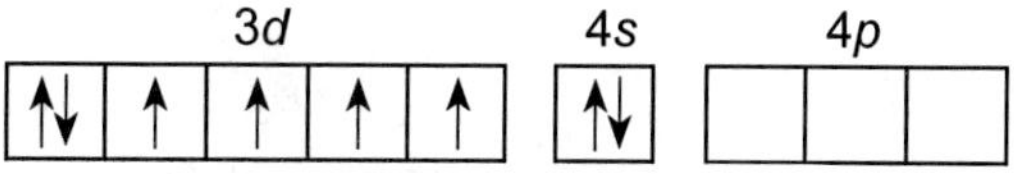

'4*f*' orbital is too far removed and, therefore, does not participate in the hybridisation.

CO is a carbonyl ligand and due to the nature of bonding between C and O surprisingly 'C' atom acts as an electron pair donor and not the 'O' atom although 'O' atom is more electronegative out of the two *e.g.*

←:C≡O:

CO is considered as one of the very strong field ligands. Hence, configuration of Fe° becomes:

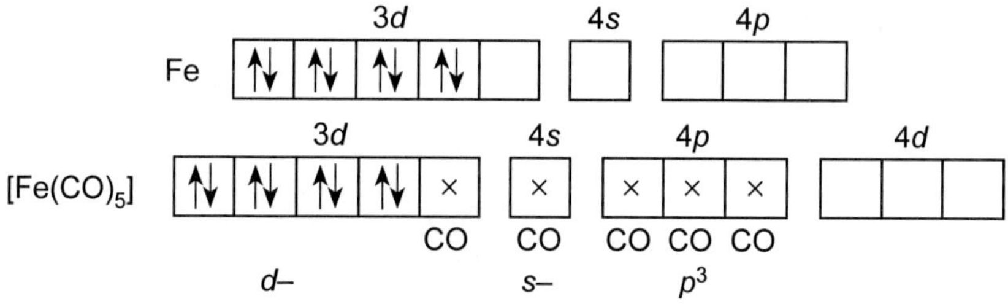

The dsp^3 hybridisation suggests trigonal bipyramidal geometry, The complex is obviously diamagnetic having zero dipole moment value.

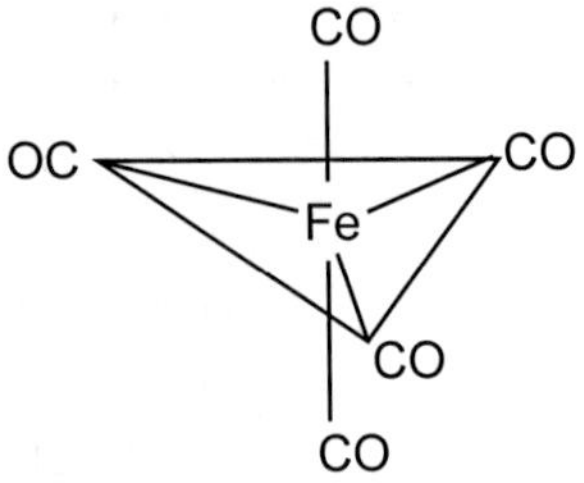

Another example is:

$3^{+} + 3^{-} + 0 = 0$

$[Ni\,Br_3(PEt_3)_2]$

In this complex, Ni is trivalent. Br^- is a weak field ligand but triethyl phosphine on the other hand is a strong field ligand due to the electron releasing power of the ethyl group which enhances the electron donor capacity of 'P' atom.

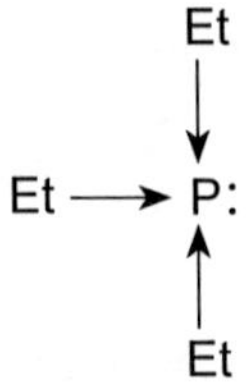

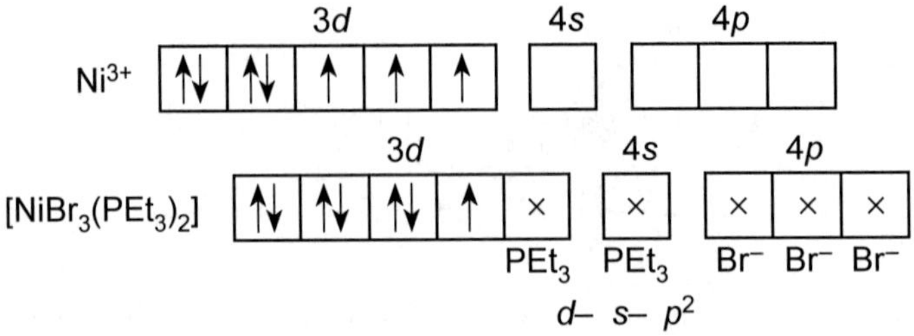

The trigonal bipyramidal geometry in this complex will acquire a symmetrical disposition of the ligands. Three Br ligands occupy the triangular base position while the two PEt_3 ligands occupy the apex or perpendicular positions. The complex is paramagnetic and has a dipole moment value of 1.83 BM which suggests one unpaired electron. Further, the complex is stabilised by resonance between the ligand positions.

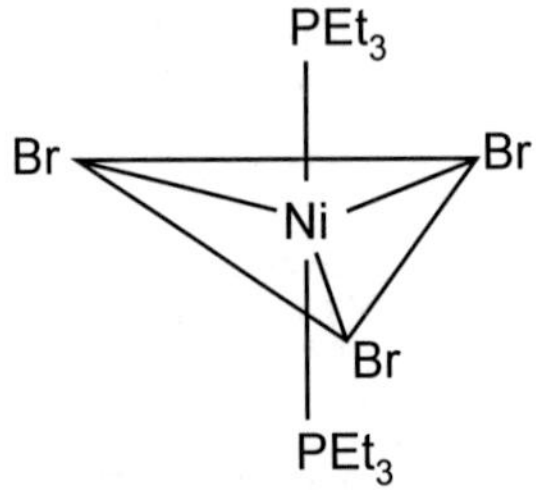

Trigonal bipyramidal

Limitations of VBT (Valency Bond Theory)

1. It cannot explain the relative stabilities for different shapes and coordination number in metal complexes. Octahedral (sp^3d^2 or d^2sp^3) and tetrahedral sp^3 and square planar (dsp^2) cannot be distinguished from one another merely on the basis of unpaired electrons.
2. VBT cannot explain the theory of colour associated with the complex ions. *e.g.,* aqua complexes of Co^{2+}, Ti^{3+}, Fe^{3+}, Cr^{+3} show varying colours.
3. VBT is unable to explain the finer details of the magnetic properties and the magnitude of the orbital contribution to the magnetic moments. For example, both octahedral (sp^2d^2) and tetrahedral complexes of $Co^{2+}(d^7)$ system have three unpaired electrons therefore p should be around 3.87 BM but in tetrahedral complexes p is between 4.4 to 4.8 BM and in octahedral complex it is around 4.7 and 5.2 BM. This is attributed to orbital contribution which is not explained by VBT.
4. VBT does not explain why the square planar complex of (d^9) system like $[Cu(NH_3)_4]^{2+}$ ion are not reducing agents like inner orbital octahedral complexes of Co^{2+} (d^7) when in both cases promotion of electron to a higher energy is required.
5. VBT lays too much emphasis on the role of the central metal ion while the importance of ligands is underplayed.
6. VBT cannot explain the rates and mechanism of reactions. The differences in the energies of activation are not explained.

6.4 CRYSTAL FIELD THEORY (CFT)

This theory was advanced by Hans Bethe and van Vleck and was initially applied to ionic crystals. It is widely accepted over valence bond theory. It assumes the ionic model of bonding between the central metal ion and the ligands.

The salient features of the crystal field theory are:

1. The central metal ion is regarded as a positive ion with charge equal to the oxidation state.
2. Ligands are treated as negative point charges, *e.g.,* F^-, CN^-, Cl^- etc. If the ligand is neutral such as NH_3 molecule then the negative end of the dipole in the molecule is facing towards the metal ion.
3. The metal orbitals and the ligand orbitals have no interaction. It is purely electrostatic interaction.
4. The electron on the central metal ion are under repulsive forces from those on the ligands.
5. The '*d*' orbitals in the metal ion have the same energy, *i.e.,* degenerate when in the free state. However, when a complex is formed the

ligands destroy the degeneracy of these orbitals *i.e.,* the orbitals now have different energies. The five degenerate '*d*' orbitals are oriented in space but the ligands attack in one direction, therefore, it is impossible for all the negative charges to stay spherically symmetrical. If the ligands surrounding the metal ion are spherically symmetrical then the '*d*' orbitals remain degenerate. In the process of ligands approaching the central metal ion some '*d*' orbitals fall into the plane of ligand attack . The energy of these orbitals is raised due to repulsive forces between the ligands and the electrons on the metal ion. The remaining '*d*' orbitals which go between the ligands and are not influenced by the ligands go to a lower energy state. This leads to the splitting of degenerate '*d*' orbitals. The pattern of '*d*' orbital splitting is different in different crystalline field. This is attributed to the fact that in different crystalline fields, different '*d*' orbitals are influenced and hence amount and kind of splitting is different. Most of the complexes are four and six coordinate complexes.

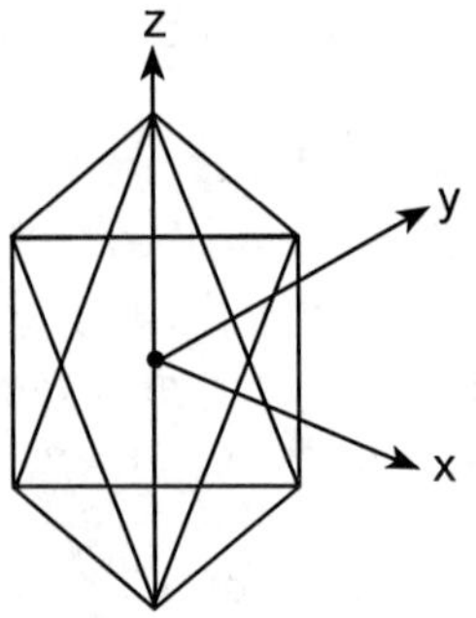

In an octahedral complex, the metal ion is at the centre of an octahedron and the ligands are at the six corners. The directions x, *y* and z are the three adjacent corners of the octahedron.

(*i*) The lobes of degenerate '*d*' orbitals that point towards *x*, *y* and *z* axes ($dx^2–y^2$ and dz^2) are called e_g orbitals. Due to electrostatic repulsion they stay at higher energy state.

(*ii*) The lobes of *dxy, dxz* and *dyz* orbitals point in between the axes, *i.e.,* they go between the ligands and hence do not experience any electrostatic force. They are symbolised by t_{2g} and due to non-interaction by the ligands occupy lower energy state.

Therefore, degenerated '*d*' orbitals get divided into two groups (*a*) axial and (*b*) non-axial.

The pattern of octahedral splitting is drawn in figure below for easy visualisation of the concept.

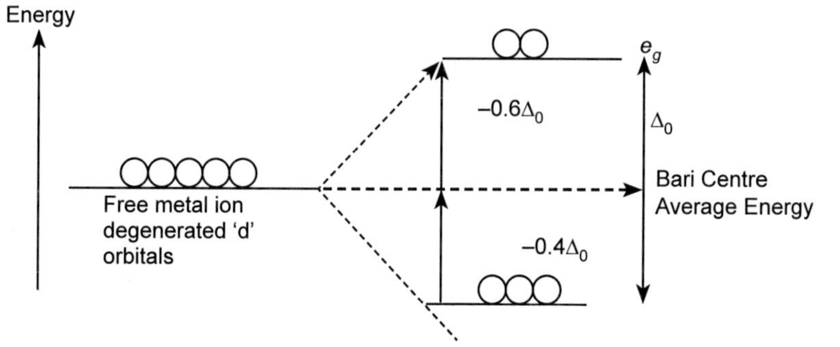

Energy levels of '*d*' orbitals e_g and t_{2g} in octahedral field.

The difference in energy between the two levels is Δ_0 or $10D_q$. The e_g levels are $+0.6\Delta_0$ above average energy and the t_{2g} orbitals are $-0.4\Delta_0$ below the average energy.

The energy gap Δ_0 between e_g and t_{2g} levels falls in the range of UV-visible spectrum. Therefore, the complexes in which the electron in the lower level (t_{2g}) has the scope to get promoted to higher level e_g (by absorbing the light of a particular wavelength from the visible light) exhibit colour.

The spectrum of $[Ti(H_2O)_6]^{3+}$ is illustrated in the diagram below:

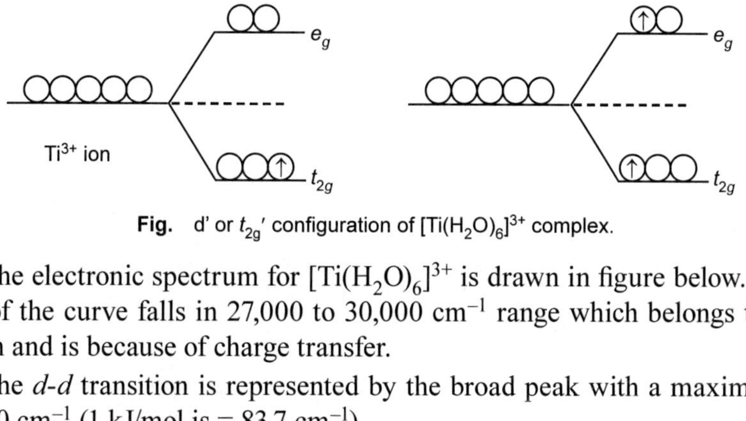

Fig. d' or t_{2g}' configuration of $[Ti(H_2O)_6]^{3+}$ complex.

The electronic spectrum for $[Ti(H_2O)_6]^{3+}$ is drawn in figure below. Steep part of the curve falls in 27,000 to 30,000 cm^{-1} range which belongs to UV region and is because of charge transfer.

The *d-d* transition is represented by the broad peak with a maximum at 20,300 cm^{-1} (1 kJ/mol is = 83.7 cm^{-1}).

$\therefore$ Δ_0 value for $[Ti(H_2O)_6]^{3+}$ is $\frac{20,300}{83.7} = 243$ kJ/mol. This value falls for the colour red-violet in the visible spectra. Here, yellow and green are absorbed to excite the electron and red-violet complementary colour is transmitted.

Due to crystal field splitting the energy acquired by t_{2g}' electron of $[Ti(H_2O)_6]^{3+}$ is $2/5\Delta_0$ below the energy of *d* orbitals (degenerate) and the complex is thus more stable.

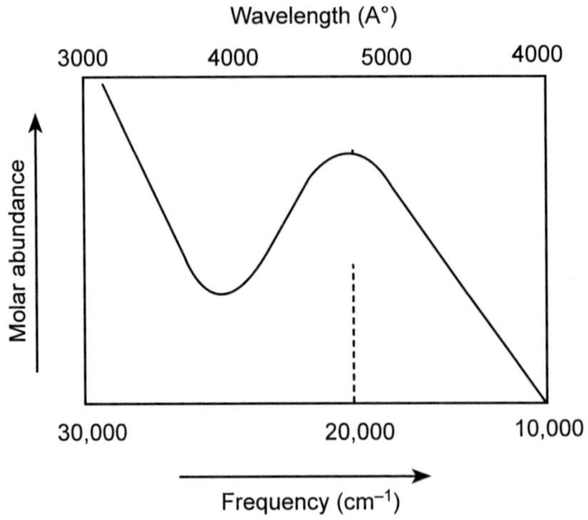

The crystal field stabilisation energy in this case is $\frac{2}{5} \times 243 = 97$ kJ Mol^{-1}

Table 6.1: Colours absorbed and colours observe

Colour absorbed	Colour observed	Wavenumber observed (cm^{-1})
yellow	red violet	24000–26000
yellow	indigo	23000–24000
orange	blue	21000–23000
red	blue-green	20000–21000
purple	green	18000–20000
red-violet	yellow-green	17300–18000
indigo	yellow	16400–17300
blue	orange	15300–16400
blue green	red	12800–15300

Table 6.2: CFSE and pairing energy for some complexes

Complex	Configuration	D_0 (cm^{-1})	P (cm^{-1})	Predicted	Bond
$[Fe(H_2O)_6]^{2+}$	d^6	10400	17600	high spin	high spin
$Fe(CN)_6]^{4+}$	d^6	32850	17600	low spin	low spin
$[CoF_6]^{3-}$	d^7	13000	21000	high spin	high spin
$[Co(NH_3)_6]^{3+}$	d^7	23000	2100	low spin	low spin

The magnitude of Δ_0 varies depending upon the interaction between the metal ion and the ligand. If the interaction is strong then Δ_0 is high and otherwise less. Due to variation in the value of Δ_0 of each complex gives different Δ_0 values hence absorption of colours and observed colours are different.

Magnitude of Δ_0

It depends upon the factors:

(*i*) The nature of the ligands

(*ii*) The charge on the metal ion

(*iii*) Metal belongingto 3*d*, 4*d* or 5*d* transition series. Those ligands which cause a small degree of splitting are called weak field ligands those with large Δ_0 are named strong field ligands. Most of the Δ_0 values are ranging from 7000 cm^{-1} to 30,000 cm^{-1}. The order of the strength of the ligands is:

Weak Field Strong Field

$I^- < Br^- < S^{2-} < Cl^- < NO_3^- < F^- < OH^- <$ EtOH < Oxalate $< H_2O <$ edta $< NH_3 <$ en < dipyridyl < o-phenanthroline $< NO_2^- < CN^- <$ CO

It is ambiguous to explain both σ and π bonding. The halides are in order expected from electrostatic force. In other cases covalent bonding aspect of σ-donation is followed:

Halide < O donors < N donors < C donors.

The CF splitting by CN' ion is double that of halide ions. This is attributed to the concept of two way bonding in which first the ligand donates electrons to vacant metal orbitals and then t_{2g} electrons in metal donate to the vacant orbital on the ligand. Similarly, N and C donors act as π acceptors.

Magnitude of Δ_0 for M^{3+} complexes in 3*d* series are 50% bigger than M^{2+} complexes.

Table 6.3: Crystal field splittings by various ligands

Complex	Absorption peak	
	(cm^{-1})	(cm^{-1})
$[Cr^{III}Cl_6]^{3-}$	13640	163
$[Cr(H_2O)_6]^{3+}$	17830	213
$[Cr^{III}(NH_3)_6]^{3+}$	21680	259
$[Cr^{III}(CN)_6]^{3-}$	26280	314

In M^{3+} octahedral complexes in case of weak field ligands, Hund's rule is followed while filling the electrons in the splitted '*d*' orbitals because the Δ_0 is greater. In strong field ligands energy provided by the ligand is greater than Δ_0 therefore electrons prefer to pair rather than go to the excited '*d*' (e_g). The total stabilisation energy is $\Delta_0 + \rho$. The complex with most unpaired electrons is called high spin or 'spin free' and the other 'low spin' or 'spin paired'.

In a way weak field Cl^- ion causes a small splitting of Δ_0. Therefore, it is energetically fovourable to occupy upper *E* level.

For a strong field CN^-, Δ_0 value is large. Less energy is required to pair the electrons to form a low spin complex. These effects ot high spin and complex are summarised in Table overleaf.

Table 6.4: CFSE and electronic arrangements in octahedral complexes

Number of *d* electrons	Arrangement in weak ligand field				Arrangement in strong ligand field			
	b_1	e_g	CFSE Δ_0	Spin only magnetic moment μ_s (*D*)	t_{2g}	e_g	CFSE Δ_0	Spin only magnetic moment μ_s (*D*)
d^1	[↑][ ][ ]	[ ][ ]	−0.4	1.73	[↑][ ][ ]	[ ][ ]	−0.4	1.73
d^2	[↑][↑][ ]	[ ][ ]	−0.8	2.83	[↑][↑][ ]	[ ][ ]	−0.8	2.83
d^3	[↑][↑][↑]	[ ][ ]	−1.2	3.87	[↑][↑][↑]	[ ][ ]	−1.2	3.87
d^4	[↑][↑][↑]	[↑][ ]	−1.2 +0.6 = −0.6	4.90	[↑][↑][↑]	[↑][ ]	−1.6	2.83
d^5	[↑][↑][↑]	[↑][↑]	−1.2 +1.2 = −0.0	5.92	[↑][↑][↑]	[↑][↑]	−2.0	1.73
d^6	[↑↓][↑][↑]	[↑][↑]	−1.6 +1.2 = −0.4	4.90	[↑↓][↑][↑]	[↑][↑]	−2.4	0.00
d^7	[↑↓][↑↓][↑]	[↑][↑]	−2.0 +1.2 = −0.8	3.87	[↑↓][↑↓][↑]	[↑][↑]	−2.4 +0.6 = −1.8	1.73
d^8	[↑↓][↑↓][↑↓]	[↑][↑]	−2.4 +1.2 = −1.2	2.83	[↑↓][↑↓][↑↓]	[↑][↑]	−2 4 +1.2 = −1.2	2.83
d^9	[↑↓][↑↓][↑↓]	[↑↓][↑]	−2.4 +1.8 = −0.6	1.73	[↑↓][↑↓][↑↓]	[↑↓][↑]	−2 4 +1.8 = −0.6	1.73
d^{10}	[↑↓][↑↓][↑↓]	[↑↓][↑↓]	−2.4 +2.4 = −0.0	0.00	[↑↓][↑↓][↑↓]	[↑↓][↑↓]	−2.4 +2.4 = −0.0	0.00

Effects of Crystal Field Splitting

In octahedral completes, filling of t_{2g} orbitals lowers the energy by $-0.4\Delta_0$ per electron. Filling of e_g orbitals increases the energy by $+0.6\Delta_0$ per electron.

$$\therefore \quad \text{CFSE}_{(oct)} = -0.4\, n_{(t2g)} + 0.6\, n_{(eg)}$$

where '*n*' stands for the number of electrons CFSE is zero for d^0, d^5 and d^{10} configuration in a weak field, *e.g.*, $[Mn(H_2O)_6]^{2+}$

This proves that unfilled, half-filled and completely filled '*d*' orbitals have extra stability. Hence, the theory of splitting of '*d*' orbitals in octahedral environment is on firm footing. The configurations of d° and d^{10}, *i.e.*, Ca^{2+} and Zn^{2+} respective also exhibit zero CFSE.

Ionic radii for M^{2+} ions in 3*d* series are expected to decrease smoothly because of poor shielding effect of '*d*' electrons and the increasing nuclear charge.

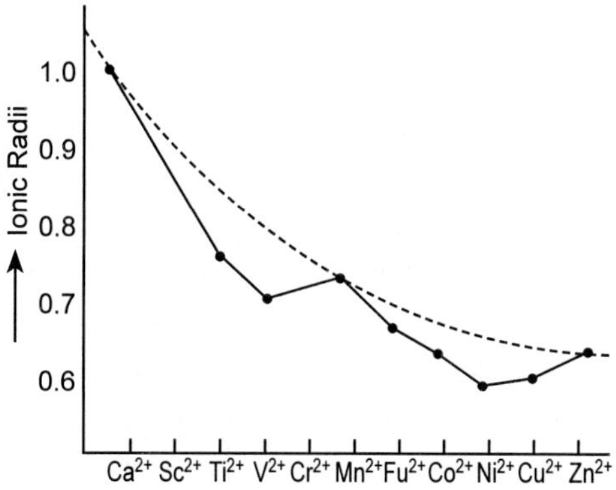

Octahedral ionic radii of M^{2+} first transition elements.

According to the plot there is decrease of radii when the electrons are in t_{2g} and the slight increase where in e_g orbitals. In t_{2g} the metal orbitals and ligand orbitals do not interact effectively therefore, the size of M^{2+} ions having electrons in t_{2g} show a steady decrease. Whenever electrons in M^{2+} ions go to orbital the e_g electrons interfere with ligand electrons. Nuclear charge thus has a lesser hold on electrons which results in slight increase in the ionic radii.

Tetragonal Distortion or Jahn-Teller Distortion

The shapes of the octahedral metal complexes is affected by the electron pairs which try to stay as far away to avoid repulsion. This distortion depends upon the '*d*' orbitals. If the '*d*' orbitals are symmetrical then the shape is regular octahedron. If it is asymmetrical then the regular octahedron may get distorted to tetragonal or square planar shape.

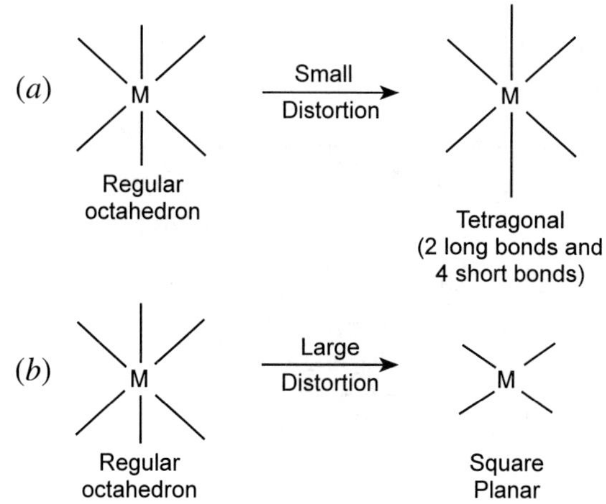

If the *d* electrons are symmetrically arranged, they will repel all the six ligands equally thus the resulting structure is a regular octahedron. If the *d* electrons are asymmetrically arranged then they will repel some ligands more than the others. This leads to distortion of the octahedron, *i.e.,* some ligands are prevented to approach the metal as closely as others.

The e_g orbitals point directly at the ligands. Therefore, these ligands are repelled more than the others. This causes great distortion in the octahedral shape. The t_{2g} orbitals do not directly point at the ligands. These orbitals affect the stereochemistry to a very small extent that is not observable.

The two e_g orbitals $d_x^2 - y^2$, d_7^2 are doubly degenerate but if filled asymmetrically then the degeneracy is destroyed.

If the d_z^2 orbital contains one more electron than the $d_x^2 - y^2$ orbital then the ligands approaching along +2 and –2 will experience greater repulsion than the remaining four ligands. This is called tetragonal distortion which is generally observed.

Table 6.5: The Distortions of Octahedral Structures that Result from the Presence of d Electrons

System	Predicted structure	Comments
High spin:		
d^1, d^6	Tetragonal distortion	Not observed
d^2, d^7	Tetragonal distortion	Not observed
d^3, d^8	No. distortion	Experimentally verified
d^4, d^9	Large tetragonal distortion	Experimentally verified
d^5, d^{10}	No. distortion	Experimentally verified

If the $d_x^2 - y^2$ orbital contains the extra electron then elongation will take place along *x* andy axis. This implies that ligands approach more closely along z axis, *i.e.,* four long bonds and two short bonds. This is like compressing the octahedron along z axis and is called tetragonal compression. Tetragonal elongation is common as regards tetragonal compression but it is not possible to foretell which will occur. The two possibilities have been elaborated in the following Tables.

Table 6.6: Symmetric electronic arrangement

Electronic configuration	t_{2g}	Nature of ligand field	Examples
d^0	[ ][ ][ ] [ ][ ]	Strong or weak	$Ti^{IV}O_2$, $[Ti^{IV}F_6]^{2-}$ $[Ti^{IV}Cl_6]^2$
d^3	[↑][↑][↑] [ ][ ]	Strong or weak	$[Cr^{III}$ (oxalate)$_3]^{3-}$ $[Cr^{III}(H_2O)_6]^{3+}$
d^5	[↑][↑][↑] [↑][↑]	Weak	$[Mn^{II}F_6]^{4-}$ $[F^{III}F_6]^{3-}$
d^6	[↑↓][↑↓][↑↓] [↑↓][↑↓]	Strong	$[Fe^{II}(CN)_6]^{4-}$ $[Co^{III}(NH_3)_6]^{3+}$
d^8	[↑↓][↑↓][↑↓] [ ][ ]	Weak	$[Ni^{II}F_6]^4$ $[Ni^{II}(H_2O)_6]^{2+}$
d^{10}	[↑↓][↑↓][↑↓] [↑][↑]	Strong or weak	$[Zn^{II}(NH_6)_6]^{2+}$ $[Zn^{II}(H_2O)_6]^{2+}$

Table: 6.7: Asymmetrical electronic arrangements

Electronic configuration	t_{2g}	e_g	Nature of ligand field	Examples
d^4	↑ ↑ ↑	↑	Weak field (high-spin complex)	Cr(+ II), Mn(+ III)
d^7	↑↓ ↑↓ ↑↓	↑	Strong field (low-spin complex)	Co(+ II), Ni(+ III)
d^9	↑↓ ↑↓ ↑↓	↑↓ ↑	Either strong or weak	Cu(+ II)

For example, CrF_2 is a distorted rutile (TiO_2) structure. Cr^{2+} is a d^4 configuration, F^- is a weak field ligand, so t_{2g} has three electrons while e_g contains one electron. The e_g orbital has $d_x^2 - y^2$ and d_z^2 orbitals, $d_x^2 - y^2$ has four lobes and d_z^2 has two lobes. To have minimum repulsion, single electron will prefer to go to d_z^2 orbital. This causes the splitting of degenerate e_g Orbitals. Due to greater repulsion $d_x^2 - y^2$ is of higher energy whereas less repulsion gives lower energy to d_z^2 orbital and greater stability. Thus, two ligands approaching along z-axis experience greater repulsion while the four ligands along x and y axis ($+x$, $-x$, $+y$, $-y$) experience less repulsion. This brings about tetragonal distortion of octahedral structure, *i.e.*, there are four short bonds and two long bonds.

In Ni^{2+}, d^8 configuration in octahedral environment in weak field ligand arrangement causes a split in e_g orbitals and gives tetragonally distorted octahedral geometry while d^8 configuration in strong field octahedral arrangement also causes a split in t_{2g} orbital. As the ligand attack is along xy plane, d_{xy} occupies higher energy than d_{xz} or d_{yz}.

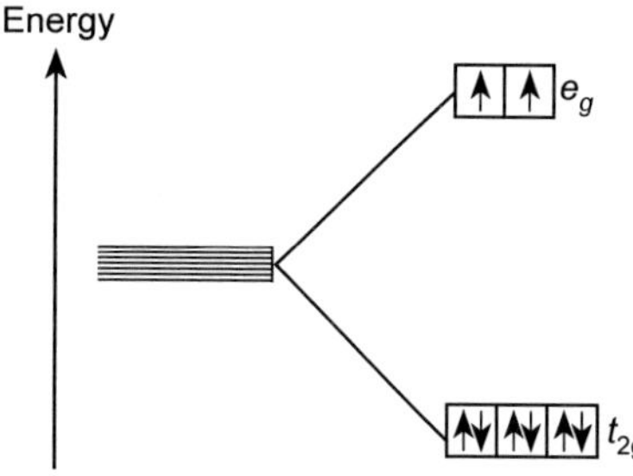

Fig. d^8 arrangement in weak octahedral field.

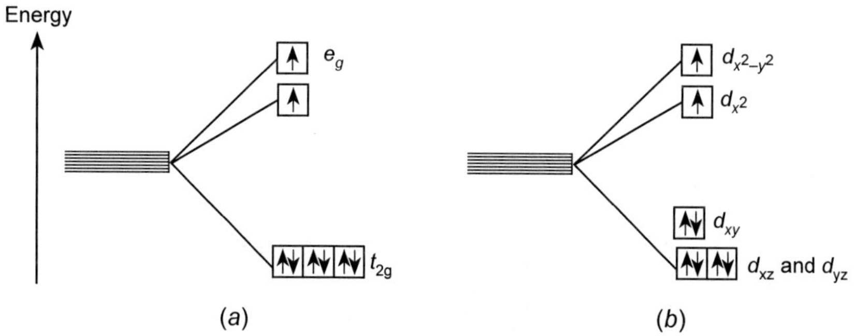

(a) and (b): d^8 arrangement in very strong octahedral field. Tetragonal distortion splits (a) the e_g level; and (*b*) also splits the t_{2g} level. The d_{xy} orbital is higher in energy than the d_n or d_{yz}. (For simplicity this is sometimes ignored)

Square Planar

In octahedrally disposed d^8 metal ion both in weak and strong field ligands, six electrons occupy t_{2g} orbitals and two electrons occupy e_g orbitals as illustrated in the Figure, *e.g.*, $[Ni(H_2O)_6]^{2-}$ and $[Ni(NH_3)_6]^{2+}$.

As the attack by the ligand is focussed along *xy* plane $dx^2 - y^2$ occupy higher energy than d_z^2 among e_g orbitals and d_{xy} occupies higher energy than d_{xz} and d_{yz}. If the interaction between the metal ion and ligand is very strong then the tetragonal distortion may become large so that d_z^2 occupies lower energy then d_{xy}.

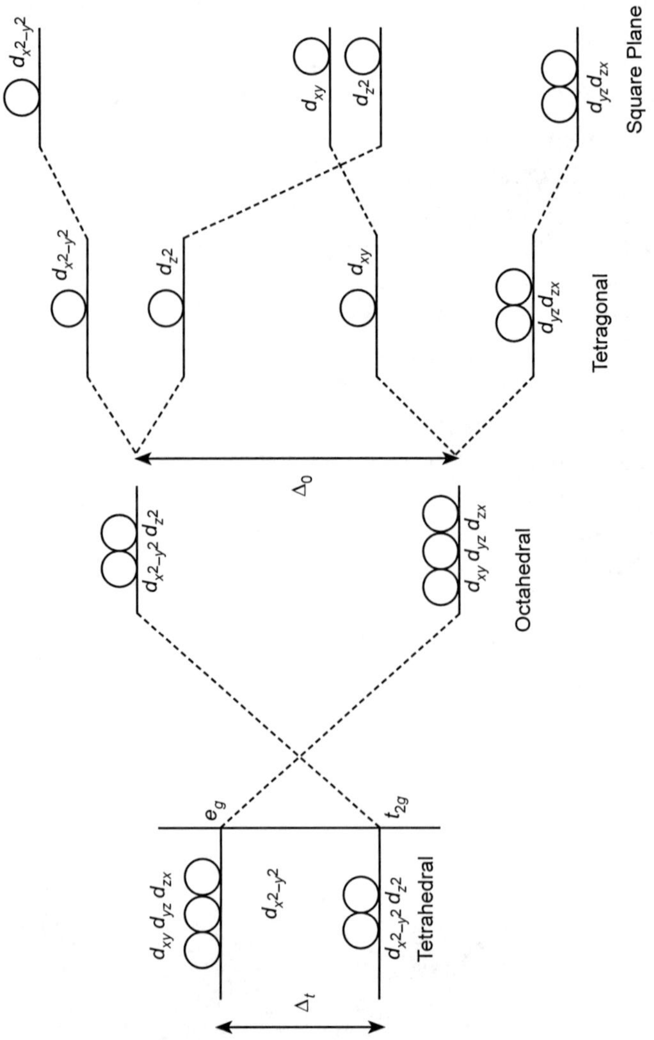

Crystal field splitting of *d* orbitals of central ion in complexes having different geometries.

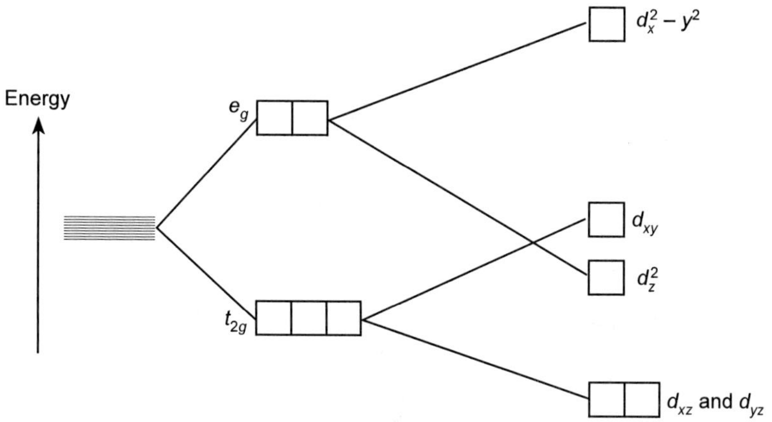

Tetragonal distortion

In square planar Co^{2+} , Ni^{2+} and Cu^{2+}, d_z^2 has nearly the same energy as d_{xz} and d_{yz} orbitals. In $[PtCl_4]^{2-}$, $d_z{}^2$ is lower than d_{xz} and d_{yz} orbitals. In Ni (d^8), strong crystal field splitting (Δ_0) is very large for 4*d* and 5*d* transition elements.

Table 6.8: Ions that form square planar complexes

Electronic configuration	Ions	Type of field	Number of unpaired electrons
d^4	Cr(+ II)	Weak	4
d^5	Fe(+ II)	(Haem)	2
d^7	Co(+II)	Strong	1
d^8	Ni(+II), Rh(I), Ir(+I)	Strong	0
d^8	Pt(+II), Pt(+II), Au(+III)	Strong and weak	0
d^9	Cu(+H), Ag(+II)	Strong and weak	1

Tetrahedral Complexes

Regular tetrahedron can be visualised as a cube in which one atom is at the centre surrounded by four ligands at the four of the corners of a cube, *x, y* and *z* point to the centre of the faces of the cube. e_g orbitals point along *x, y* and z faces whereas t_{2g} point between *x, y* and z towards the centres of the edges of the cube.

In tetrahedral environment orientation of the e_g and t_{2g} orbitals is reverse to that of octahedral, *i.e.,* t_{2g} orbitals are $4\Delta t$ above the weighted mean energy (Bari Centre) and the e_g orbitals are $6\Delta t$ below the mean.

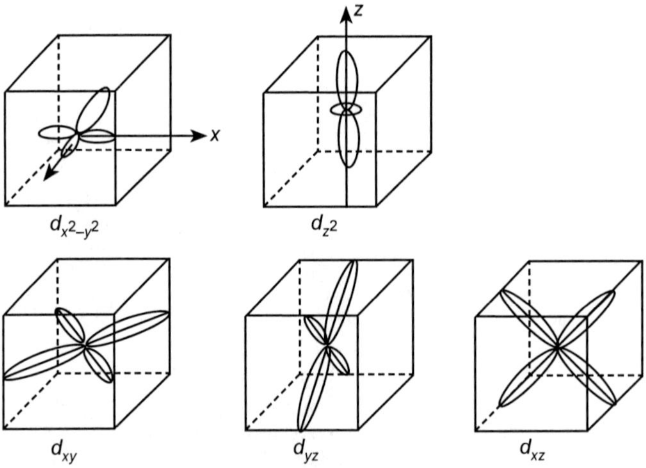

Orientation of '*d*' orbitals in a cube

In a tetrahedron the t_{2g} orbitals are not fully focussed at the metal ion therefore the magnitude of splitting is nearly half of that of octahedral fields.

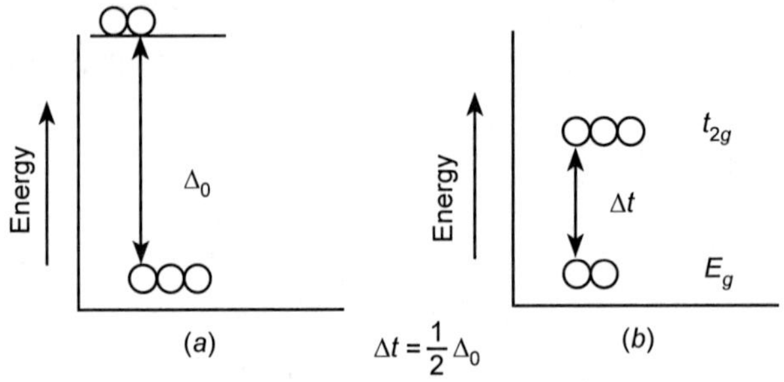

(*a*) Octahedral splitting (*b*) Tetrahedral splitting

This is attributed to:

1. There are four ligands instead of six in octahedral hence CFT is 2/3 of the size.
2. The direction of the orbitals does not coincide with the direction of the ligand orbitals. This reduces the CF splitting further by 2/3. Thus, Δt is $\approx \frac{2}{3} \times \frac{2}{3} = \frac{4}{9}$ of octahedral. As *At* is smaller than Δ_0, therefore, it is not energetically favourable to pair electrons. Hence, all tetrahedral complexes are high spin. For all electronic configurations CFSE of octahedral is greater than CFSE of tetrahedral. It shows that octahedral complexes are more stable than tetrahedral complexes. Despite this some tetrahedral complexes are stable and are formed. This is due to the following reasons:

(*i*) If ligands are bulky and crowding is caused in octahedral complex.

(*ii*) Tetrahedral complexes d°, d^{2}, d^{5}, d^{7} and d^{10} are regular *e.g.*, $[Ti^{IV}Cl_4]$, $[MnO_4]^-$, $[Fe^{VI}O_4]^{2-}$ $[Fe^{III}Cl_4]^-$, $[CoCl_4]^{2-}$ and $[ZnCl_4]^{2-}$

(*iii*) Ligands are weak field thus loss in CFSE is not the deciding factor.

(*iv*) Central metal has low oxidation state which reduces Δt.

(*v*) In configuration d°, d^{5}, d^{10} there is no CFSE gain in energy.

(*vi*) In d^1 and d^6 loss in CFSE is $0.27\Delta_0$ or d^2 and d^1 loss is $0.53\Delta_o$ which is very small.

Transition metal chlorides, bromides and iodides prefer to form tetrahedral complexes because of bulky ligands.

Table 6.9: CFSE and electronic arrangements in tetrahedral complexes.

Number of *d* electrons	Arrangement of electrons $^{e}9\ ^{f}2_{0}$	SPM only magnetic moment	Tetrahedral CFSE	Tetrahedral CFSE scaled for comparison with octahedral values, assuming	Octahedral CFSE Δ_0	
		e_g	t_{2g}	Δ_t $\frac{4}{9}\Delta_0$	Weak field	Weak Field
d^1	[↑][ ] [ ][ ][ ]	1.73	−0.6	−0.27	−0.4	−0.4
d^2	[↑][↑] [ ][ ][ ]	2.83	−1.2	−0.53	−0.8	−0.8
d^3	[↑][↑] [↑][ ][ ]	3.87	−1.2 + 0.4 = −0.8	−0.36	−1.2	−1.2
d^4	[↑][↑] [↑][↑][ ]	4.90	1.2 + 0.8 = −0.4	−0.18	−0.6	−1.6
d^5	[↑][↑] [↑][↑][↑]	5.92	−1.2 + 1.2 = 0.0	−0.00	−0.0	−2.0
d^6	[↑↓][↑] [↑][↑][↑]	4.90	−1.8 + 1.2 = −0.6	−0.27	−0.4	−2.4
d^7	[↑↓][↑↓] [↑][↑][↑]	3.87	−2.4 + 1.2 = −1.2	−0.53	−0.8	−1.8
d^8	[↑↓][↑↓] [↑↓][↑][↑]	2.83	−2.4 + 1.6 = −0.8	−0.36	−1.2	−1.2
d^9	[↑↓][↑↓] [↑↓][↑↓][↑]	1.73	−2.4 + 2.0 = −0.4	−0.18	−0.6	−0.6
d^{10}	[↑↓][↑↓] [↑↓][↑↓][↑↓]	0.00	−2.4 + 2.4 = 0.0	−0.00	−0.0	−0.0

6.5 LIGAND FIELD THEORY (LFT)

The crystal field theory is based on electrostatic model but it successfully explains shapes, spectra and magnetic properties of complexes. This theory ignores the existence of some covalent bonding that evidently occurs in transition metal complexes. It is referred to as modified CFT or ligand field theory.

The concept of some element of covalent bonding is supported by:

(*i*) In metal carbonyls and nitrosyls, metal and ligand are in the zero oxidation state. Therefore, electrostatic attraction between metal and ligand is ruled out, thus the bonding is covalent.

e.g., The model of bonding in the ligand molecule CO is:

$:\!C \leftrightarrows O$ The molecular orbital model shows filled *sp* hybrid orbital on carbon atom. The donation of the pair of electrons from CO molecule to the vacant orbital of the metal give a bond.

$$\underset{\text{neutral}}{\ominus M \oplus} + :\!C \leftrightarrows O \longrightarrow M \longleftarrow :CO \quad \sigma \text{ bond}$$

Ligand to metal.

Thus, metal acquires negative charge which is unnatural. According to Pauling's electroneutrality principle back donation of the pair of electrons from metal to CO takes place is shown:

$$\overset{\delta-}{M:} + C \leftrightarrows O \longrightarrow M \longrightarrow CO \quad \sigma \text{ bond}$$

Metal to ligand.

This results in some amount of π bonding. Both the types of bonds strengthen M—CO bond. Hence, the stability of carbonyls is increased.

In the former bond (σ bond), CO molecule is acting as a donor and metal atom as acceptor. In the later case, metal atom acts as a donor and CO molecule as an acceptor. In metal carbonyls both metal and the ligand act as donors and acceptors.

(*ii*) The order of ligands in the spectrochemical series cannot be explained on the basis of electrostatic model.

(*iii*) There is evidence from NMR studies that there exists some unpaired electron density on the ligands. This indicates sharing of electrons and some covalency. In interpreting spectra it shows inter-electronic repulsion, the '*d*' electrons from the metal go onto the ligand and are delocalised, thereby, generating some covalency. Similarly, electron delocalisation factor *k* can be used in interpreting NMR.

The LFT is used as an intermediate which includes the change from CFT to molecular orbital theory. Cotton and Wilkinson have called LFT as ACFT (Adjusted Crystal Field Theory). It is applicable in case of ligands such as CO, CN^-, phen and NO_2^- which provide largest crystal fields and thus are able to form π-bonds. The π-bonding markedly increases the magnitude of crystal field splitting.

Crystal field splitting ability of the ligands has been observed to decrease in the order.

Strong field ligands *Intermediate field ligands*

CO, CN^- > phen > NO_2^- > en > NH_3 > NCS^- > H_2O > F^- > $RCOO^-$ > OH^- > Cl^- > Br^- > I^-.

Complex Stability

The electrostatic model of a complex points to increase in energy of *d* orbitals when it is approached by the ligand. But inspite of this the complexes are formed.

This suggests that a complex is of lower energy configuration. It implies that increase in energy of the *d* orbital of metal is more than compensated by the metal ligand bonding.

In octahedral arrangement of ligand the t_{2g} and e_g sets of orbitals have different energies. The difference in their energies is given by Δ_0. Electron has a tendency to go to lower energy orbital first and if it is filled then it will go to the orbital of higher energy. The fact that half-filled configuration has extra stability implies that in half-filled configuration whether complex, or no complex is formed has gained energy towards stability and is zero. Therefore, hypothetically e_g orbital has a gain of +0.4 Δ_0 for each electron added to it and −0.6 A„ for each electron in addition to t_{2g}. This is called CFSE (crystal field stabilisation energy). To an extent, it can determine the relative stability of complexes.

Tabl 6.10: Crystal Field Stabilization Energies for Metal Ions in Octahedral Complexes

d-electrons in metal ions	Stabilization,			Stabilization,		
	t_{2g}	e_g	Δ_0	t_{2g}	e_g	Δ_0
1	↑ ○ ○	○ ○	0.4			
2	↑ ↑ ○	○ ○	0.8			
3	↑ ↑ ↑	○ ○	1.2			
4	↑ ↑ ↑	↑ ○	0.6	↑↓ ↑ ↑	○ ○	1.6
5	↑ ↑ ↑	↑ ↑	0.0	↑↓ ↑↓ ↑	○ ○	2.0
6	↑↓ ↑ ↑	↑ ↑	0.4	↑↓ ↑↓ ↑↓	○ ○	2.4
7	↑↓ ↑↓ ↑	↑ ↑	08	↑↓ ↑↓ ↑↓	↑ ○	1.8
8	↑↓ ↑↓ ↑↓	↑ ↑	1.2			
9	↑↓ ↑↓ ↑↓	↑↓ ↑	0.6			
10	↑↓ ↑↓ ↑↓	↑↓ ↑↓				

If Δ_0 is less than the pairing energy of electrons in *d*-orbitals, then Hund's rule of maximum multiplicity is obeyed resulting in high spin complexes. If Δ_0 is greater than the pairing energy then low spin complexes will result. In nut shell electrons will follow a system that is convenient and of low energy.

The energy of separation Δ_0 also depends upon the geometry. In tetrahedral geometry the value $\Delta t = \frac{1}{2}\Delta_0$.

Hence, in tetrahedral disposition there are no low spin complexes because the energy of pairing is always more than energy of excitation.

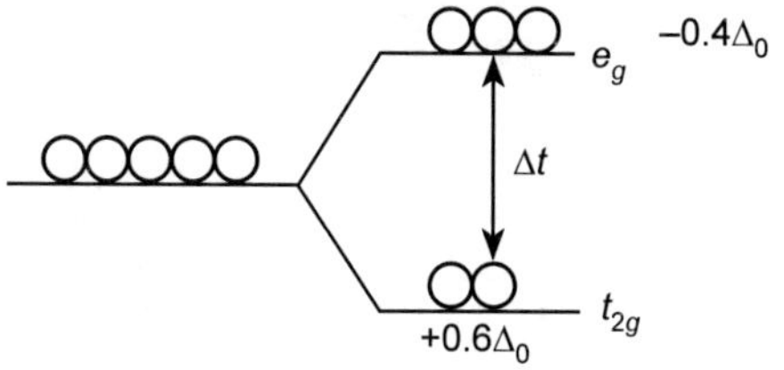

For each electron in t_{2g} gain of $+0.6\Delta_0$ and for each electron in $-0.4\Delta_t$ is observed so that at half-filled state, Δ_t is zero.

6.6 MOLECULAR ORBITAL THEORY (MOT)

The acceptability of molecular orbital theory (MOT) to chemists has increased over the years. It indicates both covalent and ionic character of chemical bonds. The MOT treats the distribution of electrons in the molecules whereas atomic theory relates to electron distribution in atoms. In most books, the molecular orbital treatment of simplest molecules is discussed. Calculation of shapes of molecular orbitals is based on linear combination of atomic orbitals method (LCAO).

It is natural that molecular orbitals should resemble atomic orbitals of atoms of which molecule is made. The LCAO method applied to two atomic orbitals gives two molecular orbitals, one molecular orbital is formed by addition of parts of atomic orbitals that overlap and the other on subtraction.

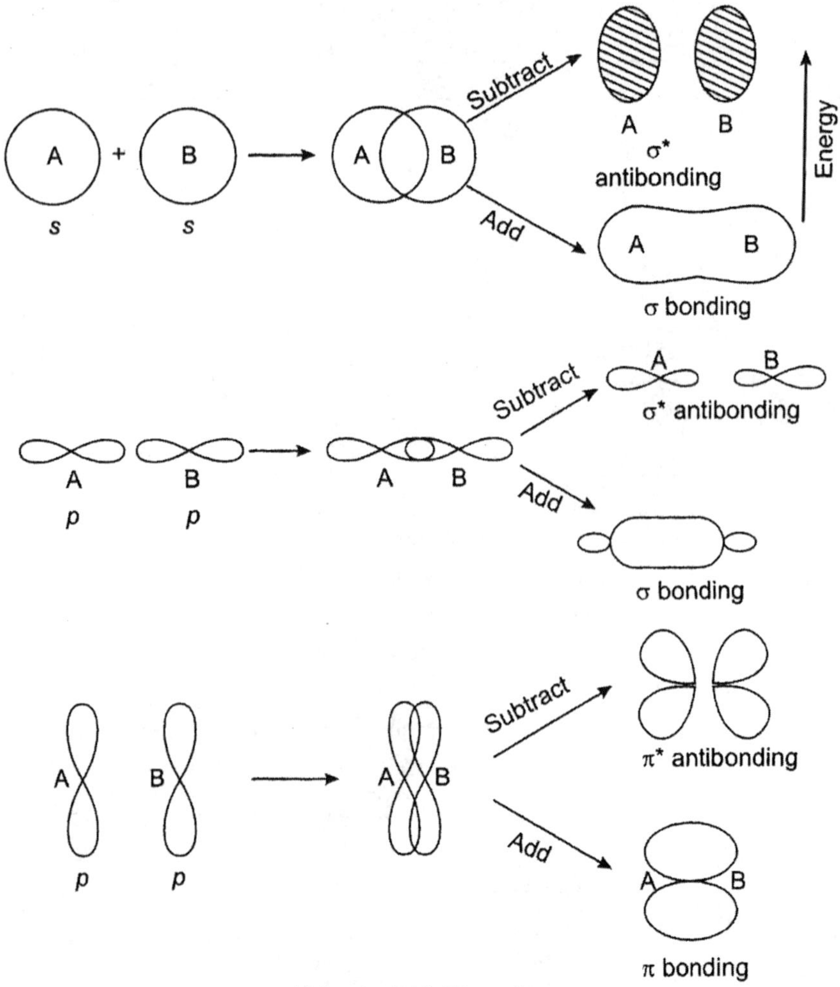

Molecular Orbital Formation

The difference between the energy of atomic orbitals and the bonding molecular orbitals depends upon the extent of overlap; large overlap results in small difference and thus molecules will be of lower energy than the separated atoms. The orbitals of interest are low-energy orbitals where the electrons reside.

If different types of orbitals are present then energies of atomic orbitals are different. The more electronegative atom will have the lowest energy atomic orbitals. The larger the difference in energy between two atomic orbitals that give a molecular orbital, the more ionic the bond, the amount of overlap between the atomic orbitals is a measure of covalent bonding.

The MO energy level diagram for metal complexes are complicated than that of simple diatomic molecules.

The MO diagrams for $[Co(NH_3)_6]^{3+}$ and $[CoF_6]^{3-}$ are given in figure underneath:

3+ 6–
(*i*) **$[CoF_6]^{3-}$**

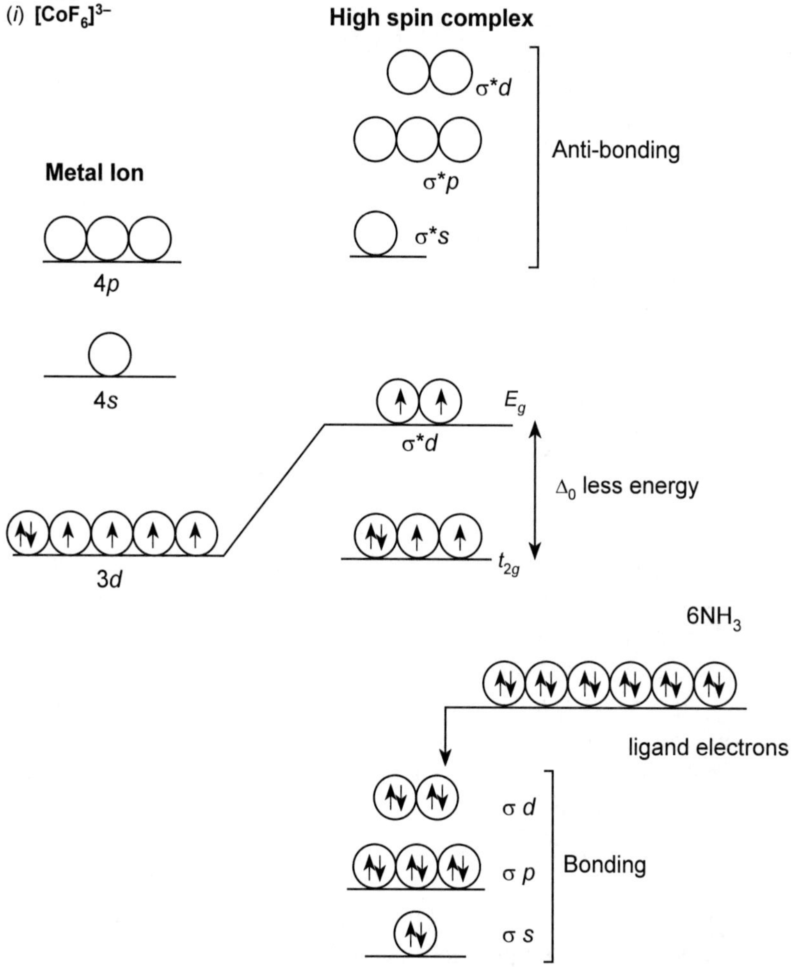

MO picture of bonding between Co^{3+} and $6F^-$ ions

(*ii*) $[Co(NH_3)_6]^{3+}$

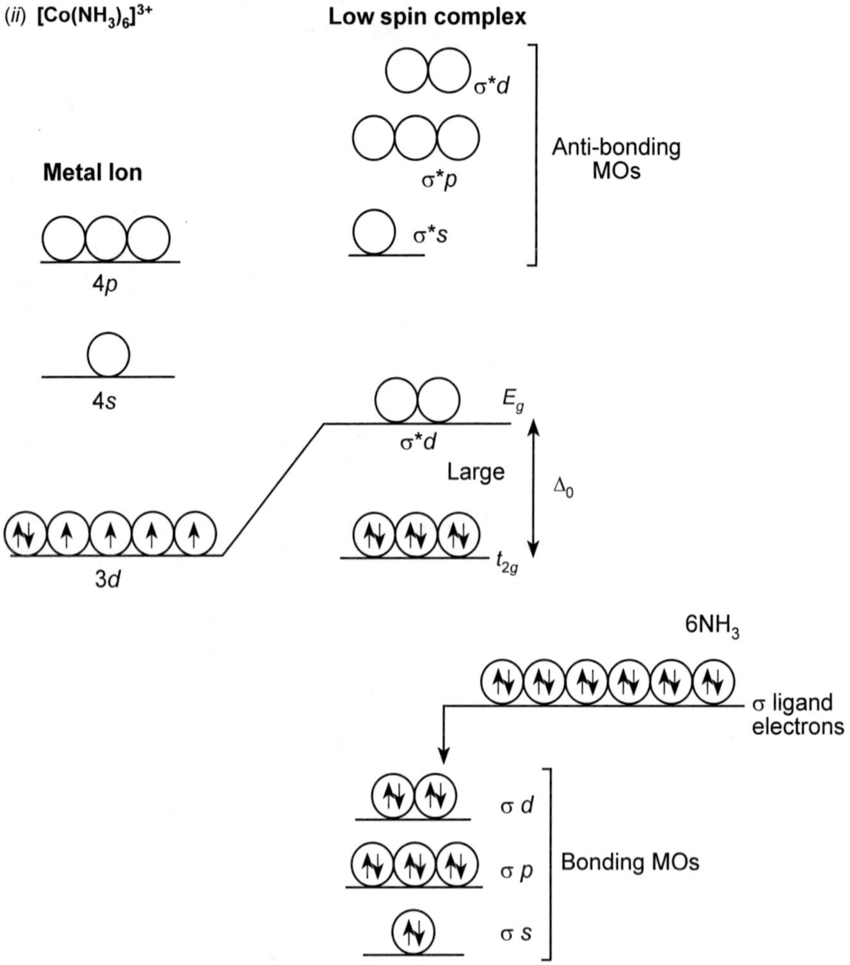

MO picture of bonding between Co^{3+} and $6NH_3$ molecules

On the left are *3d,* 4*s* and 4*p* orbitals of Co^{3+}. The AO's of lower or higher energy are immaterial. On the right, six ligands are placed at one energy level. The ligand orbitals are lower in energy than the metal orbitals. This gives some ionic character to the bonds. Bonding orbitals behave more like ligand orbitals than metal orbitals and placing metal electrons to these ligand orbitals transfers electronic charge from metal to ligand. Two *d* orbitals namely e_g (d_z^2 and $d_x^2{-}y^2$), 4s and three *p* orbitals are oriented where the ligands are located. This presents a fit picture of atomic orbital overlap with ligands resultings in six bonding and six, antibonding MO's $\sigma_s(1)$, $\sigma_p(3)$, $\sigma_d(2)$ and $\sigma_s^*(1)$, $\sigma_p^*(3)$, $\sigma_d^*(2)$. The t_{2g} (d_{xy}, d_{yz}, d_{zx}) orbitals do not point towards ligand and hence do not participate in *a* bonding and are called non-bonding orbitals.

The ligand electrons are placed in the complex MO's, *i.e.,* the six bonding MOs are filled as the complex is 6 coordinate forming six metal-to-ligand bonds. The remaining electrons are distributed among the non-bonding MO's (t_{2g} orbitals and σ^*d orbitals).

If the splitting energy Δ between non-bonding t_{2g} and σ^*d MO is small then Hund's Rule is obeyed as in [CoF_6] the d electrons are distributed as $t_{2g}^2\, e_g^2$. The presence of two electrons in σ^*d orbitals and hence Co—F is weakened.

Ammonia (NH_3) is a strong field ligand and give large A therefore all the Co(Ill) electrons fill the lower energy t_{2g} orbitals, *i.e.,* $t_{2g}^6\, e_g^0$.

CFT attributes to the CF splitting essentially to the electrostatic repulsion while MOT to the covalent bonding, *i.e.,* the extent of overlap between e_g metal orbitals with ligand orbitals.

Bonding in $[Fe(CN)_6]^{4-}$ ion

In case of $[CoF_6]^{3-}$ and $(Co(NH_3)_6]^{3+}$ and other similar metal complexes only a o overlap of AO's was considered. In $[Fe(CN)_6]^{4-}$ ion both *a* and *n* bonding occurs.

In the σ bond, ligands act as Lewis bases and hence donate a pair of electrons o empty e_g (d_z^2 and d_x^2–y^2) orbitals. In the π bond CN^- ion acts as a Lewis acid and accepts electrons from the totally filled t_{2g} (d_{xy}, d_{yz}, d_{2x}) orbitals of the metal.

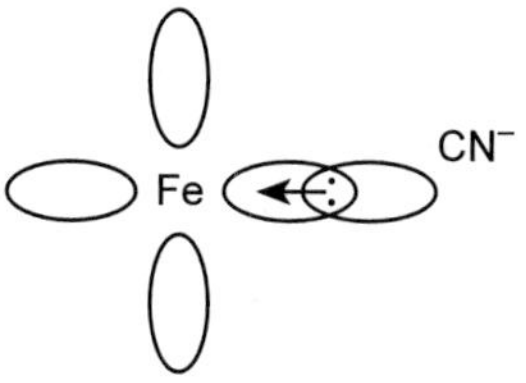

Fe ← CN⁻ σ bond

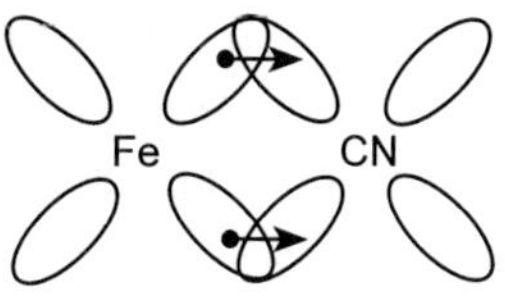

Fe → CN⁻ π bond

The formation of both σ and π bond strengthens the metal ligand bonding. In reality it is a two way bond formation and hence leads to stability of the $[Fe(CN)_6]^{4-}$ complex ion. In MnO_4^- both *a* and *n* bonds are observed, ligand oxygen provides the electrons for a *n* bond.

The large CF splitting provided by CN^- and CO and other π bonding ligands can be explained on this basis. The t_{2g} orbitals of metal in octahedral environment are oriented for π-bonding. The t_{2g} metal orbitals point between the ligands and hence no o bond can be formed.

In a π bond in case of CN^-, low energy electrons are partially transferred to a low energy t_{2g} orbitals. Due to this t_{2g} orbitals occupy lower energy. This explains that MOT is very useful in contributing to both covalent and ionic nature of metal-ligand bond. According to the both covalent and ionic model oi metal-ligand bond, MOT is most correct and acceptable theory to date.

SUMMARY

Metal-ligand bonds in complexes are not as simple as they appear. They are partly covalent and partly ionic. Covalent bonding is explained by Lewis acid-base and valence bond theory whereas ionic bonding is supported by Crystal Field Theory (CFT) and Ligand Field Theory (LFT). In reality the bond is neither covalent nor ionic therefore Molecular Orbital Theory (MOT) is invoked to cover both the aspects. Valence bond theory, depending on the strength of the ligand gives inner and outer orbital complexes in octahedral geometry. The six bonds between metal and ligands are of equal strength. Therefore, this may be due to Hybridisation of bonding orbitals. Crystal field theory based on ionic model takes care of bonding where degenerated orbitals belonging to metal ions split under the influence of ligands into high energy e_g and low energy t_{2g} orbitals.

The energy gained in octahedral environment for every electron in e_g orbital is hypothetically 0.4 A_0 and energy loss for every electron in t_{2g} is

0.6 A_0 so that in case of unfilled, half-filled and completely filled *d* orbital the gain in energy is zero or indirectly they are extra stable. This is crystal field stabilisation energy, CFSE. It gives useful data for relative stability of a complex. In octahedral splitting there are high spin and low spin complexes.

In tetrahedral environment splitting is $\Delta t = ½\ \Delta_0$. Therefore, low spin complexes are not favoured. Both VBT and CFT do not satisfy all the aspects of bonding in complexes both singly or in a combined way therefore one theory which takes care of all the aspects of bonding in complexes is Molecular Orbital Theory (MOT). This theory envisages both covalent and ionic aspect of the complex. In this after *d* orbital splitting, the low energy electron in the ligands pick up vacant low energy orbitals of the metal and accordingly choose two *s*, three *p* and two *d* orbitals for octahedral complex exemplified by $[CoF_6]^{2-}$ and $[Co(NH_3)_6]^{3+}$. The role of e_g and t_{2g} orbitals is further explained in $[Fe(CN)_6]^{4-}$ complex.

EXERCISE

1. What is EAN rule for complex stability?
2. All carbonyls except $[V(CO)_6]$ obey LAN rule. Justify.
3. Bring out the examples of complexes which do not follow EAN rule.
4. Calculate EAN of the central atom:
   $[Ca(NH_3)_4]^{2+}$, $[No(Co)_4]$, $[Co(NH_3)_6]^{3+}$, $[Fe(CN)_6]^{4-}$, $[Fe(CN)_6]^{3-}$, $[Pt(NH_3)_2Cl_2]$, $[PdCl_4]^{2-}$
5. Complex formation is simply a Lewis acid-base reaction where metal ion acts as Lewis acid and ligand is a Lewis base. Illustrate with example.
6. Give reason why:

(*a*) Inner orbital octahedral complex of Co^2 is unstable.
(*b*) Inner orbital octahedral complex of Ni^{2+} is not possible.
(*c*) Low spin tetrahedral complexes are not known.
(*d*) Cu^{2+} ion prefers square planar complexes over tetrahedral or octahedral.
(*e*) Ni^{2+} ion forms 4 coordinate square planar rather than tetrahedral.
(*f*) $[Fe(CN)_6]^{4-}$ is a low spin complex while $[Fe(H_2O)_6]^{3-}$ is a high spin complex.
(*g*) $[Sc(H_2O)_6]^{3+}$ ion is colourless and $[Ti(H_2O)_6]^{3-}$ ion is coloured.
(*h*) $[Co(NH_3)_6]^{3-}$ ion is diamagnetic and $[CoF_6]^{3-}$ ion is paramagnetic.
(*i*) $[Ni(Co)_4]$ complex is tetrahedral while $[Ni(CN)_4]^{2-}$ is square planar.
(*j*) Aqueous solution of most of the transition metal ions is coloured.
(*k*) CN^- forms 6-coordinate complex with Cr(III) and Co(III) but with Ni(II) complex formed is 4-coordinate.
(*l*) On electronegativity scale halide ions are arranged $I^- < Br^- < CF^- < F^-$ and on spectra chemical scales the arrangement is $F^- < Cl^- < Br^- < I^-$.
(*m*) Cu^{2+} complexes are coloured but Zn^{2+} are colourless.
(*n*) Fe forms low spin octahedral complexes.
(*o*) Square planar complexes are more stable than octahedral complexes.
(*p*) The values of magnetic moment (BM) for the following complexes are:
(*i*) $K_3[FeF_6] = 5.9$; $[Co(NH_3)_6]Cl_3 = 0.0$; $[Ni(H_2O)_6]Cl = 2.9$
(*ii*) $[Fe(H_2O)_6]^{2+} = 4.9$ and $[Fe(CN)_6]^{4-} = 0.0$

7. Define 'inner and outer orbital complexes'.

8. Apply VBT to explain geometry in:
$[Cu(CN)_4]^{2-}$ and $[Cu(NH_3)_4]^{2+}$ ions

9. Explain hybridisation of Cu in $[Cu(NH_3)_4]SO_4$ and Fe in $[Fe(H_2O)_6]Cl_3$,

10. Enumerate the limitations of VBT.

11. Discuss the geometrical shapes associated with hybrid orbitals
(*a*) sp^3 (*b*) dsp^2 and (c) d^2sp^3 with examples.

12. Explain crystal field stabilisation energy and its significance.

13. Draw the shapes of various '*d*' orbitals and explain why they are split into two classes t_{2g} and e_g in octahedral environment.

14. The *d* orbitals are split into groups with different energy in an octahedral field. On the basis of crystal field splitting draw the CFSE for high spin and low spin arrangements. ($-0.6\Delta_0$ for every electron in E_g and 0.4 for each electron in t_{2g}).

15. Draw the energy level diagram of 3d orbitals in a tetrahedral environment.

16. Draw the energy level diagram for the following complexes and calculate the CFSE:
    (*a*) d^6octahedral low spin
    (*b*) d^9 octahedral with tetragonal elagation
    (*c*) d^8 square planar
    (*d*) d^6 tetrahedral
17. Calculate CFSE value for d^6 ion both in octahedral and tetrahedral complexes.
18. Draw the diagram to show the pattern of *d* orbital splitting changes in octahedral, tetragonal and square planar complex.
19. What are *d–d* transitions? In some complexes they are forbidden and they do occur in some complexes.
20. Describe the Jahn-Teller effect in octahedral complexes of Cr^{2+} and Cu^{2+}.
21. Usually Cu^{2+} complexes are four coordinate but with fluoride ligand it is six coordinate with four Cu—F distance 1.93Å and two F^- at 2.27Å. Explain.
22. Describe Δ_0 changes when M^{2+} changes to M^{3+} and how it changes in a vertical triad between 3*d,* 4*d* and 5*d* elements.
23. What are CFSE values for:
    (*a*) $[CoF_5]^{3-}$ (*b*) $[Co(NH_3)_6]^{3+}$
    (*c*) $[Fe(H_20)_6]^{2+}$ (*d*) $[Fe(CN)_6]^{4-}$
    (*e*) $[Fe(CN)_6]^{3-}$
24. Complex $[Ni(CN)_4]^{2-}$ is diamagnetic and square planar while $[NiCI_4]^{2-}$ is paramagnetic and tetrahedral. Explain the observation.
25. What is the molecular orbital theory? How far do$s it meet the shortcomings of valency bond theory and crystal field theory?
26. Draw the molecular orbitals diagram for the following complexes:
    (*i*) $[CoF,]^{3-}$ (*ii*) $[Co(NH_3)_6]^{3+}$
27. How do we explain the bonding in $[Fe(CN)_6]^{4-}$ on the basis of molecular Orbital Theory (MOT).
28. Compare VBT, CFT and MOT in the following complexes:
    $[Fe(H_2O)_6]^{2+}$, $[Ni(NH_3)_6]^{2+}$ high spin and $[Co(C_2O_4)_3]^{3-}$ low spin.
29. What is ligand field theory? Give example of bonding in metal carbonyls.

■ ■ ■

7

Preparation and Isolation of Complexes

7.1 PREPARATION

Preparation of (*i*) complexes, (*ii*) metal carbonyls and organometallic compounds is just an addition of a slight excess of ligands to soluble salt of the metal but lot of problems are encountered in its isolation. In the conductivity measurement of the complex we come to know whether the complex is salt-like or covalent. Covalent complexes are soluble in non-polar solvents. These are associated with low melting and boiling points and generally have soft crystals. Metal carbonyls and organometallic compounds belong to this category, *e.g.*, $K[(PtCl_3C_2H_4)]$ and $Fe(C_5H_5)_2$.

Preparation: The experimental methods generally used for the preparation of complexes are of very simple nature.

Step I: A soluble salt solution of the metal ion is taken in a solvent (H_2O) or non-aqueous medium in which it is highly soluble.

Step II: A slight excess of the ligand is added to the salt solution. It is generally observed by the change in colour of the salt solution. Change in colour of the salt solution indicates the formation of the complex compound in solution.

Trans Effect

Regarding the preparation of square planar complexes of Pt(II) it was observed that certain ligands easily replace the groups in the trans-position. This is called Trans Effect. Therefore, two square planar complexes can be prepared for $[Pt(NH_3)_2Cl_2]$ *i.e., cis* and *trans.*

1. If this complex is prepared by adding NH_4OH ($NH_3.H_2O$) to a solution of $[PtCl_4]^{2-}$ then the dipole moment value suggests *cis*-isomer.

2. If the complex is prepared by adding HCl to $[Pt(NH_3)_4]^{2+}$, then the dipole value indicates *trans* form.

Example,

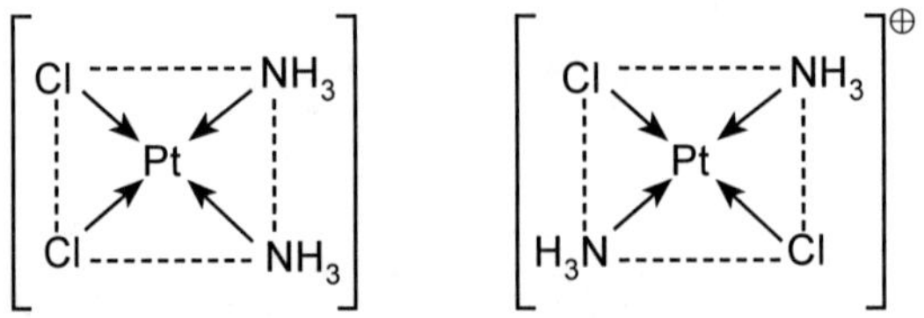

Cis *Trans* form chelate

In the case of bidentate chelate ligands in which the two donor atoms are different *cis* and *trans* may exist

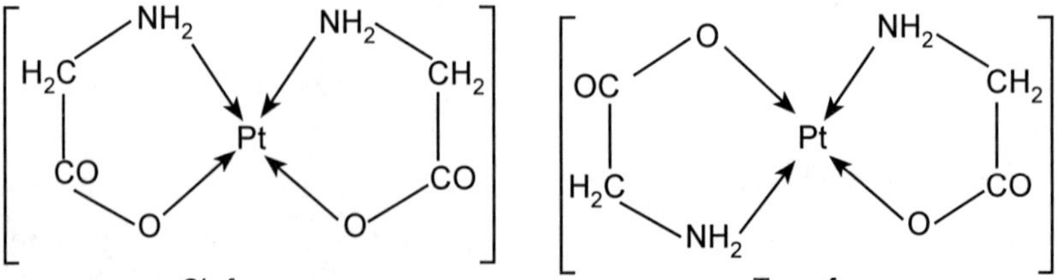

Cis form *Trans* form

Trans effect of a number of ligands decreases in the order:

$CN^- \sim CO \sim C_2H_4 > PH_3 \sim HS > NO_2^- > I^- > Br^- > Cl^- > NH_3 \sim Py > OH^- > H_2O$

In complexes other than Pt(II), trans effect is not as effective.

Preparation of Metal Carbonyls

(*a*) Carbonyls were first prepared by Mond. Nickel iron and cobalt carbonyls can be prepared by passing gaseous carbon monoxide over the metal at 1 atm and 25°C.

$$Ni + 4CO \xrightarrow[25°C]{1\ atm} [Ni(CO)_4] \text{ colourless}$$

$[Ni(CO)_4]$ boils at 43°C $[Co_2(CO)_8]$ can be prepared at high pressure and temperature. On strong heating they can be separated into metal and CO. Simple monomeric carbonyls obey EAN rule. Black $[V(CO)_6^-]$ is the only carbonyl which does not obey EAN rule. Some important carbonyls are $[Fe(CO)_5]$, $Mn_2(CO)_{10}$ and $[Cr(CO)_6]$.

(*b*) Preparation of olefins of transition metals was done by W.C. Zeise. According to him ethylene reacts with $[PtCl_4]^{2-}$ in dilute HCl solution to yield, complex containing both ethylene and platinum. This complex acquires a square planar geometry with ethylene pointing between the two carbon atoms.

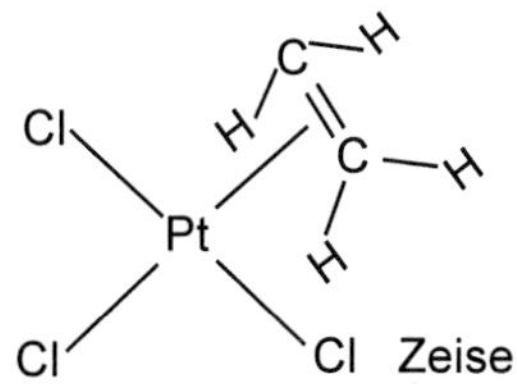

Zeise salt

Additional stability is provided by π bonding between filled metal orbitals and vacant antibonding ethylene molecular orbitals.

Preparation of Organometallic (Sandwich Compounds)

The first sandwich that came to known was *bis* (7i-cyclopentadienyl) iron (II) which was later called Ferrocene. It is an orange coloured compound having B.P. 249°C. It does not decompose at its B.P. It is neither attacked by the addition of NaOH or HC1. The compound is diamagnetic and non-polar. C_5H_5 acts as a six electron donor like benzene. The two ligands donate correct number of electrons so that it obeys EAN rule, hence highly stable $Fe(C_5H_5)_2$, $Cr(C_6H_6)_2$, and $Mn(C_5H_5)(C_6H_5)$.

Ferrocene can react like any other aromatic compound without any harm to metal-ligand bond. These compounds are prepared by taking the metal halide and sodium or potassium cyclopentadienide.

$$FeCl_2 + 2K(C_5H_5) \longrightarrow 2KCl + \underset{\text{Orange}}{(C_5H_5)_2Fe}$$

Dibenzene chromium can be prepared by the following reaction:

$$CrCl_3 + 2Al + AlCl_3 + 6C_6H_6 \longrightarrow \underset{\text{Yellow}}{3[(C_6H_6)_2Cr]} + AlCl_4^-$$

Dibenzene chromium sandwich compounds are less stable than the corresponding cyclopentadienyl compounds. They are readily oxidised and decomposed. To further stabilise these compounds they are converted into their respective salts.

$$(C_6H_6)_2Cr + ClO_4^- \longrightarrow \underset{\text{Yellow}}{[(C_6H_6)_2Cr]^+ClO_4^-}$$

Preparation of Transition Metal σ Bonded Complexes

Quite recently these compounds have gained importance. They contain a bonded alkyl or aryl groups. The ligands such as CO, $C_5H_5^-$ or phosphines increase the possibility of o bonded compound formation.

Example,

$$(CO)_5MnNa + CH_3I \longrightarrow NaI + \underset{\text{Colourless}}{(CO)_5Mn(CH_3)}$$

$$Cis[\{P(C_2H_5)_3\}_2PtBr_2] + 2CH_3MgBr \xrightarrow{\text{Ether}} \underset{\text{Colourless}}{Cis\text{—}[\{P(C_2H_5)_3\}_2Pt(CH_3)_2]} + 2MgBr_2$$

7.2 ISOLATION

Common techniques are as under:

1. Evaporate the solvent and cool the concentrated reaction mixture in an ice bath to obtain the crystals of the complex.
2. If crystals are not obtained by evaporation of the solvent then some crystals of the sample are added to induce seeding by scratching the inner wall of the container below the surface of the liquid. This technique is called 'nucleation' or 'seeding'.
3. A suitable solvent is added that is miscible with the solvent of the reaction mixture but the desired complex is insoluble. To cause precipitation of the complex scratching and screening may be desired.
4. If complex is a cation then suitable anion is added to the reaction mixture with which it forms an insoluble salt or a suitable cation is added to the reaction mixture to precipitate the salt.

For example, in the preparation of the tetraammine copper (II) sulphate, the steps involved are as follows:

1. Aqueous copper sulphate solution is taken.
2. To the solution in step 1. measured amount of ammonia (1:1) is added. Remember ammonia is in slight excess. Deep blue colouration indicates that complex is formed.
3. To isolate the complex a quantitative amount of ethyl alcohol is added Ethyl alcohol is chosen because it is miscible with water but the complex is sparingly soluble. The solution is warmed slightly so that the complex in alcohol solution is dissolved completely.
4. The reaction is allowed to cool so that the crystals appear out of the saturated solution.
5. The crystals so obtained are finally dried and preserved.

Purification may be done by distillation, sublimation or chromatographically.

SUMMARY

Preparation of complex is very simple. A metal salt solution is taken in aqueous medium and a slight excess of the ligand is added. The complex is highly soluble in aqueous medium, therefore it is extracted by the following methods:

(*i*) Evaporation of the solvent if the complex does not decompose at high temperature.

(*ii*) Add a suitable solvent to the reaction mixture. The solvent should be miscible with the reaction medium but the complex must be sparingly soluble in it *e.g.*, C_2H_5OH (water loving solvent).

(*iii*) The non-aqueous solvent in which the preparation is carried out must be volatile *e.g.*, ether.

$$\underset{\text{salt solution}}{[Cu(H_2O)_4^{2+}]\,SO_4} + \underset{\text{ligand}}{4NH_3} \xrightarrow{H_2NOH} [Cu(NH_3)_4]\,SO_4$$

For preparation of organometallic compounds cyclopentadiene is treated with alkali metal to form the salt which in turn is reacted with transition metal salt in a suitable solvent. The solvents used may be $CHCl_3$, THF, DMF, alcohol, acetone etc.

$$C_5H_6 + \underset{\text{metal}}{Na} \xrightarrow{CHCl_3} C_5H_5Na + \frac{1}{2}H_2$$

$$2C_5H_5Na + TiCl_4 \xrightarrow{THF} (C_5H_5)_2TiCl_2 + 2NaCl$$

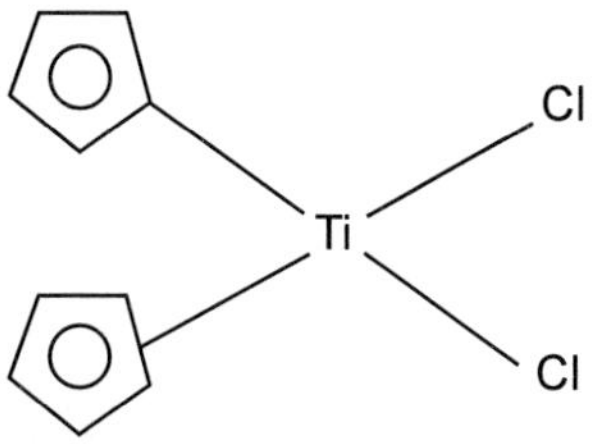

bis n^5–(cyclopentadienyl)titanium (*iv*) dichloride orange red crystals. The main plank of preparation is choosing a suitable solvent in which the ingredients are soluble but the complex obtained is insoluble. The complex is later isolated from the solvent by the methods mentioned in the extraction.

EXERCISE

1. What are the basic points to look into the preparation and isolation of complexes?
2. How will you prepare tetraamminecopper (II) sulphate from aqueous solution of copper sulphate? How will you isolate or extract the complex?
3. Prepare potassium trioxalatochromate (III) complex from $K_2Cr_2O_7$ and extract the crystals.
4. How do we choose an appropriate solvent to isolate the complex compound?
5. List out the complexes where it can be isolated by evaporation or distillation.
6. What are the conditions for the preparation of $[Ni(CO)_4]$ complex by Mond's Process?
7. Why ether is chosen as a suitable solvent for the isolation of metallocenes?

8. One H-atom is easily replaceable by the more electropositive Na and K metals in cyclopentadiene. Comment

9. Give preparation of Zeise salt.

10. Write the chemical equations involved in the preparation of η^5–$(C_5H_5)_2Fe$ and η^5–$(C_5H_5)_2TiCl_2$ orange red crystals.

11. Give two examples of metal carbonyls which are dimeric.

12. How *cis* and *trans* isomers of $[Pt(NH_3)_4Cl_2]^{2+}$ are isolated?

13. What is *trans* effect? Explain with example.

14. In the extraction of which type of complexes Ethyl alcohol or DMF (dimethyl formamide) is used.

15. What are the essentials for a complex formation?

16. What are the different steps in the isolation of a desired complex compound?

■ ■ ■

8

Stability of Complex Ion

8.1 INTRODUCTION

Metal ions can form complexes with solvent molecules and with ligands in solution. For the metal ions, it is important to combine with a preferred ligand which is able to impart higher stability.

In a copper (II) salt solution, if ammonia is added then Cu(II) aqua complex is replaced by NH_3 as given in the reaction:

$$[Cu(H_2O)_4]^{2+} + NH_3 \rightleftharpoons [Cu(NH_3)(H_2O)_3]^{2+}$$

$$[Cu(NH_3)_2(H_2O)_2]^{2+} + NH_3 \rightleftharpoons [Cu(NH_3)_2(H_2O)_2]^{2+}$$

$$[CUNH_3(H_2O)_3]^{2+} \rightleftharpoons [Cu(NH_3)_3H_2O]^{2+}$$

$$[Cu(NH_3)_3H_2O]^{2+} + NH_3 \rightleftharpoons [Cu(NH_3)_4]^{2+}$$

Therefore, cupric ions have preference for a certain ligand over the other. This may be attributed to the fact that the positive metal ions prefer anionic ligands to neutral and positive ligands. The stability of a complex is indicated by stability constants. Example,

$$H^+ + F^+ \rightleftharpoons HF$$

$$K_{\text{stability constant}} = \frac{[HF]}{[H^+][F^-]}$$

In the context of stability constant generally solvent molecules are water but non-aqueous solvents are preferred.

Metal complexes are formed stepwise and equilibrium constant can be studied for each step.

The water molecules which surround the metal ions to form hydration spheres are not taken into account even in equilibrium constant. In dilute solutions the activity of water is close to unity.

Metal complexes are formed due to stepwise exchange of ligand *e.g.*

$$Ag^+ + NH_3 \longrightarrow [Ag(NH_3{}^+)]\ K_1$$

$$[AgNH_3]^+ + NH_3 \longrightarrow [Ag(NH_3)_2]^+\ K_2$$

The equilibrium constants are given by:

$$K_1 = \frac{[AgNH_3]^+}{[Ag^+]\,[NH_3]} \text{ and } K_2 = \frac{[Ag(NH_3)_2]^+}{[AgNH_3] + [NH_3]}$$

The overall stability constant is expressed as:

$$K_1K_2 = \frac{[Ag(NH_3)^+]\,[Ag(NH_3)_2]^+}{[Ag^+]\,[NH_3]\,[AgNH_3]^+\,[NH_3]}$$

Large equilibrium constant values indicate higher concentration of complex ions with respect to the compounds. Only the exchange of water ligand is fast while others are slow due to resistance offered to the exchange.

8.2 FACTORS INFLUENCING STABILITY

In general there are number of factors controlling stability of complexes. A complex comprises of a metal ion and a ligand therefore it is pertinent to discuss the kind of metal ions and ligands which together contribute to the stability.

On the basis of energy consideration, if heat is evolved during the complexation then stability of complex is higher because the complex itself goes to the lower energy state. Secondly, if the entropy or disorder is more in products than the reactants, then greater entropy yields greater stability.

On the basis of electrostatic model, charge on the metal ions should be greater and the size or radius should be smaller. This will make the metal ions hold the ligands strongly. In fact there is a relationship between charge-to-radius, *i.e.*, $\frac{e \text{ charge}}{r \text{ radius}}$ should be high.

Another factor worth considering is crystal field stabilisation energy (CFSE). If electrons in the metal ions reside in t_{2g} orbitals according to crystal field theory, hypothetically every single electron in t_{2g} stabilises the complex by 0.4 Δ_0 in octahedral complexes.

On the basis of quality of ligands, they have an important say in complex stability. In ionic model, if the ligand is electronegative then its attraction to metal ion is strong and thus the stability. The polar ligands yield greater binding force. Also, the base strength of the ligand gives stable complexes.

The metals are classified as:

(*i*) 'X' metals are alkali, alkaline earth, 3*d* elements, lanthanides ant actinides.

(*ii*) 'Y' metals are 4*d*, 5*d*.

X metals form stable complexes with ligands having donor atoms N, O or F on the basis of Ionic Model Picture while class 'Y' metals contribute covalent bonding to metal-ligand bond. They have a π bond character. There is transfer of electron density from metal to the ligand (Back donation). Their stability depends upon (*a*) base strength (*b*) chelate effects. Class 'Y' metals prefer ligands in which *d* orbital exists such as $P(C_2H_5)_3$, S^{2-} and I^- which help in π bond formation and electrons get delocalised. In short, they form stable complexes with higher members of N, O and F families. Chelating ligands because of the ring structure impart higher stability to metal chelates. Edta forms very stable complexes with Ca^{2+} ions, Mg^{2+} ions and Zn^{2+} ions. Five and six membered chelates are more stable than the rest. Besides, ligands containing multiple bonds form stable complexes, *e.g.*, acetyl acetone.

(*i*)

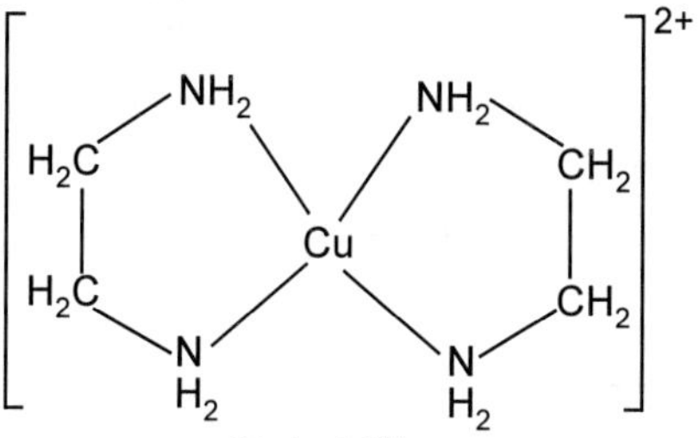

$[Cu(en)_2]^{2+}$

(*ii*)

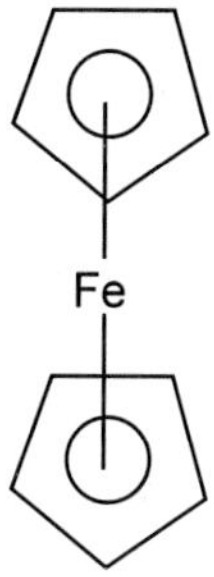

Ferrocene (orange)

(*iii*)

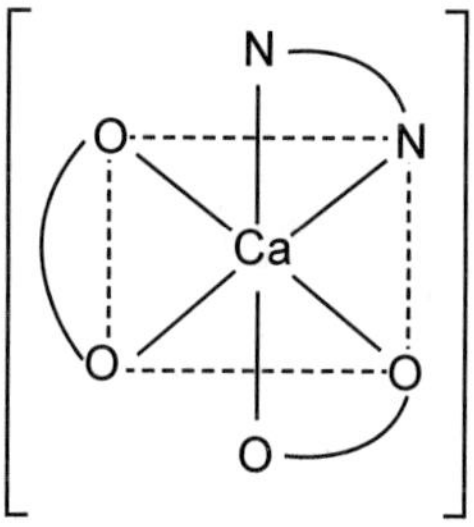

[Ca edta]

Stability of $[Ni(en)_3]^{2+}$ is greater than $[Ni(NH_3)_6]^{2+}$ ion, due to chelating effects.

SUMMARY

Stability of complex ion is measured by the stability constant *e.g.*,

$$A + B \rightarrow C$$

$$K = \frac{[C]}{[A][B]}$$

There are two factors on which stability of a complex depends: (*i*) Type of metal ion, (*ii*) type of ligand.

The factors that influence the complex formation are:

1. Size and charge of the metal ion: Smaller size and larger charge on the metal ion imparts greater stability to the complex.
2. CFSE plays an important role in the stability of the complexes in the first row transition elements.
3. Class 'X' metals (3*d* elements) form stable complexes with ligands having donor atoms N, O and F, while class *Y* metals (4*d* and 5*d* elements) form stable complexes with higher member of N, O and F families, *i.e.*, P, S and Cl etc.
4. (Base Strength of the Ligand): Greater the base strength, greater the stability of the complex.
5. Chelated complexes are stable due to greater entropy. More extensive chelation provides still greater stability.
6. Five or six-membered ring complexes are more stable. Five-membered ring ligand should be o bonded and six-numbered 7i-bonded.
7. Branched chain chelating ring complex give steric strain and make the complexes less stable than the simple unbranched ligands. Steric factor may also be due to geometry sometimes.

EXERCISE

1. Write the equilibrium constant value 'C and stability constant P in the following reactions:

$$Ni^{2+} + 4CN^- \rightleftharpoons [Ni(CN)_4]^{2-}$$

$$Ag^+ + 2NH_3 \rightleftharpoons [Ag(NH_3)_2]^+$$

$$Cr^{3+} + 3en \rightleftharpoons [Cr(en)_3]^{3+}$$

$$Fe^{3+} + 4Cl^- \rightleftharpoons [FeCl_4]^-$$

Which of the above values is greatest and smallest?

2. Calculate whether PbS will get precipitated with the following data:

Cation	$[K\{M(edta)\}]^{2-}$	$K_{sp}[MS]$
Pb^{2+}	2×10^{18}	4×10^{-26}
Ni^{2+}	2×10^{-18}	1×10^{-22}
Co^{2+}	1.6×10^{16}	5×10^{-22}
Zn^{2+}	3.9×10^{16}	6×10^{-27}
Cd^{2+}	2.6×10^{16}	1×10^{-20}

3. In qualitative analysis we come to know that AgCl dissolves in excess NH_3 whereas Agl does not. This inference is based on stability constant value of $[Ag(NH_3)_2]^+$. On the basis of this value AgCl dissolves but not Agl. This value is large to dissolve AgCl and not large enough to dissolve Agl.

$$K_{sp}(AgCl) = 1.7 \times 10^{-10}$$
$$K_{sp}(Agl) = 8.5 \times 10^{-17}$$
$$\beta_2[Ag(NH_3)_2] = 1.5 \times 10^7$$

Calculate whether AgCl or Agl will precipitate.

4. State the important factors that play a key role in the stability of a complex.

5. What is the contribution of ligands in determination of the stability of metal complexes?

6. Discuss the effect of chelation in the stability of complexes.

7. An important criterion for the stability is charge to radius ratio of the metal ions. Support your answer with suitable examples.

8. The stability of a complex can be gauged by (*i*) increase in evolution of heat (*ii*) increase in entropy as the reaction proceeds. How far do you agree?

9. Class 'X' metals (3*d* series) prefer ligands O and N-containing ligands and F^- while class 'Y' metals (4*d* and 5*d* series) prefer heavier element in N, O and F families, *i.e.*, P, S^{2-} and I^-. Why?

10. Ligands that do not have double bonds prefer 5-membered ring chelates and those with π-bonds prefer 6-membered ring chelates. Explain.

11. What role is played by steric strain in the chelates?

■ ■ ■

9

Kinetics or Inert and Labile Complexes

9.1 INTRODUCTION

Inert and labile complexes have been classified on the basis of the rate of reaction. The rate of reaction is described as decrease in number of moles of reactants or increase in the number of product molecules.

Example,

$$[Co(NH_3)_5Cl]^{2+} + H_2O \longrightarrow [Co(NH_3)_5(H_2O)]^{2+} + Cl^-$$

The convenient way of expressing rate of reaction quantitatively is half-life. The standard condition in which salt containing $[Co(NH_3)_5Cl]^{2+}$ is reduced to one-half concentration is 25°C and 113 hrs.

According to the rate law, dissociation of a molecule in which product molecule is obtained through a variety of intermediates is given by:

$$A \longrightarrow A'$$

$$\text{rate} \propto [A]$$

$$\text{rate} = k[A] \qquad \text{'P'...}(i)$$

where k is the rate constant. For reactions taking place rapidly, k is large and for a slow reactions k is small. This may also apply to the conversion of *cis*-isomer to *trans*-isomer *e.g.*,

$$cis\text{-}[Co(en)_2Cl_2]^+ \xrightarrow{k} trans\text{-}[Co(en)_2Cl_2]^+$$

$$\text{rate} = k[cis\text{-}Co(en)_2Cl_2]^+$$

In case the reaction proceeds through an intermediate *e.g.*

1. $A + D \xrightarrow{\text{Slow}} E$
2. $E \xrightarrow{\text{Fast}} A' + D$

In the above reactions product E is formed as a result of collision between A and D.

In reaction (*i*)

rate of formation of E $\propto$ [A][D]

rate $E = k[A][D]$ 'Q' ...(*ii*)

This is a slow reaction as indicated and a rate determining step. In reaction (*ii*) A′ is formed as fast as E is formed. Rate of A′ is equal to rate of E formation and D is not consumed in the reaction but the rate is dependent on its concentration, therefore, D acts as a catalyst.

$$\text{rate} = k[\text{A}][\text{D}] \quad ...(iii)$$

Interconversion of the optical isomers of $[Co(en)_3]^{3+}$ is catalysed by $[Co(en)_3]^{2+}$

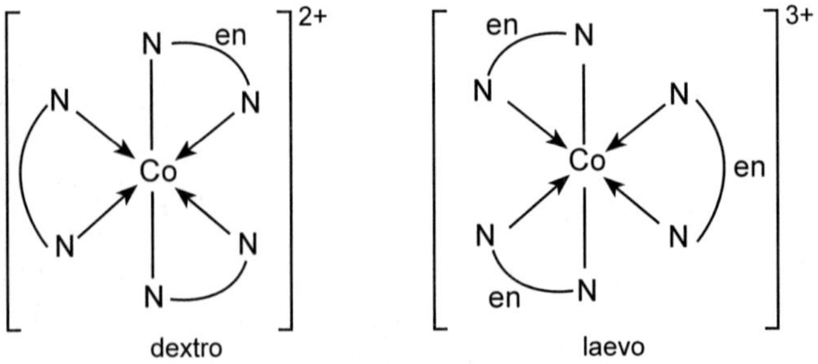

Hence, rate $= k[Co(en)_3]^{2+}[Co(en)_3]^{3+}$

Another possibility is a three body collision to form the product. This is rare and unlikely.

The equation 'P' is a first order reaction, 'Q' is a second order reaction and lastly the three body collision is a third order reaction.

Effective Collision

The rate of a reaction depends upon the rate law and rate constant

$$\text{Rate} = k[\text{reactants}]$$

The product molecules can be formed either rapidly or at a slow rate. This is related to the collision between the reactant species as collisions provide a certain amount of collision energy. If the collisions are effective then the product is formed. All collisions are not successful, *i.e.,* they do not provide enough energy to form the product.

Therefore, it is important that the collisions are effective. This depends upon geometry and orientation of the colliding molecules *e.g.,* a cyanide ion must approach the metal ion with carbon facing the metal to form metal carbon bond

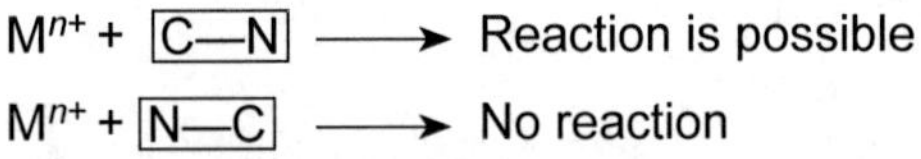

This applies more in the gas phase because in solution metal ion is captive of surrounding solvent molecules. All collisions are not having the same speed, some are fast while some are slow. The slowest step is the rate determining step. For a successful reaction, the colliding molecules get additional collision energy which helps to form an **activated complex**. The energy necessary to form activated complex is called activation energy.

Small activation energy will make the reaction proceed faster and high activation energy will make it slow. With the use of a catalyst, activated complex of lower energy is formed and the reaction can be made rapid.

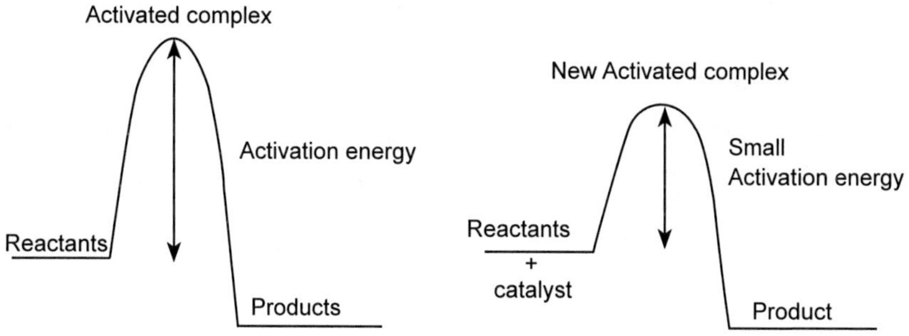

(*a*) Reaction proceeding without a catalyst

(*b*) Reaction with the help of a catalyst

Therefore, with the help of a catalyst a new activated complex is formed having lower energy. The rate constant *k* indicates how often a molecule absorbs enough energy to form an activated complex.

9.2 INERT AND LABILE COMPLEXES

In general, complexes in which ligands are rapidly replaced by other ligands are labile and those in which replacement is slow are called Inert.

To be more accurate and quantitative, lability and inertness is gauged at 25°C and 0.1M concentration of reactants. It is not essential that a stable complex may be inert and unstable complex may be labile.

The stability indicates that equilibrium of the reaction is far to the right.

$$[Ni(H_2O)_6]^{2+} + 4CN^- \longrightarrow [Ni(CN)_4]^{2-} + 6H_2O \quad ...(i)$$

$$[Ni(CN)_4]^{2+} + 4CN^- \longrightarrow [Ni(CN)_4]^{2-} + 2H_2O \quad ...(ii)$$

In equation (*ii*) if labelled ^{14}C cyanide ligand is added to common CN^- with ^{12}C, the replacement is instantaneous. Hence, it proves that the complex is stable but still labile.

Similarly Co(III) ammine complexes are unstable in acid solution *e.g.*

$$4[Co(NH_3)_6]^{3+} + 2H^+ + 26H_2 \longrightarrow [Cc(H_2O)_6]^{2+} + 24NH_3^+ + O_2$$

However, $[Co(NH_3)_6]3+$ can be stored in acid solution and remains undissociatcd at room temperature for days together.

It should be properly understood that stability depends upon the difference in energy of products and reactants and lability is the difference in energy of the reactants and activated complex; if this is large the reaction is slow and if less, it is fast. Taube stressed that electronic structure of the complex plays an important role. In six coordinated complexes, lability and inertness can be predicted with some amount of correctness. Here, the number of electrons in *d* orbitals play a significant role.

9.2.1 Classification of Inert and Labile Complexes on the Basis of Electrons in *d* Orbitals

Labile Complexes

(*a*) Labile complexes must have electrons in e_g ($d_z^2, d_x^2 - y^2$) according to the crystal field theory *e.g.,* $[Co(NH_3)_6]^{3+}$, d^7 ($t_{2g}5E_g^2$); $[Cu(H_2O)_6]^{2+}$, $d^9(t_{2g}6e_g^3)$; $[Fe(H_2O)_6]^{3+}$ $d^5(t_{2g}3e_g^2)$.

(*b*) Complexes which have less than three elcctrons in '*d*' *e.g.*,$[Ti(H_2O)_6]^{3+}$, d^1; $[Ca\ EDTA]^{2+}$ d° and $[V\ (phen)_3]^{3+}$ d^2.

Inert Complexes

d^3 configuration and low spin d^4, d^5, d^6

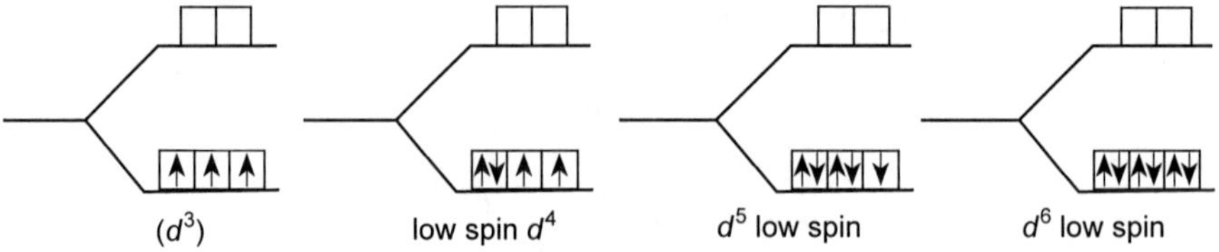

9.2.2 Inert and Labile Complexes on the Basis of CFSE

If the CFSE of a compound is larger than CFSE of activated complex then it reacts slowly and is termed inert. If the difference is small, it is rapid and is called labile. The change in CFSE of the compound and activated complex affects the rate of reaction. The CFSE of the compound is added to activation energy.

If the activated complex is less CFSE stabilised than the original compound, then activation energy increases and reaction becomes slow (Inert). If CFSE of the activated complex is more CFSE stabilised then activation energy decreases and reaction becomes faster (labile).

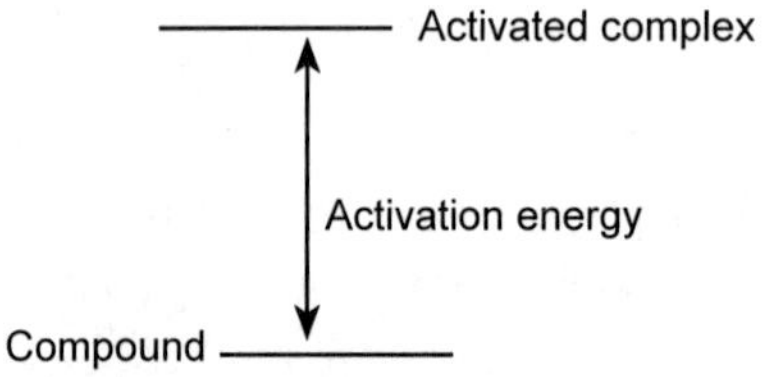

Less Stabilised Activated Complex

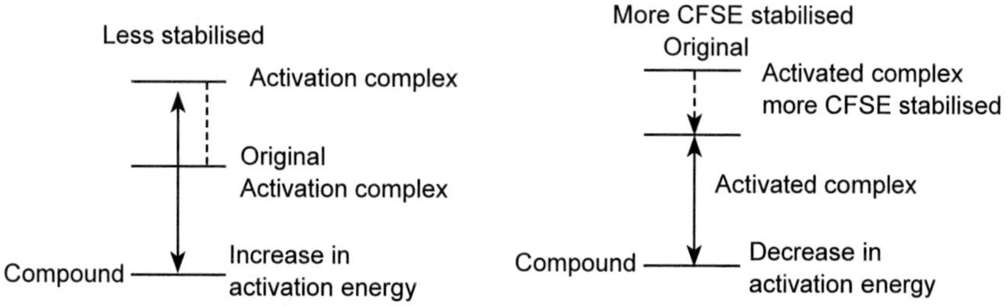

The above two factors give a crude approximation of the rate of reaction since in a number of cases they do not hold. Finally, it is the mechanism of reaction which has a greater say. The octahedral complexes react through an intermediate 5-coordinate geometry, *i.e.,* square pyramidal.

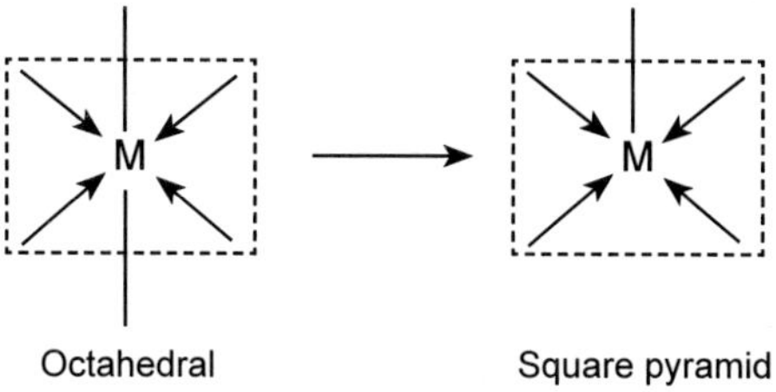

9.2.3 Role of a Catalyst (Kinetics)

The slow reaction may be speeded up with the employment of elevated temperature or pressure Alternatively the catalyst may successfully prepare metal complexes. Prolonged heating may destroy the complex. Catalysts speed up the reactions through the formation of an alternative activated complex of lower activation energy.

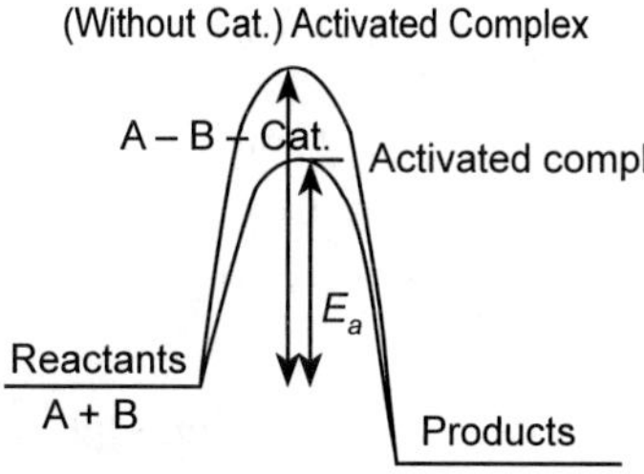

e.g.,

$$\underset{\text{Pink}}{[Co(H_2O)_6]^{2+}} \xrightarrow[NH_4Cl]{NH_3—H_2O—O_2} \underset{\text{Purple}}{[Co(NH_3)_5Cl]Cl_2}$$

In the presence of charcoal under the same conditions;

$$\underset{\text{Pink}}{[Co(H_2O)_6]Cl_2} \xrightarrow[NH_4Cl\ +\ charcoal]{NH_3—H_2O—O_2} \underset{\text{Orange}}{[Co(NH_3)_6]Cl_3}$$

Catalyst is incapable of changing the oxidation state of Co(II) but this happens due to the formation of an activated bridged intermediate which helps in air oxidation of Co(ll) to Co(III) through peroxobridged cobalt (III) complex *e.g.*,

$$\underset{\text{Orange}}{[Co(NH_3)_6]^{2+}} + O_2 \longrightarrow \underset{\text{Pink}}{[(H_3N)_5Co—O—O—Co(NH_3)_5]^{4+}} + 2NH_3$$

The bridged complex in turn reacts with NH_3 to form $[Co(NH_3)_6]^{3+}$. This may proceed through cleavage of O—O bond. Without the catalyst, ammonia may alternatively form $[Co(NH_3)_5OH]^{2+}$

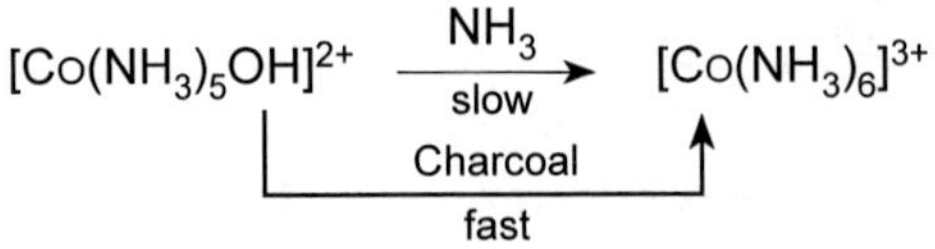

Recently homogeneous catalysts were observed to find role in the preparation of Pt(IV) complexes. Pt(II) acts as a catalyst in the preparation of Pt(IV) complexes.

$$\underset{\text{Yellow}}{trans\text{-}[Pt(NH_3)_4Cl_2]} + Br^- \xrightarrow[\text{Cat.}]{[Pt(NH_3)_4]^{2+}} \underset{\text{Orange}}{trans\text{-}[Pt(NH_3)_4Br_2]} + 2Cl^-$$

Pt(II) catalysis proceeds through an activated bridged complex and a two electron redox reaction.

This change is represented by:

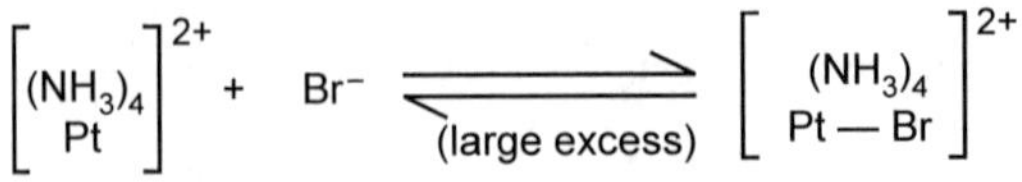

...(*i*)

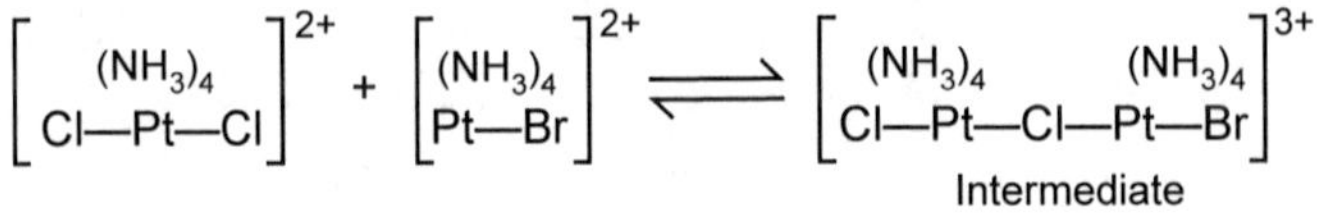

...(*ii*)

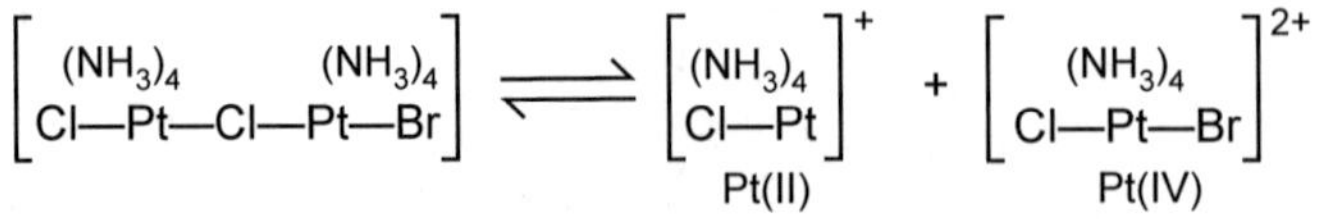

...(*iii*)

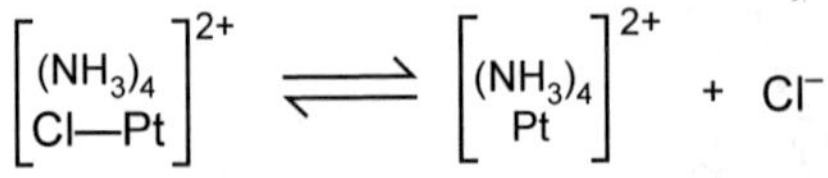

...(*iv*)

The catalyst $[Pt(NH_3)_4]^{2+}$ is weakly bonded with Br which is in large excess. In equation (*i*) in the formation of a bridged complex there is transfer of two electrons from Pt(II) to Pt(IV) through Cl bridging atom resulting in original Pt(II) to become Pt(IV). New Pt(IV) contains Br^- ion to give trans-$[Pt(NH_3)_4ClBr]^{2+}$. The catalyst is regenerated in equations (*iii*) and (*iv*).

SUMMARY

The rate of reaction is related to the concentration of the reactant molecule

$$A \rightarrow B \qquad ...(i)$$

$$\text{rate} \propto [A]$$

$$\text{rate} = k[A]$$

'*k*' is called the rate constant. For faster reaction '*k*' is large and small for slow reaction. The slow step is called the rate determining step *e.g.*,

(*i*) $$A + D \xrightarrow{\text{Slow}} E$$

(*ii*) $$E \xrightarrow{\text{Fast}} A' + D \quad ..(ii)$$

The step (*i*) is rate determining step. D is not consumed in the above steps therefore D acts as a catalyst, catalyst alters the rate of reaction but is left unused after the reaction.

The third mechanism is unlikely as it involves formation of slow intermediate E and E decomposes fast. This is quite rare *e.g.*.

$$A + D + D \xrightarrow{\text{Slow}} E \quad ...(iii)$$

$$E \xrightarrow{\text{Slow}} A' + 2D$$

$$\text{rate} = k[A'] [D]^2$$

Mechanism (*i*) is first order ($A \rightarrow A'$), mechanism (*ii*) is second order and mechanism (*iii*) is the third order reaction and has a first and second order dependence. The order of reaction depends upon the number of species colliding to form a product.

$$\text{rate} = k[A] [D] [D] = k[A] [D]^2$$

$$\text{order} = 1 + 2 = 3$$

Order is sum of index orientation of metal ion and ligand. Orientation should be to give successful collision.

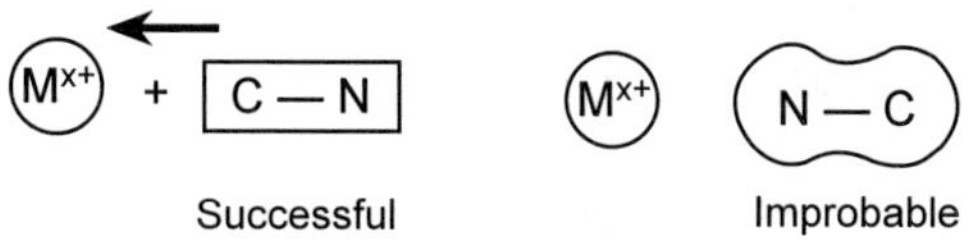

The collisions provide the energy required for successful collision. The additional collision energy gives activated complex. The energy necessary to form activated complex is called activation energy. If the activation energy is large, then the reaction is slow but the slow reaction can be made fast with the help of a catalyst which helps to form activated complex of lower energy.

If ligands are rapidly exchanged then it is called labile and if slow it is inert. The standard condition for inert and labile are : 25°C and 0.1 M concentration of reactant and the time limit is one minute.

Inert and labile can be classified on the basis of:

(*i*) Metal ion configuration

(*ii*) CFSE of complex compound and that of activated complex

(*iii*) Mechanism or route of the reaction and geometry of the intermediate complex (*iv*) Steric factors regarding the geometry

EXERCISE

1. Define rate law by taking example of a suitable reaction.
2. Rate of a reaction depends upon the slow step in the multi-step reaction. Justify.
3. How does effective collision affect the rate of reaction?
4. 'Catalyst increases the rate of reaction by forming an alternative activated complex of lower energy'. Explain.
5. Specify the standard condition of temperature, concentration and time for the classification of inert and labile complexes.
6. Indicate which of the following complexes are inert or labile and give reason for your choice:
   (*i*) $[A1(C_2O_4)_3]^{3-}$ (*ii*) $[Cr(C_2O_4)_3]^{3-}$ (*iii*) $[CoF_6]^{3-}$ high spin.
   (*iv*) $[Fe(CN)_6]^{4-}$ low spin (*v*) $[V(H_2O)_6]^{3+}$
   (*vi*) $[V(H_2O)_6]^{2+}$ (*vii*) $[Ni(NH_3)_6]^{2+}$ (*viii*) $[PtCl_6]^{2+}$ low spin.
7. Discuss how the CFSE values of complex compound and the activated complex, determine the inert and labile complexes.
8. Why d^2 and d^8 high spin metal ion configuration form inert octahedral complexes, while all other high spin complexes are labile?
9. Indicate the decreasing order of lability in hydroxides of (*i*) Mg^{2+}, Ca^{2+}, Sr^{2+} and Ba^{2+}; Mg^{2+}, $A1^{3+}$ and Si^{4+} and explain your answer.
10. Explain why $[Co(NH_3)_6]^{3+}$ is reduced much more slowly than $[Co(NH_3)_5Cl]^{2+}$ by $[Cr(H_2O)_6]^{3+}$. Write the equation in each case.
11. The complex $[Co(NH_3)_5SO_3]^+$ reacts in acid solution to give SO_2.
    $\therefore$ rate = $k[Co(NH_3)_5SO_3]^+[H^+]$
    $$[Co(NH_3)_5OSO_2]^+ + 2H^+ \xrightarrow{H_2{}^{18}O} [Co(NH_3)_5(H_2O)] + SO_2$$
    The reaction occurs without the incorporation of ^{18}O. Propose a mechanism that agrees with this change.
12. Complexes $[AuCl_4]^-$ and $[Au(dien)Cl]^{2+}$ undergo radioactive *Cl exchange
    $$[Au(dien)Cl]^{2+} + {}^*Cl^- \longrightarrow [Au(dien){}^*Cl]^{2+} + Cl^-$$
    rate = k[complex] + k'[complex] $[Cl^-]$
    Propose a mechanism.
13. The [M bipy]^{2f} complexes of 3*d* elements dissociate in H_2O as:
    $$[M\,bipy_3]^{2+} + 2H_2O \longrightarrow [M(bipy)_2(H_2O)_2]^{2+} + bipy$$
    The rates of these reactions increase in the order:
    $Fe^{2+} < Ni^{2+} < Co^{2+} \sim Mn^{2+} \sim Cu^{2+} \sim Zn^{2+}$
    Give reason.

■ ■ ■

10

Substitution Reactions in Complexes

10.1 INTRODUCTION

There are three different ways in which substitution reactions in complexes can occur:

(*i*) substition reactions in aqueous solutions.

(*ii*) substition reactions in non-aqueous solvents.

(*iii*) substition reactions where no solvent is required.

Aqueous solution: Most of the metal ions form aqua complexes in water solution. Water is regarded as a very good ligand. For preparation of complex, a metal salt solution is taken in aqueous medium and coordinating ligand is added in excess as no reaction in this category is quantitative. For example,

$$\underset{\text{blue}}{[Cu(H_2O)_4]^{2+}} + 4NH_3 \longrightarrow \underset{\text{dark blue}}{[Cu(NH_3)_4]^{2+}} + 4H_2O \qquad ...(1)$$

H_2O is replaced by ammonia molecules as NH_3 is a strong field ligand. In brief in all substitution reactions a strong field ligand replaces a ligand which is less strong. In the above exchange it is important to consider CFSE of octahedral and square pyramid geometries as the exchange of ligand is through 5-coordinate intermediate.

In d^3 or d^8 complexes there occurs a loss in CFSE in going to square pyramid geometry hence they react slowly and are termed inert. All other high spin complexes actually react fast and are labile. On the basis of high-spin and low-spin calculations the rates of reactions decrease in the order:

$d^5 > d^4 > d^8 \sim d^3 > d^6$ (all are low spin configurations)

Table 10.1: Change in CFSE

System	Octahedral	Square Pyramid	Change in CFSE
$d_1°$	0	0	0
d^1, d^6	0.40	0.45	–0.05
d^2, d^7	0.80	0.91	–0.011
d^3, d^8	1.20	1.00	+0.20
d^4, d^9	0.60	0.91	–0.31
d^5, d^{10}	0	0	0

The above sequence has been supported by the experimental result. The above values also confirm that half-filled has extra stability.

The rate of reaction is strongly guided by charge and size of the atom. Highly charged ions form stable complexes and so they react slowly decrease in lability is given by

$$[AlF_6]^{3-} > [SiF_6]^{2-} > [PF_6]^{-} > SF_6$$

If ligand is same then size of the metal ion is important. Central atoms with small ionic radii react slowly compared to those with large ionic radii

$$[Mg(H_2O)_6]^{2+} < [Ca(H_2O)_6]^{2+} < [Sr(H_2O)_6]^{2+}$$

Thus, it is generalised that central metal ion with large $\frac{\text{charge}}{\text{radius}}$ ratio reacts slowest.

As a corollary to the above statement, it is found that 3*d* row of transition metal ions due to small size react slow, the slowest being $[Ni(H_2O)_6]^{2+}$ *i.e.*, d^8 system compared to 4*d* and 5*d* metal ions. Generally four coordinated complexes (square planar and tetrahedral) react rapidly than six coordinated complexes. In four coordinated the fifth group has a lot of space to enter. For square planar $\frac{e}{r}$ ratio does not work well.

In all, the rate of reaction depends upon the mechanism of the reaction which in turn involves activation energy. Substitution reactions in complexes in case of the octahedral complex proceed through breaking of metal-ligand bonds therefore large positive charge will retard the loss of a ligand and the rate of reaction will be slow (inert complex). In four coordinated systems formation of new metal-ligand bond is important, therefore, large positive charge on the central atom favours the extra metal-ligand bond formation.

10.2 MECHANISM OF SUBSTITUTION

6-Coordinate Complexes

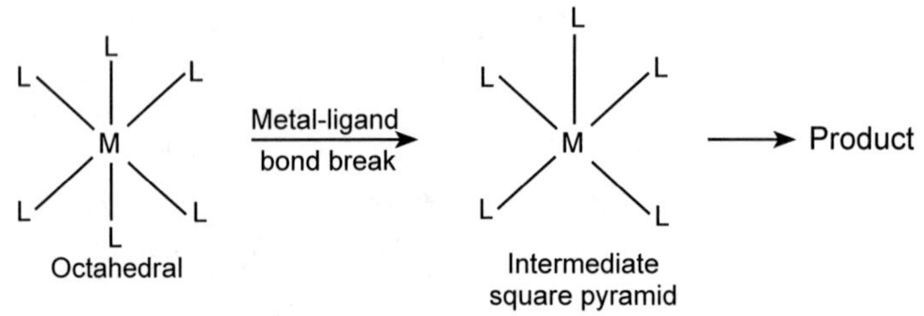

4-Coordinate Complexes

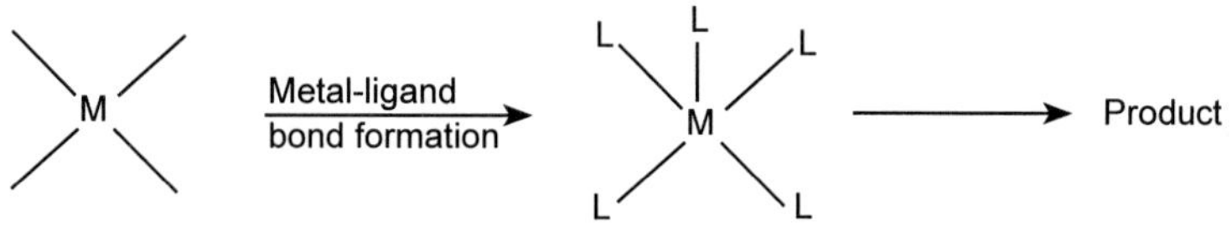

5-coordinate intermediate is a must for substitution in octahedral and square planar complexes.

As observed, of reaction conditions for 6-coordinated systems is opposite to that in 4-coordinated systems. Since the reaction can proceed in a number of different paths therefore it is not possible to draw generalisation in the rates of six or four coordinated complexes. Similarly, a chelating ligand easily replaces a non-chelating ligand. The dark blue complex in case of Cu^{2+} is extracted by ethanol. Isolation of complexes is a very important aspect of complex formation. In reality the replacement of H_2O by NH_3 is a step wise change but is difficult to isolate desired mixed complex. For example,

$[Cu(H_2O)_3NH_3]^2$, $[Cu(H_2O)_2(NH_3)_2]^{2+}$, $[Cu(H_2O)(NH_3)_3]^{2+}$ and $[Cu(NH_3)_4]^{2+}$.

Under drastic experimental conditions of temperature and pressure some complexes may be isolated. For example,

$$K_3[RhCl_6] + 3K_2C_2O_4 \xrightarrow[100°C]{2\ hrs} K_3[Rh(C_2O_4)_3] + 6KCl$$

In some reactions more than one type of ligand may get substituted.

$$\underset{\text{purple}}{[Co(NH_3)_5Cl]Cl_2} + 3en \xrightarrow{\text{Steam}} \underset{\text{orange}}{[Co(en)_3]Cl_3} + 5NH_3 \quad ...(2)$$

Steam bath is used for the above reaction because the reaction is slow.

Mixed complexes can be achieved by limiting the concentration of the ligand *e.g.*,

$$\underset{\text{red}}{K_2[PtCl_6]} + en \longrightarrow \underset{\text{yellow}}{[Pt(en)Cl_2]} + 2KCl \quad ...(4)$$

$$\underset{\text{yellow}}{[Pt(en)Cl_2]} + 2NH_3 \longrightarrow \underset{\text{colourless}}{[Pt(en)(NH_3)_2]Cl_2} \quad ...(5)$$

Dichloro product is non-ionic and separates from the aqueous reaction mixture.

10.3 SUBSTITUTION IN NON-AQUEOUS SOLVENTS

Non-aqueous solvents are used in the preparation of metal complexes because (*a*) there is strong metal-oxygen affinity (*b*) ligand is insoluble in water. Strong metal-oxygen bonds are observed in Al(III), Fe(III) and Cr(III) metal complexes.

On addition of basic ligand like ethylene diamine to Cr(III) aqua complex, metal-oxygen bond being strong remains intact but oxygen-hydrogen bonds are broken and instead the following reaction takes place:

$$[Cr(H_2O)_6]^{3+} + 3en \longrightarrow \underset{\text{green}}{[Cr(H_2O)_3(OH)_3]\downarrow} + 3enH^+$$

If anhydrous Cr(III) salt is taken in non-aqueous solvent then substitution of ethylene diamine (en) is observed:

$$\underset{\text{(anhydrous) purple}}{CrCl_3 + 3en} \xrightarrow{\text{ether}} \underset{\text{yellow}}{[Cr(en)_3]Cl_3}$$

Conclusively, Cr(III) complexes are prepared directly without the aqueous solution. The common non-aqueous solvent is DMF (dimethyl formamide) or THF (tetrahydrofuran). For example,

$$\underset{\text{blue-grey}}{[Cr(DMF)_3Cl_3]} + 2en \xrightarrow{\text{DMF}} \underset{\text{violet}}{Cls\text{-}[Cr(en)_2Cl_2]Cl}$$

DMF is represented as $HCON(CH_3)_2$

When ligand is not soluble in water then a water miscible non-aqueous solvent is used which just enough to dissolve the ligand. Then this solution is added to the saturated aqueous solution of the metal ion. Metal complexes of bipyridyl (bipy) and phenanthroline (phen) are prepared by this route *e.g.*,

$$\underset{\text{Colourless}}{[Fe(H_2O)_6]^{2+}} + 3bipy \xrightarrow{C_2H_5OH\text{–}H_2O} \underset{\text{Intense red}}{[Fe(bipy)_3]^{2+}} + 6H_20$$

10.4 SUBSTITUTION WHEN NO SOLVENT IS USED

This involves direct reaction between anhydrous salt and a liquid ligand which also acts as a medium for the reaction. In this preparation liquid ligand is used in large excess to serve as a reaction mixture.

$$\underset{\text{yellow}}{NiCl_2} + 6NH_3\ \text{(liquid)} \longrightarrow \underset{\text{violet}}{[Ni(NH_3)_6]Cl_2}$$

Evaporation starts at room temperature as NH_3 boils at –33°C and the dry residue obtained is the pure ammine complex. This type of preparation is most suitable for metal ammine complexes. Liquor ammonia is preferred to avoid $Cr(OH)_3$ formation in Cr (III) complexes.

Another direct reaction method involves $PtCl_2$ or $PtCl_4$ and liquid ethylene diamine. $PtCl_2$ and $PtCl_4$ act as Lewis acids and ethylenediamine as a Lewis base. Due to strong acid-base reaction it is strongly exothermic. Quite recently metal dimethyl sulphoxide complexes have been characterised by this method:

$$\underset{\text{pink}}{Co(ClO_4)_2} + 6(CH_3)_2\,SO \longrightarrow \underset{\text{pink}}{[Co\{(CH_3)_2SO\}_6](ClO_4)_2}$$

10.5 MECHANISMS OF SUBSTITUTION

There are two basic mechanisms in substitution reactions:

(*i*) Dissociation of a metal-ligand bond

(*ii*) Formation of a metal-ligand bond. (For octahedral substitution reaction)

$$[MA_5B] + C \longrightarrow [MA_5C] + B \quad ...(1)$$

Dissociation of a bond is slow and hence rate determing step is loss of B. The subsequent displacement by C to the intermediate is rapid.

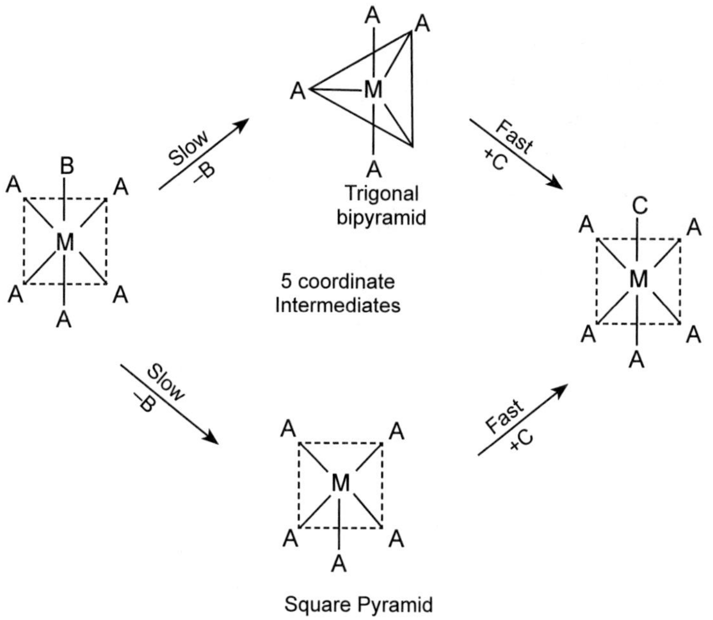

This is called SN_1 substitution, *i.e.,* substitution nucleophilic first order because the adding ligand searches a positive metal ion.

The SN_2 mechanism involves a 7-coordinate intermediate as shown:

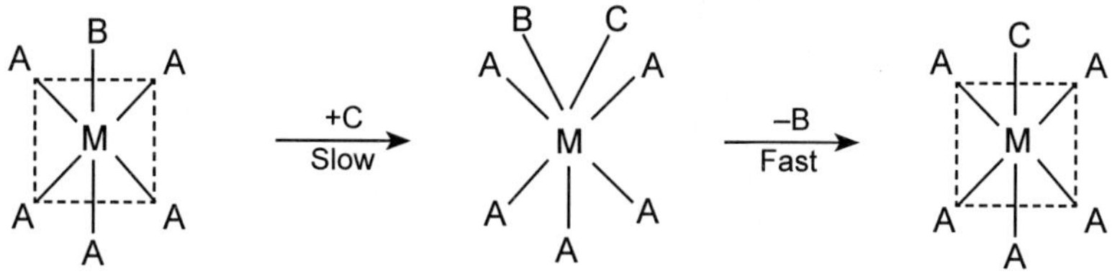

7-Coordinate Intermediate

The difference in SN_1 and SN_2 reactions is that in former, bond breaking is important and in SN_2 formation of an additional metal-ligand is important.

In SN_1 mechanism, If metal-ligand bond is strong then dissociation will be slow. SN_2 mechanism has no plausible reason to proceed through additional bond formation which can lead to congestion or crowding of ligands (steric hinderance). Therefore, the dissociative mechanism seems to be more acceptable.

In square planar complexes SN_2 mechanism finds favour. The breaking of $M \leftarrow L$ bond is very difficult as the charge on the complex will become more positive so the new $M \leftarrow L$ bond formation is more acceptable. It is observed that the charge on the complex has a small effect on the rate of the reaction. Therefore bond making and bond breaking are both slow and thus important, which is a characteristic of SN_2 process (substitution nucleophilic second order).

In general, substitution in 4 coordinate complex is easier as there is lot of vacant space above and below the square plane for the attack of the ligand to form 5 coordinate intermediate.

SUMMARY

In metal complexes, a strong field ligand replaces a ligand which is relatively less strong. The strength of the ligand is based on the focus of electronic charge in one direction *e.g.,*

$$\underset{\text{blue}}{[Cu\,(H_2O)_4]\,SO_4} + 4NH_3 \longrightarrow \underset{\text{dark blue}}{[Cu\,(NH_3)_4]\,S0_4}$$

Substitution reactions may take place in aqueous, non-aqueous and in the absence of a solvent. The above substitution reaction is carried out in aqueous medium. Other examples include,

$$\underset{\text{purple}}{[Cu\,(NH_3)_5\,Cl]\,Cl_2} + 3en \longrightarrow \underset{\text{orange}}{[Co\,(en)_3]\,Cl_3} + 5NH_3$$

If the metal ion is highly soluble in H_2O and the ligand is insoluble in water then substitution is carried out in non-aqueous solvent. In case of anhydrous Cr(III) salt the substitution of ethylene diamine is smooth in non-aqeous solvent to yield $[Cr(en)_3]3+$ *e.g.*,

$$\underset{\text{(anhydrous) purple}}{Cr\,Cl_3 + 3en} \xrightarrow{\text{ether}} \underset{\text{yellow}}{[Cr(en)_3]Cl_3}$$

Similarly the reaction between an anhydrous salt and liquid ligand in excess, serves as a solvent *e.g.,*

$$\underset{\text{yellow}}{NiCl_2} + \underset{\text{liquid}}{6NH_3} \longrightarrow \underset{\text{violet}}{[Ni\,(NH_3)_6]\,Cl}$$

Liquid ammonia boils at –33°C and the dry residue is pure nickel ammine complex. In this case, use of liquid ammonia also prevents $Cr(OH)_3$ precipitate formation. Trans effect depends upon the geometry of the complex. The mechanism of substitution reactions ultimately lies in the geometry. There are two basics in the mechanism of substitution reactions (a) dissociative and *(b)* displacement reactions.

The dissociative mechanism proceeds through loss of *i-x* to give 5-coordinate intermediate *e.g.,*

$$[MX_5Y] + Z \longrightarrow [MX_5Z] + Y$$

This is called SN_1 mechanism

In SN_2 mechanism displacement is through 7-coordinate intermediate in a slow step and then dissociation in fast step.

It involves two reactants in the slow step hence, it is called SN_2. In SN_1 main plank is bond breaking and in SN_2 it is additional bond formation that is stronger than bond breaking. Therefore bond, making and bond breaking is the rate determining step.

EXERCISE

1. What is the underlying principle of substitution reactions in complexes?

2. The substitution of H_2O by NH_3 molecule in a Cu(II) salt solution in water is instant even at room temperature. Numerically it should be:

$$[Cu(H_2O)_4]^{2+} + 4NH_3 \longrightarrow [Cu(NH_3)_2(H_2O)_2]^{2+} + 2NH_3$$

but it takes place step wise and all the $4H_2O$ molecules are substituted by NH_3. Give reason.

3. Balance the equation and show how the complex is isolated:

$$\underset{\text{purple}}{[Co(NH_3)_5Cl]C!_2} + 3en \longrightarrow$$

4. Why is it necessary to use non-aqueous solvent for the extraction of a complex compound? How do we choose a suitable solvent for their extraction?

5. $NiCl_2 + 6NH_3$ (liquid) $\longrightarrow [Ni(NH_3)_6]Cl_2$

In the above substitution why is solvent not used?

6. Elaborate how stereochemistry in the different stages of a reaction is the guiding force for the substitution reaction.

7. Explain SN, and SN_2 reactions in the context of substitution reactions of the complexes. What is the rate determining step in the multi-step substitution reaction ?

8. Discuss the substitution reactions in a 6-coordinate complex. Give pictorial representation to the reaction mechanisms that are involved in the change. Illustrate with the help of a suitable example.

9. Square planar complexes are more likely to react through SN_2 mechanism. Explain.

10. The substitution reaction is a two step reaction, *i.e.,* an intermediate complex is formed and the reaction mechanism is SN, (substitution nucleophilic, 1st order). Discuss.

■ ■ ■

SELECT TOPICS

11

Coordination Complexes and Biological Systems

11.1 INTRODUCTION

Metals are an essential part of the coordination complexes. It is the presence of the metals which provide the surface area for the ligands to coordinate with them. An average adult who has a medium built with a body weight 70 kg contain 43 kg water and 4 kg minerals. About 40 naturally occurring minerals are present in the living bodies such as plants and animals.

The minerals that are essential for the healthy human life must contain.

11.1.1 Essential Metals

1. Na, K, Mg, Ca, Mn, Fe, Co, Cu, Zn, Mo.
2. Some other *beneficial metals* include : Li, V, Cr, Ni, Sn, and W
3. *Non-metals*: H, C, Si, N, O, P, S, Se, F, Cl, Br and I.
   (*a*) Out of the above mentioned elements some elements are available in *abundance* such as : H, O, C, N, Na, K, Ca, Mg, P, S and Cl
   (*b*) the less abundant elements are: Mn, Fe, Co, Cu, Zn, and I.
   (*c*) The elements occurring in trace amount are: Si, V, Cr, Ni, Sn, F, Br, and Se.

In plants the role of boron, aluminum and barium is also important. The most important function of the metals is their ability to form complexes. Sodium and potassium play the role of charge carries and help to maintain osmotic pressure on either side of cell wall. These ions also maintain the sensitivity of nerves and muscle. Sodium ions are present in the blood plasma of verteberates while the fuid contain K^+ ions in the cytoplasm.

There are 50 enzymes which are activated by charged K^+. The reactions catalyzed by K^+ are phosphoryation or elimination of—COOH group that are important in protein synthesis and respiration. Na^+ inhibits respiration and activate few enzymes.

Magnesium and calcium are required for structure formation and to start reaction Ca^{2+} maintains the correct rhythm of heart beat and conversion of fibrinogen to fibrin. Calcium is used in cell wall structures which are manifested as shell and bones.

It is also used to stabilize these structures e.g. in thermolysin where there are four Ca^{2+} ions. Mg^{2+} ions form complexes with nucleic acids and help the nerve impulse transmissions. In animals the tissues are complexed with ADP and ATP. Mg is also an essential constituent of chlorophyll in green plants.

Manganese activates all those enzymes which magnesium does—*e.g.* kinases synthases and dehydrogenases. It can activate some RNA and DNA polymerases, guanylate cyclases. It is involved in the synthesis of glycoproteins in the formation of cartelage and bone. The different valencies of manganese are observed in Mn-enzyme function.

There are innumerable iron proteins but to date 70 are isolated. They perform the same function but-differ in molecular mass due to polymerisation.

Iron proteins contain porphyrin, flavins or ptyrins and sometimes other metals such as Mo, Cu or Ni. Iron proteins have been divided into two categories.

(*a*) Proteins that are having sulphide groups. They may be labile or non-labile.

(*b*) Proteins that do not have sulfur. The later groups may be sub divided into (i) Iron complexed by porphyrin (heme) (ii) those group which do not contain heme and are oxo-bridged with dinuclear iron with bridging corboxylates. The Fe-S proteins contain Fe S_4, Fe_2S_2 or Fe_4S_4 or with mixed metals like Fe_7MoS_8 active sites which accept electrons to reduce Fe(III) to Fe(II) in the Fe-S Cluster.

Some common examples of iron proteins are given below:

(*i*) Hame proteins are : Myoglobine, hemeglobin, cytochromes, catalases.

(*ii*) Fe-S proteins and flavin proteins are: Ferredoxine, rabredoxin, hydrogenase, reductase, *e.g.* nitrite to sulphide reductase, sulphide oxidase, xanthine oxidase, dehydrogenase, molybdeo ferredoxin etc.

There are several kinds of hemes characteristic of hemo proteins.

Heme (A) Hemoglobin, myoglobin and catalase.

Heme (B) includes cytochromes and proto heme (Heme B), Fe-proto-porphyrin IX. It is bonded through ionic and hydrophobic interactions.

The α-type cytochromes contain Heme A and the e-type cytochromes contain heme C.

In the case of Heme C there are two thio ethers bonds. Their structures are:

The iron in all hemoproteins is either five or six coordinate etc. Becuase protein donates at least one ligand. The donation of sixth ligand from protein has an important bearing on the chemical reactivity of central iron. In hemoglobin and myoglobin the iron is five coordinate. This gives way for oxygen molecule to add reversibly and make the iron six coordinate. If the goblin protein is absent then ferro heme Fe(II) is oxidized to Fe(III) from heme.

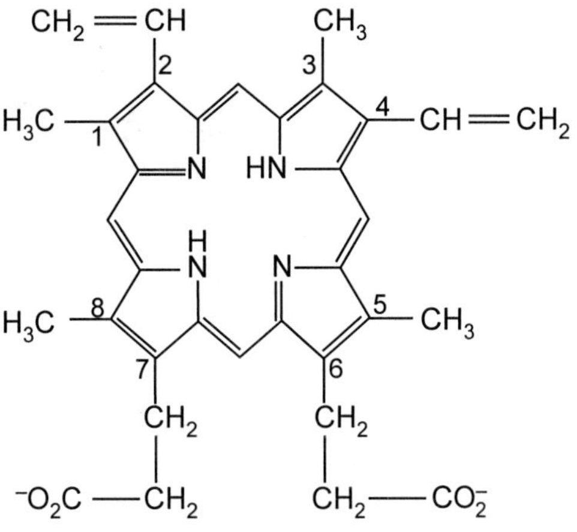

Fig. 11.1: Protoporphyrin IX

11.1.2 Cyanide and CO Poisoning

Both hemoglobin and α-cytochrome, α can bond CN^- and CO to the Fe(II) as the sixth ligand.

The fatal effect of cyanide poisoning are linked to inhibition not of hemoglobin but α-cytochrome.

Cyanide is strongly bound to met-hemoglobin than α-cytochrome and very weakly to hemoglobin. It is removed as blood when filtered through spleen where erythrocytes containing met hemoglobin are destroyed. The fatal effects of cyanide poisoning are linked to inhibition not of hemoglobin but α-cytochrome. Therefore in cyanide poisoning the treatment is inhalation of amyl nitrite or injection of $NaNO_2$ which oxidises part of hemoglobin to met hemoglobin Fe(III). Met hemoglobin does not bind CO. Therefore CO poisoning the best remedy is to administer O_2 for inhalation

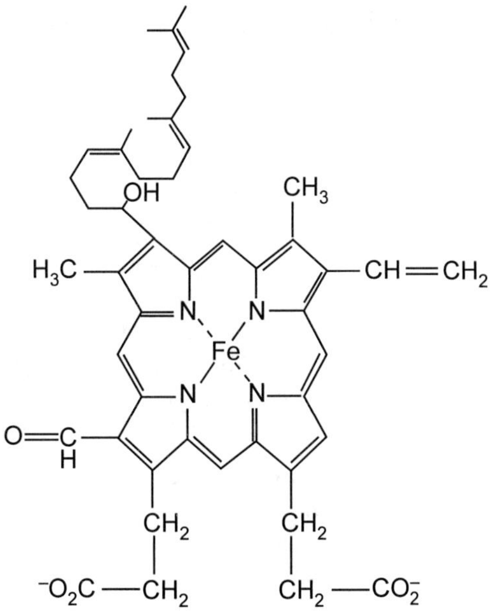

Fig. 11.2: Heme A

to displace CO. In hemoglobin and α-cytochromes the iron center of the respiratory β and γ cytochromes are six coordinate. These cytochrome do not interact with cyanide or CO and hence are not autoxidised by oxygen.

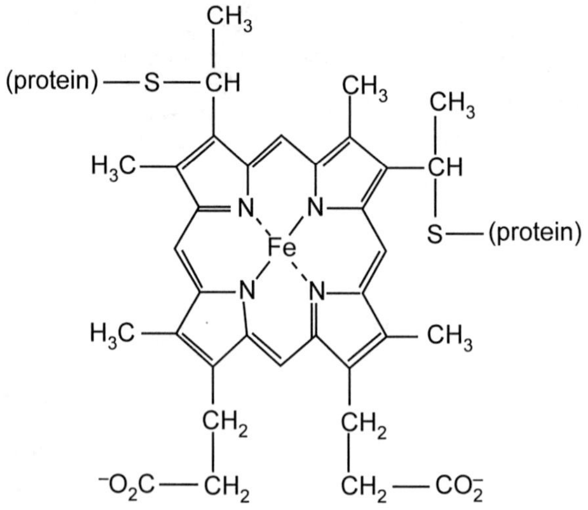

Fig. 11.3: Heme C

11.2 HEMOGLOBIN THE OXYGEN CARRIER

Iron porphyrin, proteins and enzymes are collectively known as heme proteins or hemoproteins .these proteins exhibit distinctly different functions

(*i*) reversibly oxygen (O_2) binding for its transport as in hemoglobins or

(*ii*) storage of oxygen (O_2)as in myoglobins oxidation by O_2 with reduction of O_2 to H_2O as in the case of cytochrome c oxidase, decomposition of H_2O_2 as it catalyses and peroxidases. Hemoglobin is the red colour of erythrocytes that account for the 95%of the dry weight of the red blood cell. Its main function is carrier of oxygen but it is also related with the transport of CO_2 and acid –base balance of the body. Hemoglobin also absorbs a large volume of oxygen in the lungs and carries it to the tissues.

$$\underset{\text{Hemoglobin}}{Hb} + O_2 \rightleftharpoons \underset{\text{Oxy hemoglobin}}{Hb\text{—}O_2}$$

Hemoglobin has an oxygen absorbing capability e.g. 1g hemoglobin when fully oxygenated contains 1.39 ml of oxygen. Besides, hemoglobin contains 50 mg of Fe for 100 ml of blood.

11.2.1 Transport of Oxygen

Oxygen transport is very important for respiration. Its production in photosynthesis provides examples of biologically important redox reaction which includes electron transfer, transfer of atom and photosyntheses processes.

11.2.2 Oxygen Carriers

The distribution of oxygen to the various parts of the body is carried by cell which contain iron where as carriers contain copper which is present in extra cellular fluids. Cell-interior is a reducing medium that contains Fe as Fe (II) besides this,cell gives a porphyrin ring which acts as a ligand to form non-labile Fe(II) complex.

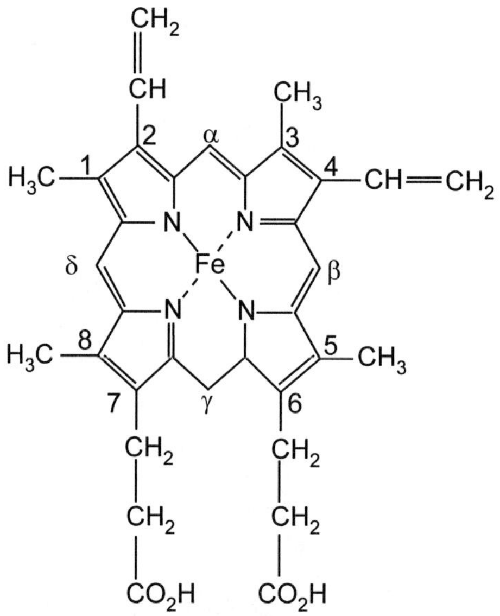

Fig. 11.4: Iron(II)-protoporphyrin-IX (Heme-B); (Fe----N═Fe←N)

The porphyrin ring is susceptible to oxidation and protein the reductases in the cell. On the other hand both Cu(I) and Cu(II) are able to form strong complexes with imidazole side chains of poly peptide residue. Therefore Cu-leased carrier can survive unaltered in the blood iron and copper carrier are able to play distinct role in the same organism.

11.2.3 Metal Ions Strong and Transportation in the Human System

Iron(II)from the stomach is oxidised ,bound and transported in the blood stream as a complex of Fe(III) with a protein called transferrin. It can remain in the system (stored)only as Ferritin which is Fe (III)-protein complex with 15-20% iron. Feritin helps to store iron in the form of Fe (III) complex with proteins. It can be understood as a protein shell surrounding a core which is ferric hydroxy phosphate of composition $–FeO(OH)_8(FeO.H_2PO_4)$ containing 57% iron. Ferritin is present in the organs of all mammals specially in liver, spleen and bone marrow. The iron stored in such a way is released as and when needed in the synthesis of iron—containing hemoglobin, myoglobin and the cytochromes.

As a generalization it should be understood that the metal ions in the human body can be bound, stored and transported in the from of metal ion complexes. 70%of Fe metal is present as hemoglobin and myoglobin while the remaining is stored as ferritin.

Transferrin is a protein with a higher molecular weight .this can bind two iron atoms per molecule. The two binding sites are similar but not identical in their binding strength. The iron is bound so firmly that no other protein can take away iron. Except those at the site of hemoglobin synthesis. In transferrin iron (III)is bound to at least five groups in the protein.

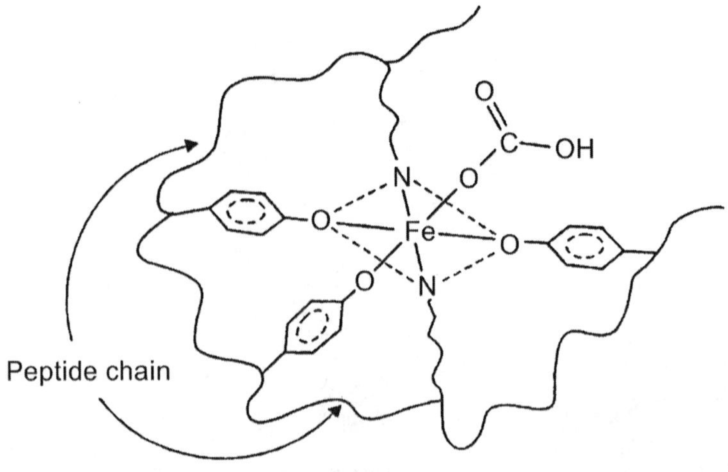

Fig. 11.5: Fe(III) Transferring Complex

Copper in trace amount is essential for the synthesis of hemoglobin. It occurs in the form of metallo protein, hemo cuprein in the blood cells (0.34% copper).

11.3 ROLE OF SOME METALS IN THE HUMAN BODY

11.3.1 Selenium

Selenium is toxic when its concentration is above 10 ppm but it is also essential for life when the concentration is below 10 ppm. Average man requires < 0.2 mg per day. The deficiency of selenium is caused when the concentration becomes less than 0.1 ppm. the deficiency of Se result in liver necorsis and muscle dystrophy.

A good member of seleno-organic compounds are derived from bacteria fungi and green plants. Some naturally occurring organo-selenium compounds are listed below

Some naturally occurring organo-selenium compounds

Name	*Formula*	*Sources*
Dimethyl diselenide	MeSeSeMe	Fungi; *Astragalus*
Se-methylselenocysteine	CySeMe	Se accumulators, *e.g. Astragalus, Stanleya*
Selenocysteine seleninic acid	CySeOH	*Lolium, Trifolium*
Selenomethionine	$CyCH_2SeMe$ }	Bacteria; fungi; *Lemna, Lolium, Trifolium*
Selenocystine	CySeCy }	
Selenocystathionine	$CyCH_2SeCy$	Se accumulators, e.g. *Astragalus, Morinda*
Selenohomocystine	$CyCH_2SeSeCH_2Cy$	One *Astragalus* spp.
Selenomethionine selenoxide	$CyCH_2Se(Me)$=0	*Lolium, Trifolium*
Se-propenylselenocysteine selenoxide	CySeOCH=CHMe	*Allium*

Most of the above mentioned compounds are amino acids. Amino cystine seems to be largely distributed. Other seleno-compounds include peptides, adenosine phosphoselenate obtained from bacteria and yeasts. Still other seleno-compounds may be storage compounds of this highly toxic selenium.

11.3.2 Boron

At present ,the only naturally occurring boron is boromycin having the formula $C_{45}H_{74}O_{15}NB$ obtained from streptomycin which is easily hydroysed to boric acid. It seems boron has role to play in RNA or carbohydrate transport. Boron deficiency in plants inhibits ATPase activity in the cell membrane, and uptake of PO_4^{3-} ions.

11.3.3 Molybdenum

Molybdenum (Mo) is quite abundant in the earth‘s crust. In natural water concentration of Mo is higher than those of Fe, Cu and Zn. Aluminum is abundant but not very essential to the humans although an adult contains 60 mg of Al in the lungs for inhalation. Al is reported to give strength to enzyme activity but K^+ and Zn^{2+} are required for the full enzyme activity. Silicon is also abundant on the earth's crust but its ultra trace amount is essential for the human system. Ti and Zn are abundantly available but its trace amount are required to get extremely insoluble oxides at the relevant Ph values. They do not form stable complexes with the complexing agents present in the biological system. The essential element may be harmful at high level of concentration. In reality, many of the essential elements belong to the *d*-block metals which have the ability to form coordination complexes and chelote compounds in the most characteristic aspect of chemistry.

11.4 SODIUM POTASSIUM EQUILIBRIUM

In the animal cells higher amount of K^+ are present in the intra cellular fluid than outside (extra cellular fluid) and the reverse is true of Na^+ ions. The adinosine tri phosphate (ATP) provides energy to maintain equilibrium in case the equilibrium between Na^+/K^+ is disturbed. Large amount of ATP is consumed in such an eventuality. Therefore it acts as a pump to maintain the Na^+ and K^+ ions equilibrium. If Na^+ and K^+ ions are present in excess of Mg^{2+} required by all ATPases (the anzyme which catalyses the hydrolysis of ATP to ADP). The enzyme activity is related to quantitative extent of the ion transport. The alternative view is that ATPase is phosphorylated *i.e.* the enzyme phosphate bond is established at an asparate site. This can happen only in the presence of Na+ and Mg^{2+} ions. The phosphorylated product is hydrolysed if K^+ ions are present. It is further emphasized that the enzyme undergoes conformational change before phosphorylation.

Looking at these changes a mechanism as outlined below was proposed.

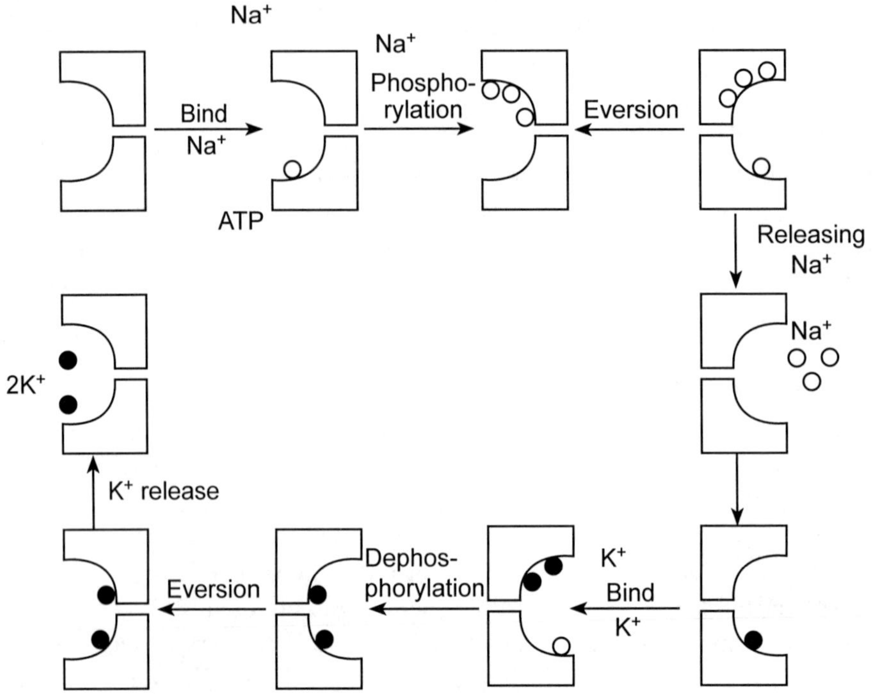

Fig. 11.6

At first ATP and $3Na^+$ ions bind to the inside of the membrane and the enzyme is phosphorylated. Then it undergoes a conformational change called *eversion*, which brings the release of $3Na^+$ ions outside the membrane. These three Na^+ ions are replaced by the two K^+ ions. The attachment of the K^+ ions induces dephosphorylation and hydrolysis of ATP to ADP which brings about a conformational change and carries the two K^+ ions into the inside cell where they are finally released. Now these Na^+ ions are expelled by two K^+ ions and the outside of the cell become positively charged.

11.4.1 Selectivity

Here ,we need to understand how an enzyme can be selective to Na^+ and then to K^+ ions in different conformations. The helical conformations form the walls of the ion channels. The difference between channels' selectivity towards Na^+ and K^+ ions is remarkable.

Typical conformations create specific geometrical arrangements of a number of metal–ligand binding atoms. The steric factor should also be accounted for. It is well known that crown-etherse that have antibiotic activity increase the permeability of cell membrane to cations. The chelating ligand may bind both Na^+ and K^+ ions but 100 times strongly to K^+ ions. This selectivity is stereo chemical as a K^+ ion neatly fits into the cavity or hollow created by the crown while Na^+ being smaller does not. Rather Na^+ ion comes out of the cavity easily.

11.5 BEHAVIOR OF CALCIUM AND MAGNESIUM

Out of these calcium plays an important role in the structure of bone and shell because of the insolubility of its carbonate and phosphate. Ca^{2+} ions play diverse role in the human system. Ca^{2+} ions also play non-structural functions. They function as messengers for hormonal action in inactivating muscled contraction and nerve signals in the blood clothing and stabilization of protein structures. Ca^{2+} ions may be needed as a messenger as well as to reduce the concentration of Ca^{2+} ions in the cytoplasm.

The diversity in the roles of Ca^{2+} ions is surprising as it does not form strong complexes with simple ligands in solution. The differences between Ca^{2+} and Mg^{2+} ions are interesting and arise due to three factors:

1. Because of low selectivity, Ca^{2+} ions, can bind neutral oxygen donor ligands (carbonyl and alcohols) with regard to water. Ca^{2+} ion can also bind with N-donor ligands.
2. Ca^{2+} ions resemble Na^+ and K+ ions in favouring high coordination complexes with irregular geometry. Due to 2+ charge it can attach an ion which alkali metal ions do not favour.
3. The rates of binding and dissociation of Ca^{2+} are high. In contrast the binding rate of Mg^{2+} is three times slower than of Ca^{2+}.

There exists a concentration gradient for Mg^{2+} (High concentration inside cell) and Ca^{2+} (low concentration out of the cell) as in Na^+/K^{-1} case.

It has been chacterised that one calcium ion binds with four carboxylate o atoms with a structure of distorted tetrahedral.

Similarly magnesium metal activates the most enzymes, including phosphates, carbonates and kinases but not enzymes involved in redox processes. It is observed in glycolysis, tri carboxylte acid cycle and lipid and protein metabolism.

Generally in biological systems metal complexes play the following roles:

1. They act as catalysts in many metabolic processes of the living beings which involve
   (*a*) Redox reaction with change in valency of metal in the complex.
   (*b*) Acid-base type reactions where the metal acts as a Lewis acid and can combine with subtrate to accelerate the reaction. In these metal functions as a labile coordination complex intermediate. The completion of reaction depends upon the decomposition of this intermediate complex.
2. Metal complex serves as a storage and transfer of either the metal ions or donor molecules and thus act as an agent for transmission of energy.

11.5.1 Other Metalloenzymes

Molybdenum is involved in the normal growth of many plants and humans. Studies show that many molybdenum containing enzymes in the oxidized state MoO_2^{2+} are bound to (—COO—) group of the protein and to aromatic nitrogen and phenolic oxygen of the substrate. There is no evidence of molybdenum deficient disease or symptonic in animals. The catalytic function of the molybdenum is because of the following redox reaction .

$$\text{Mo(VI)} + e \rightleftarrows \text{Mo (V)}$$

W and Mo belong to the same group in the periodic chart. They possess very similar chemistry. The have similar atomic radi, ionic radii and electro-negativity values. In the enzyme activation in the reactions Mo is preferred to W. This is due to the fact that lower oxidation states 3+, +4 and +5 are less accessible for W compared to Mo.

11.5.2 Vanadium in Biological Systems

Vanadium in trace amounts is a nutrition for many organisms *e.g.* higher animals. Marine organisms and mushrooms accumulated vanadium. Sea worm ascidian has vanadium in the blood level up 1900 ppm. This represents more than a million fold concentration with respect to vanadium in sea water. The function of vanadium in mushrooms like Aminite is not certain but it is present to the extent of 250 mg V/kg.

Vanadium (III) occurs as green complex called hemo vanadium in the blood cells of some ascidians where it functions as a transporter of O_2. Vanadium is known to be present in active sites of some enzymes. In higher amounts vanadium was found in fossil matter of animals and plant origin.

Fixing bacteria contain vanadium and in one of these three distinct nitrogenase systems based on Mo, V and Fe. This led to the discovery of some model compounds of vanadium and the known coloured V(–1) compound.

11.5.3 Metallo Enzymes as Acid Catalysts

A metallo enzyme can act as an acid catalyst either by Lewis acidity of metal ion or Bronstead acidity which is enhanced by a metal ion.

The Zn^{2+} ion is a common Lewis acid in biological systems . Zn^{2+}, Mg^{2+} and Cu^{2+} are small dipositive ion but why zinc gets preference over the other two. Zinc is different from Mg on account of the electron affinity. The Values are Zn = 27.4 eV and Mg = 22.7 eV which means that zinc is stronger Lewis acid. EA value for Cu is 28.0 eV Which is similar. Another feature which is different between zinc and copper is the variable oxidation state. Zinc has a fixed, oxidation state of +2 so redox reactions are unlikely to generate free radicals.

Another factor which is characteristic of a metal ion is that it should take up or release the ligand with ease. So, zinc complex are much more labile than the corresponding Mg^{2+} and Ni^{2+} complex. Geometrical rearrangement aspect or fluxcional behavior is facile for zinc complexes least not for Mg^{2+} and Ca^{2+} complexes. For these reasons zinc finds extensive use as metaloproteins and enzymes over Mg^{2+} and Cu^{2+}.

11.6 PHOSPHATE STORAGE AND MUSCLES

Hydrolysis of ATP to ADP in muscles provides energy needed for muscle contraction. After sometimes this energy gets depleted and very low concentration ($mol^{-1}kg^{-1}$) remains in cells and tissues. Therefore, the energy is quickly replenished with high energy phosphate which is provided by phosphagens. Phospho creatine is found in most vertebrates.

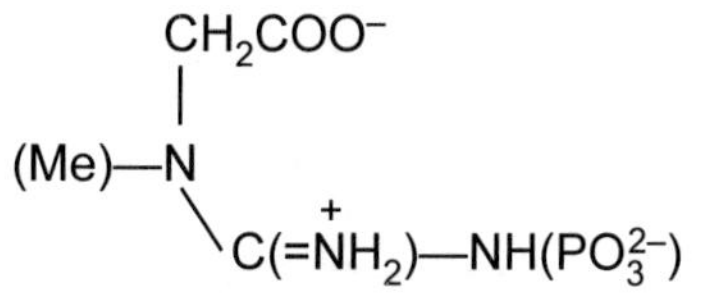

Phospho creatine

Phospho creatine in association with the Mg-enzyme creatine kinase brings about the interconversion of ADP and ATP.

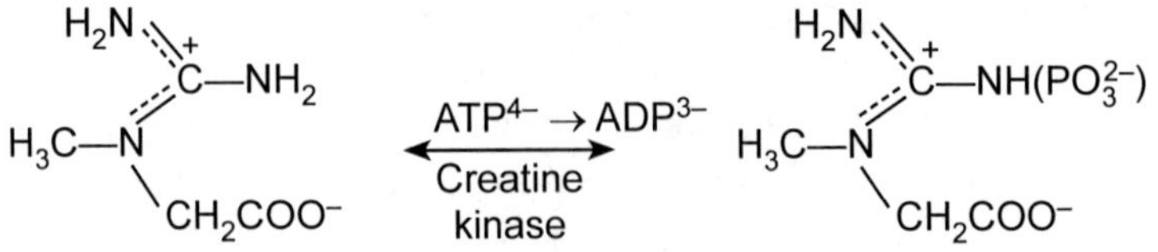

11.7 OXYGEN BINDING AND OXYGEN ENZYMES

Investigation on the interaction of O_2 with metal complexes depends upon understanding the function of oxygen carriers. The study has revealed the contribution of the biological problem as well as understanding how the molecular O_2 acts as an oxidizing agent. Regarding the d-block metals that bind O_2 cobalt provides the best picture of the oxidative reactions of oxygen. The hemoglobin molecule is a conjugated protein which consist of four sub units each of which has a poly peptide chain and an associate heme molecule. Therefore, a molecule of hemoglobin (Hb) has four atoms of iron and this can combine with 4 molecules of O_2.

The polypeptide chains fit properly in complete molecule so that heme proteins are disposed of at the corners of a tetrahedron where there occurs immediate reaction with oxygen molecule. Attachment of the heme portion to the globin causes change in the properties of the former in regard to oxygen absorption and improves the properties of globin molecule.

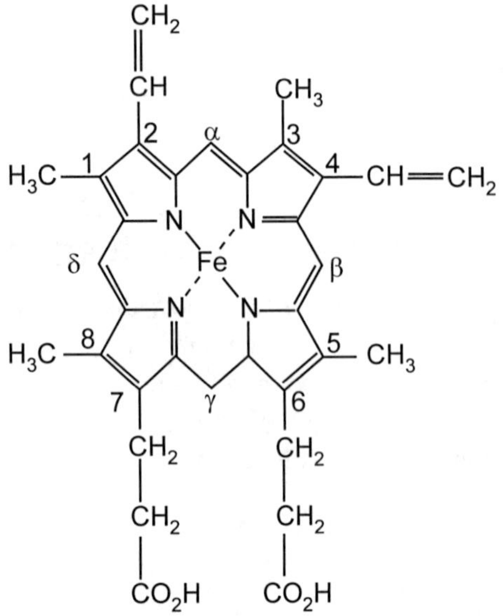

Fig. 11.7: Iron(II)-protoporphyrin-IX (Heme-B); (Fe----N=Fe←N)

11.7.1 Vitamis B_{12} or Cyanocobalamine

The vitamin B_{12} was first isolated in the year 1948 in the crystalline form and its structure was established in 1973 with the help of x-ray crystallography.

It was further established by Wood Ward that it is derivative with CN^- and cobalt (III) called cyanocabalamine. The molecule is built around a corrin ring (modified porphyrin) containing Co(III) coordinatively bound to four planar nitrogens. The fifth and sixth coordination positions of cobalt occupied by the N-atom of the imidazole ring of side chain of the corrin ring and cyanide ion respectively as shown in the figure

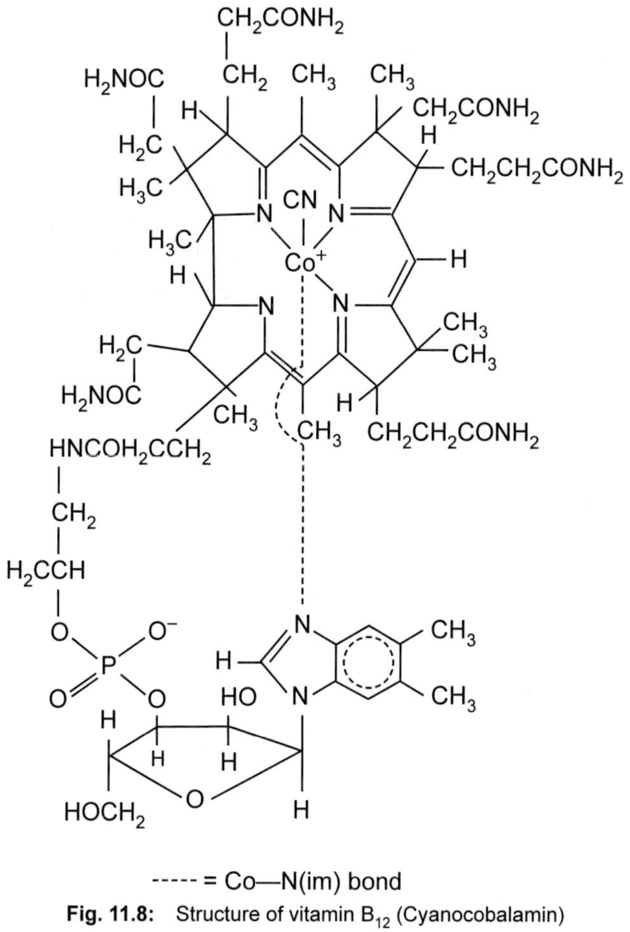

Fig. 11.8: Structure of vitamin B_{12} (Cyanocobalamin)

The above structure has been confirmed by x-ray crystallograhy.

The CN^- is the artifact of the isolation procedure and is not present in the natural system in which the sixth position of cobalt is coordinated H_2O/OH. CH_3 or adenosyl group with Co-C bond are involved in vivo reaction of vitamin B_{12}. In the case where synthetic B_{12} is administered they are formed by replacement of CN^-. The species in which H_2O is bound to Co in place of CN^- is called aqua cobalamin or hyroxocabalmin. Vitamin B_{12} which is reduced by ferredoxin in the natural system in two successive one electron steps to vitamin. B_{12r} (containing Co(II) and Vitamin B_{12} (containing Cobalt(I))

$$\underset{\text{Orange red}}{\text{Vit-B}_{12}\text{ a}} \overset{e^-}{\rightleftharpoons} \underset{\text{Brown}}{\text{Vit B}_{12}\text{ r}} \overset{e^-}{\rightleftharpoons} \underset{\text{Bluish green}}{\text{Vit B}_{12}\text{ s}}$$

Vitamin B_{12S} complex containing Co(II) behaves strongly nucleophilic and play an important role resulting in formation of Co–C bond.

The important functions of vitamin B_{12} involve the reduced Co(I) form. One of these is the reduction of organic substance and other is methylation to form Co-CH_3 in a reaction of the type:

$$B_{12}\,Co(I) + CH_3I \rightarrow [B_{12}Co(III)\text{-}CH_3]^+I^-$$

The only vitamin that contains a metal is vitamin B_{12}. It can be derived from the animal source but not found in plants. An average adult contains few mg of vitamin B_{12} in the liver. It can be synthesized by many bacteria. It is involved in the formation of methane by 'a' particular bacteria but in the process (CH_3—Hg^+) a toxic material is formed which is harmful to the environment. The reaction of B_{12} and ATP produces vitamin B_{12} active coenyme. The adenosyl group attached to cobalt(II) with a C—Co(I) band in place of CN^- of the synthetic vitamin B_{12}. Methyl cobalamin serves as an important methylating agent, aqua and hydroxo cobalamine are also biochemically active.

Photosynthesis

The green colouriing matter of the plants is called chlorophyll. Chlorophyll is a complex between porphyrin and magnesium. The main function of chrlorophyll is absorbing photons from light energy and transmitting it in the system for the purpose of photosynthesis. Photosynthesis results in the formation of carbohydrates.

$$6CO_2 + 6H_2O \rightarrow C_6H_{12}O_6 + 6O_2$$

Tracer studies show that the evolved oxygen comes from H_2O and not from CO_2 (O^{18} technique). The reaction involves the splitting of water followed by the reduction of CO_2.

$$12H_2O \rightarrow 12H_2 + 6O_2$$

$$6CO_2 + 12H_2 \rightarrow C_6H_{12}O_6 + 6H_2O$$

The overall reaction becomes:

$$6CO_2 + 12H_2O^{18} \rightarrow C_6H_{12}O_6 + 6H_2O + 6O_2^{18}$$

The reaction is endothermic and requires 469-kg mol^{-1} energy per mole CO_2 which is supplied by the sunlight. The glucose then condenses into polysaccharides like starch. Chlorophyll has many modifications which occur in the side chain of the porphyrin ring. The most abundant chlorophyll being chlorophyll as shown in figure 11.9.

Chlorophyll b has —CHO group in the place of CH_3 group bonded to carbon Chlorophyll c has —CH = COOH in the place of $—CH_2—CH_2—COOC_{20}H_{39}$ attached to carbon.

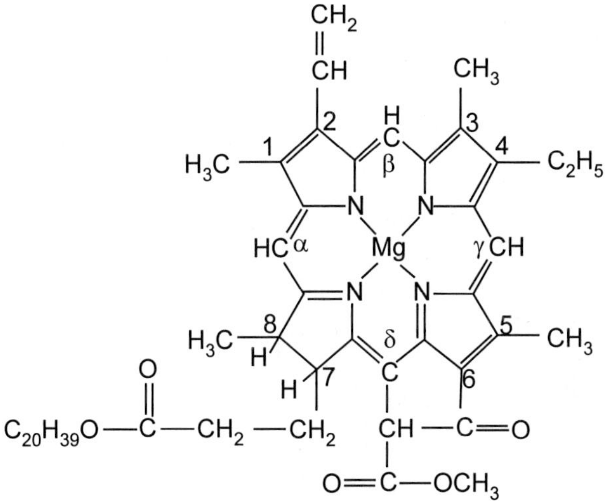

Fig. 11.9: Structure of chlorophil

Higher plants and green algae contain chlorophyll a and b in the ratio of 3:1. Brown algae contain chlorophyll c. Marine algae contain chlorophyll c and a, but red algae contain mainly chlorophyll a and d. chlorophyll a is observed in all O_2 evolving organisms.

11.7.2 Food and Trace Elements

The constituents of the food has a great effect on nutrients presents in the intake of food. Some of the trace elements for example Ca, Mg, Fe, Zn Cu, Mn etc have a marked effect in the absorption of nutrients: Non-Heme iron in a cereal based iron may be absorbed to an extent of 2% while heme iron from meat may be absorbed to an extent of 30%. Absorption of Zn from the 5 gm of four different diets are

	Diets	Zn Absorption in %
1.	Milk	30
2.	Milk+cereal	20
3.	Meta-based composite	20
4	Soya intake	14

11.8 TRACE ELEMENTS ABSORPTION

11.8.1 Inhibitors in Absorption

(*a*) The negatively charged anion of phytic acid (1,2,3,4,5,6-Cydohexane hexolphophoric acid), $C_6H_6(OPO_3H_2)_6$ is a major phosphorous compound in plants especially oil seeds, legumes and cereal grain which forms insoluble complexes with the metal ions since the human's gastro-intestinal tract is unable to hydrolyse phytates. These metal phytates complexes hamper the bioavailability of metal ions. In animal model studies negative effect of phytate on the absorption of Fe and Zn has been shown. Hence the absorption of Zn in humans is reduced to 50%; Addition of phytic acid to cow's milk reduced the zinc absorption by 50%. Similarly addition of phytic acid to white bread reduces the absorption of zinc and iron considerably. The same hold for all other essential metal ions.

(*b*) **Effect of metal ions on the absorption of trace elements:** Calcium has a direct inhibitory effect on iron absorption *e.g.* addition of Ca to human milk has a strong negative effect on the absorption of iron in rats. Similarly presence of calcium in cows milk has strong inhibitory effect on iron assimilation. Similar addition of Ca does not have any affect on the absorption of Zn.

Trace elements form metal complexes in aqueous solution which are structurally similar and therefore they compete for absorption *e.g.* copper and zinc interactions and iron-manganese. Another kind of interaction exists when trace elements compete for carrier ligands within the mucosal cell as is likely for Fe and Zn. In both these interaction high level of element may inhibit the absorption of another. This is seen relevant in food items that are artificially filled with trace elements. It has been observed that high level of iron inhibits the absorption of zinc but effect is less pronounced in items of food than the solution (medicinal tonic). In iron fortified baby foods there exists lower level of plasma zinc. High level of dietery Mn similarly impairs Fe absorption in humans. Due to phytate content dietary, fibre also affects bio-availability of trace elements. Fibre with no phyfate such as cellulose, beet pulp fibre have been shown to have no effect on the absorption of Zn.

(*c*) **Phenolic Compounds:** Tannic acid contains phenolic groups (–OH)and is present in tea , coffee, red wine and certain vegetables. It forms complexes with metal ions specially Fe (III). These complexes are highly insoluble and have a negative effect on the availability of metal ions for the biological reactions it has been reported that tannic acid inhibits the absorption of Fe. Gallic acid shows a similar effect in the terms of galloyl groups. These findings explain the adverse effect of tea,coffee and red wines (amount is high in tannic acid) on the bioavailability of iron. It was observed that when coffee was taken in the breakfast meal, Fe absorption was dependent on the tannic acid content .drinking more cups of coffee in a day during lactation and pregnancy has been show to cause anaemia or low iron content.

Coffee without caffeine has the same effect which shows that caffeine is not related to in inhibition effect.

11.8.2 Enhancing the Effect of Absorption of Metals (Even Trace Elements)

(*a*) **Proteins** can affect the bioavailability in two major ways

(*i*) A particular source of protein has a positive effect on the absorption than the other.

(*ii*) total protein content in the diet affects the absorption of the elements even in trace elements.

Protein from the meat source enhances the absorption of non-heme iron. The component is cysteine containing peptide formed during digestion than the free cysteine which is readily oxidized. The peptide bound cysteine is protected until it is released in the gut where it is the chelated iron facilitating the absorption. Meat also suggests a similar effect to the absorption of zinc. This may also be due to high content of zinc in the meat. Different proteins have been found to affect trace metal bioavailability to a different level. Casein and soya proteins reduce absorption. The negative affect could be due to phosphate. Thus apart from protein structure, inorganic and organic constituents affect the bioavailability of trace elements in the diet. The effect of protein in the meat may enhance the absorption of Zn. But this could be due to high level of Zn as Zn is bound to dietery proteins. Amino acid ,small peptides produced during digestion may form chelate with Zn and its absorption. Higher concentration of protein may increase chelation and the subsequent absorption.

(*b*) **Organic Acids:** Vitamin c or ascorbic acid enhances absorption of iron by reducing Fe (III) to Fe (II) and subsequent chelation. This affect is very strong and is able to over come the phytate effect as discussed earlier. Experimentally it has been observed addition of ascorbic acid to cow's milk enhances absorption of iron in comparison to the addition of ascorbic acid to soya milk.

Ascorbic acid has no effect on the absorption of Zn as it does not chelate strongly with Zn(II), other acids like citrate, malate and fumarate may have positive effect on the absorption of trace elements due to chelation and the subsequent absorption of the metal ion.

11.8.3 Toxic Metal Ion

Many soluble metal salts such as chromates may cause local irritation or damage of tissues or even poisoning if the metal ion from the salts is absorbed in sufficient amount. Zn an Cu are essential for life even in trace amount. They stay useful in the form of metallo-enzymes. In moderate or trace, amount Zn and Cu salts cause gastro-intestinal irritation and nemesis. The action of copper is due to hydrosulphide (—SH) groups. Many enzymes contain—SH group which are essential for the enzymatic activity. Therefore, when they get bound to copper the enzyme activity is lost. Food and water contain good amount of copper but the absorption of copper is low, hence excretory mechanism in a healthy adult is enough for the systematic poisoning. In the Wilson's disease copper control mechanism fails so that absorption of copper accumulates in various organs. This finally leads to a painful death.

All the metal ions which do not partake in circulation in sufficient amount and are not excreted rapidly are toxic depending upon the extent of metal itself. The toxic effect may occur due to electrolytic imbalance damaging organs such as kidneys or the nervous system for interfering with the enzymatic process. In fact no metal ion including those which are helping the vital functions in the biological system are tolerated at high concentration. Thus, the extra cellular concentration of K^+ is 16-20 mg per 100ml but if it is doubled, strong effects on nerve or muscular activity is observed which may cause death due to cardiac depression. If the lead concentration in blood reaches 0.08 mg in 100 ml from the normal 0.03 mg/100ml. Aluminum occurs in the blood in the range 0.05 to 0.1 mg per 100 ml which, however is low in toxicity if it is administered 5g per 100 ml, the increase being 5000 times than it becomes lethal. There are reports on relationship of Al^{3+} level in brain tissue and Alzheimer's disease.

The toxicity of metal ions such as Li^+, K^+ and Ba^{2+} may occur due to:

(*a*) Low complexing ability

(*b*) Electrolytic alterations with respect to cationic charge

(*c*) Adsorptions on the negative sites

(*d*) Displacement of H^+ or other cation's consequently altering the nature and properties of enzymes proteins.

Other metal ions form complexes with the available donor sites but some have preferences for some donor atoms. *e.g.* Hg^{2+} for—SH or sulfhydryl groups. Strong complexing ion may displace weak from the normal coordination sites in metallo enzyme. Arsenic has a toxic action in living cells which is belived to be due to its combination with thiol RSH present in protoplasm, The final result is a chelate with –SH groups that are present in essential enzymes as shown:

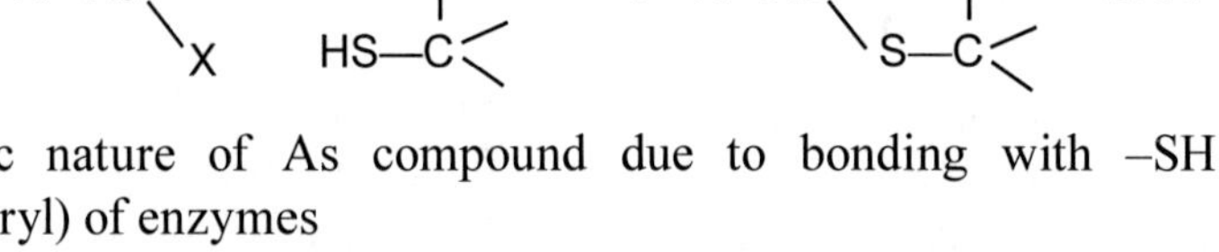

Toxic nature of As compound due to bonding with –SH Groups (Sulfhydryl) of enzymes

Similar reactions are experienced with metals such as Zn, Cd, Hg, Pb, Bi, Sb, Au all of which are preferentially bonded to sulfphur donor atoms. Hg is known to form strong bonds with –NH group of peptide by desplcing H^+. The essential metals are seen to be harmful above the threshold values.

Berylluim is now found to be carcinogenic. Inhalation of Be containing dust causes pulmonary damage and skin lesions. Beryllium prefers to bond to O but not N and S. VO_2^{2+} and V(IV) ions damage kidneys in the form of carbonate complexes. They are also toxic to enzymes present in the kidney.

In the case of transition metal ion natural stability order for complexes vis:

Mn(II) < Fe(II) < Co (II) < Ni (II) > Zin (II)

This shows that the metal ion toxic nature is related with chelation process.

11.8.4 Enzymes Inhibition

Ligands that can coordinate with the active site of metallo enzymes prevent further coordination, inhibit the enzyme activity, as CO binding to hemoglobin prevents further binding with O_2 and its function as O_2 carrier. Similarity N_3^- can occupy the space fit for CO_2, the azide ion binds to Zn displacing water and thus inhibit the functions of carbonic anhydrase. Some other inhibitors bind to Zn *e.g.* CN or H_2S inhibit the enzyme activity. Cyanide ion impairs respiration.

The first row transition metal except Mo resort to inhibition. Coordination of heavier metal ion in place of activating metal can destroy the enzyme activity. Hg^{2+} ion has a preferred affinity for sulfphur donor, being a soft lewis acid Hg^{2+} prefer S^{2-} of thiol group. Therefore Hg^{2+} tends to form extremely stable with sulphur containing amino-acids such as cysteine and methionine. The inhibition of an enzyme by Hg is an indication of the presence of thiol groups. Hence affinity of sulphur and mercury is an indication of the poisonous effect of Hg in the biological system.

On the other hand some sulfphur containing ligands have been found to be useful as antidotes for As, Pd and Hg poisoning.

11.8.5 Detoxification by Metal Ion and Chelation

Some biological sites bind the metal ions. Therefore, to dislodge these metal ions some chelating agents are administered in the form of medicines. They form stable and soluble metal chelates with metal ions thus removing the bound metal and excreting them out of the biological system. The chelating agents are of low toxicity and not metabolised. They remain unchanged in the biological system and hence act as scavenger. These chelating agents peneterate into the metal binding sites. It is essential that these cheleates should be much less toxic than free metal ions. A polydentate is considered most suitable. It picks up only metal ions so that chelated metal ion is not able to bind to any binding sites of enzymes and proteins in the system and is most suited to the elimination of the metal ion.

The most common multi dentate chelating agent is EDTA. EDTA is effectively used against circulating metal ions as it is of low toxicity, It exists as hydrophilic anions in solution and have poor power to penetrate the cells.

Generally sodium salt of calcium chelate is administered *e.g.* $Na_2[CaY]$ where YH_4 = edta by injection as the rate of absorption from the gastrointestinal tract is very low. The calcium chelate is used as a drug to remove lead poisoning since the Ca in the chelate is very easily replaced by strongly bound Pb which is thus excreted. It is also used for iron poisioning as also for excretion of Zn and radio active Sr.

Chelating agent- 2, 3-dimercapto propanol HS—CH_2—CH(SH—CH_2—OH) commonly known as BAL (British anti lewisite) was developed to protect enzymes with —SH groups used for the effect of arsenical in war gases.

It is also used in the treatment of poisoning by compounds of Hg, As, Sb, Bi and Au.

Similarly meso 2,3-dimercapto succinic acid (DMSA)

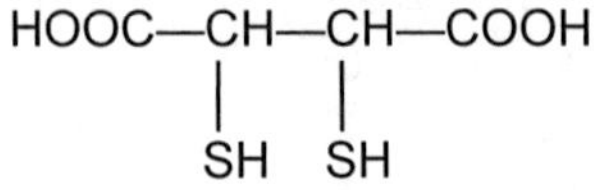

is effectively used as a drug for the treatment of Pb and Hg poisonng.

Penicillamine and and 2,2 dimethyl lithia- zolidine are bidentate ligands with low toxicity. Both of them are in use for mobilizing and excreting copper in the case of Wilson's Diseases. The dye aurine- tri carbonxlic acid is bindentate chelating agent, eradicates beryllium poisioning as it forms a stable chelate with Be(II) to facilitate excretion. Alzheimer disease is due to the accumulation of Al(III) in parts of the brain. The administration of desferrioxamine increases the excretion of aluminium through urinary excretion and decrease the increase of A.D.

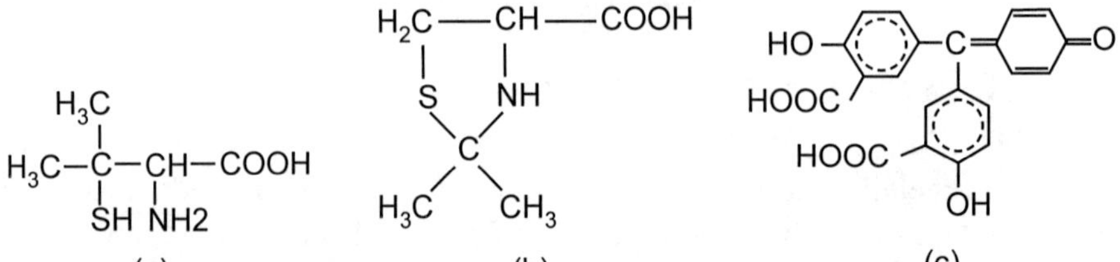

(a) (b) (c)

Fig. 11.9: (*a*) Pencillamin, (*b*) 2,2-dimethylthiozolidine-4-carboxylic acid, and (*c*) aurinetricarbodylic acid

8-Hyroxyquinoline is a famous analytical agent. It's antibacterial action is due to chelation of trace metal ions which are necessary for bacterial growth and function. Arsphenamine, 4,4′-(1,2 diarsenediyl)bis (2-amine phenol)

dihydrochloride (arrensophenolamine hydrochloride); 3,3′-diamine-4,4′ dihydroxy-arseno benzene dihydrochloride commonly known as salvarsan was prepared by B. Ehrlich was used in the treatment of syphilis and bovine pleuronuemonia.

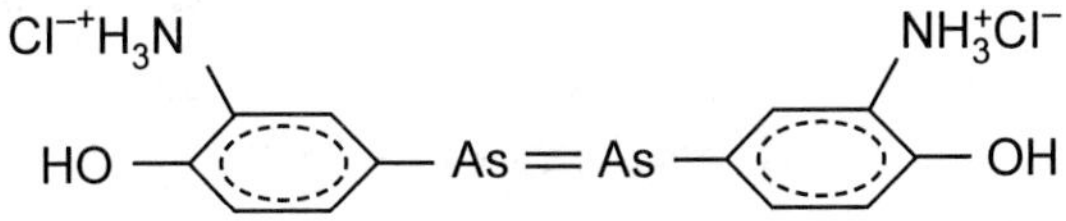

(I)

Fig. 11.10

Merbromin or mercurochrome(ii) is a well known antibacterial and antiseptic organo-mercury compound which was widely used in USA is now banned (in India it is still in use) An analogosus compound meralen is now in use.

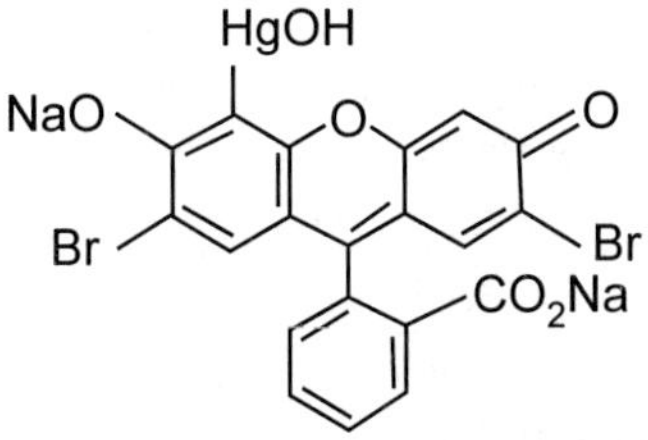

(II) Merbrowin (Mercurochrome)

Fig. 11.11

In meralein I Br and OH replaces ONa and SO_3Na in place of COONa of merbormin.

Detailed studies have shown that 8-hyroxy quinoline's antibacterial activity increases manifold in the presence of Fe(III) hence Fe(III) chelate is the active agent, Similarly anti tubercular activity of isonicotinic acid hyrazide is increased by the presence of copper (II).

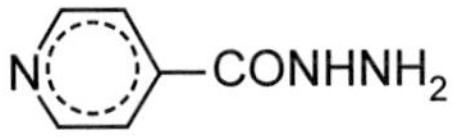

Fig. 11.12

Tetracycline has a number of o-donor atoms which form metal chelates of high stability.

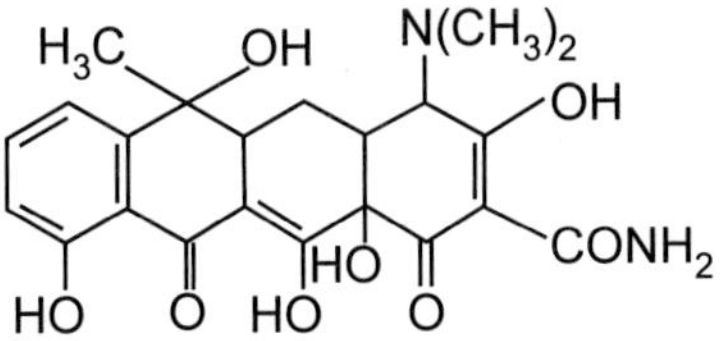

Fig. 11.13

Antipyretic action of the drugs is association with the chelated transport of Cu^{2+} ion. Metal chelates and their biological activity play an important role in cause and treatment of cancer and other diseases.

11.9 METAL COMPLEXES IN THE CURE OF CANCER

Krishschner investigated the antiviral and anti cancer of 26 inorganic compounds including metal complexes *e.g.* 6-mercaptopurine.

The metal ion is transformed with the ligand to viruses associated with cancer. This is a mechanism where the anticancer drug is released in the area of the tumour. Metal chelates of Dl- methionine and ethionine spread antibacterial activity to that of the ligand setting the metal ion fat soluble and thus capable of acting within the cell. Specially metal complexes of Pt and other group 8 to 10 metals of the Periodic Table are seen to have antibacterial power. They are potentially useful in cancer chemotherapy.

Researchers have carried out antitumour activity of some platinum compounds. The Active compoupds of Platinum are *cis*-$Pt(NH_3)_2Cl_2$, *cis*-$Pt(NH_3)_2Cl_4$, Pt(en) Cl_2, and Pt $(en)Cl_4Cl_2^{2+}$. Out of these Pt $(NH_3)_2Cl_2$ is very effective than the others. The charged $Pt(NH_3)_4$ Cl_2^{2+} is inactive in comparison to non-toxic *cis*-$Pt(NH_3)_2O_2$ which has a wide spectrum anti cancer activity.

But mer-$M(NH_3)_2$ Cl_2 where M = Rh, Ru seen to have anti tumour activity, so far about 200 complexes of different metals and different ligands have been found to have possible antitumor activity. Cis –$Pt(NH_3)_2$ Cl_2 and *cis*-Pt $(NH_3)_2$ Cl_4 have been tested against leukemia.

■ ■ ■

12

Uses and Applications of Coordination Complexes

12.1 INTRODUCTION

The coordination compounds find wide application in different aspects of daily life.

The complex compounds are used in Dyes as stated by G.T. Morgan and coworkers. They can be used as biological enzymes in the living system. Further the complexes are widely used in analytical chemistry to detect the composition of the substance by qualitative and quantitative analysis. Photography is another field where metal complex are involved in various stages of the process. The most important polymer industry uses the metal complexes as the suitable catalysts.

In the polymerization of ethylene Zeiglar natta calalyst involving titanium and aluminium complexes. The extraction of some of the metals from its over has an all important role played by metal complexes.

Yet another application of coordination complexes is in the form of artificial silk. Even in the domain of medicines complex compounds are widely used e.g. Cyanocobalamin hemoglobin etc.

These days chelation therapy is used in the treatment of heart and other diseases.

12.2 APPLICATIONS OF METAL COMPLEXES

12.2.1 Dye Industry

Schiff's bases are strong ligands and are a condensation product of amino compounds and carbonyl compounds. The metal complexes of the Schiff bases are important complexes in biological systems, polymers, dyes and medicines. They possess antibacterial, antifungal, anti-cancer and diuretic activities. They find applications in food industry, dye industry, analytical chemistry, catalysis, fungicidal, agro chemical and biological activities.

Chromium (56) azomithine (C=N) complexes, pinbage, cobalt complexes Schiff base, unsymmetrical complex 1:2 chromium dyes give fast colours to leathers, food packages, wools etc. Azo groups containing metal (57) complexes are used to make dyes poly fibers. Cobalt complexes of a Schiff base (salicyaldehyde with diamine) has excellent light resistance and storage ability and does not undergo degradation even in acidic gases(CO_2, SO_2, etc.). Zinc (II) and cadmium (II) complexes with N_2S_2-schiff base ligands give a new class of luminescent compounds.

12.2.2 Biological Enzymes

Some metals are essential for living organism while other metals are dangerous even in small doses. Metals play a pivotal role in the processes of biological systems. The metals act as a cofactor of enzymes catalyzing hydrolithical and redox processes and thus synthesize biomolecules

Low molecular weight metal complexes contribute to the structural and functional properties of metallo proteins and metallo enzymes. Structural model of metallo enzymes reveal the development of model compounds that guide us to the redox, magnetic, spectral and other properties of the molecule's active centre. The functional model helps us to understand the activity and working mechanism of metallo enzymes.

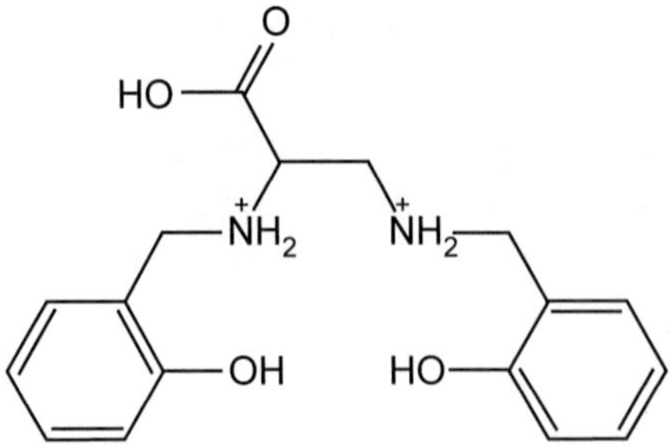

Figure 12.1: Structure of the bhbdpa ligand

Modelling the active site of metallo enzymes and metallo proteins is to establish an environment around the central metal ion which is similar to the one prevalent in the native system. For this purposes we use synthetic ligands in which donor ligands are not exactly identical to the donor set of the native ligand but their geometrical position provides similar structure to that of the native enzyme. Finally it is possible to use the same sequence as the metal binding site of the native protein.

The previously mentioned active site modelling methods can be used.

In the first instance iron(III) and copper (II) complexes of the salen derivative bhbdpa (n, n'-ethylene bis) salicylidine imine) ligand, the additional carboxylic groups may notably alter the structure and stability of the formed complexes and create a possibility for the formation of Dinuclear units. The mononuclear complexes of bhbdpa may be helpful to structural mimicking of the active centre of protocatechuate 3, 4 dioxygenase and galactose oxydase while dinuclear complexes of bhbdpa provide new examples for carboxylate bridged dimetallic cores.

In the second and third step the aim was the functional and structural modelling of copper (II) containing oxidase enzymes. For the mimicking of the native enzymes histidine rich active site, peptide ligands processing more histidine residues were used. Copper (II) and zinc (II) complexes of the multi histidine ligands are able to bind two or three metal ions.

12.3 POLYMER-METAL COMPLEXES AS CATALYSTS

In a polymeric catalyst or polymer supported catalyst, a conventional catalyst species is attached to a macromolecular backbone. Natural and synthetic polymers have been used as carriers of bio active molecules for modelling *e.g.* enzymes catalysis. They show a much lower catalytic activity than a true enzyme but they proved to be valuable in understanding some basic features of enzymes processes and in developing other type of synthetic polymers. Transition metal ions or complex ion anchored to a polymer, form an important form of these model compounds. If the matrix is polypeptide than the polymer-supported complexes are supposed to mimic the activity of metallo enzymes.

A biopolymer catalyst, such as, enzyme, shows high selectivity and activity for various reactions. The specificities are caused by intra-molecular cooperative catalysts between functional groups in the enzymes. The polymer support for chelates and catalysts, as well as synthetic reagents has grown recently since Merrifield demonstrated their use in peptide synthesis. The catalytic activities of the metal complex bound on the polymer matrices are different from the monomer analogue due to the specificities induced by the macromolecular ligands. In a polymer-metal complexes the polymers support provide a specific catalytic site and act as an effective matrix for the formation of active centres.

The catalytic activity of a polymer-metal complex is generally lower than the corresponding monomeric analogue due to steric reasons. However, the activity is sometimes be higher than the monomer due to the unsaturated coordination structure or by the strain in the formation of inactive dimeric complex due to steric effect. Polymer-metal complexes are widely used as immobilized catalysts, in which a specific catalytic behavior was induced by the polymer matrix. Besides, polymeric ligands have immense application in areas like conductivity, modification of material surface by binding of metal complexes, separation of metal ions and in bioinorganic system.

12.3.1 Catalytic Activities of Polymers-Metal Complex

Catalytically active polymers are obtained by introducing a catalytic centre to a polymer backbone and it is reasonable to assume that the catalyst bound to the polymer will show specific catalytic activity reflecting the properties of a polymer chain.

In the case of metalloenzymes such as oxidase and hemoglobin, metal complex is the active site, the macromolecular protein plays a significant role. Polymeric catalyst reduces catalyst poisoning since the catalytic site is protected by the polymered matrix. In a polymer-metal complex, aggregation is physically prevented by the rigidity of the polymer matrix.

In polymer-metal complexes, the selectivity arises out of steric hindrance or chemical environment of the polymer matrix.

Polymer-metal complexes are markedly useful as immobilized catalyst because it is more reactive than monomer analogue due to specificities of their large ligand molecules. It is mainly used in oxidation hydrogenation, hydrolysis, hydroformylation, decomposition of H_2O_2, initiation of radical polymerization and optical resolution.

1. The catalytic cycle of a metal-complex catalytic reaction is:

$$ML_{n-1} + S \rightleftharpoons L_{n-1}M\text{—}S$$

$$M'L + S^* \rightleftharpoons L_{n-1}\ M'\text{—}S^*$$

Where M stands for metal ion; L for ligand and S for substrate.

First of all substrate coordinates to a metal catalyst and form an intermediate mixed complex (LMS). The substrate is activated by the metal ion and dissociates from the catalyst and then regenerate to the original complex.

Characteristics of polymer supported catalysts are:

1. Polymers can act as insoluble carriers for catalyst. The domain in solid phase or the boundary of its surface provides a specific catalytic site. The isolation of the products from the catalyst is facilitated and the catalyst can be used repeatedly.
2. The activities of the polymer-metal complexes are sometimes lower than those of the corresponding analogues (monomeric) due to steric hindrances. They can be higher than the monomer complexes *e.g.* unsaturated coordination structures or sterical prevention of the inactive dimeric complex will increase the catalytic activity.
3. Selectivities of the reaction by polymer complexes are different from the monomeric complexes.

4. Polymer can be used as an effective matrix for the formation of active metal particles.

12.4 APPLICATION OF POLYMER-METAL COMPLEXES AS CATALYSTS

12.4.1 Oxidation

Oxidation of hyroqinone catalysed by Cu(II) complex of poly(1-vinyl imidazole), vinylamine-vinyl acetamide copolymers and polyethyleneimine were studied. The poly (1-vinyl imidazole) complex showed catalytic activity 20 times higher than cupric ions.

The oxidation mechanism is given by:

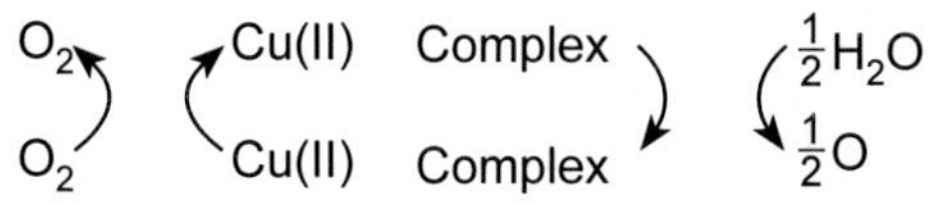

The advantages of polymer complexes were:

1. Selectivity for the substrate due to steric effect of polymer ligand
2. The metal ion are mono dispersed mono atomically giving high catalytic efficiency
3. The synthetic effect is excellent due to insolubility of the polymer catalyst.

In the auto oxidation of cumene catalysed by the metal complex of polymer supported Schiff base the activity decreased in the order Mn > CO > Ni > Cu > Zn and increased with increasing molecular weight of the Schiff base.

12.4.2 Hydrogenation

Catalytic hydrogenation proceeds through a mixed complex of catalyst and substrate. At least one coordination site of the catalyst must be free or labile so the substrate may be coordinated to the catalyst. Such a free or labile coordination site called 'unsaturation' may be released in the polymer metal complex due to steric factor. Pd complex of the diphenyl phospinated polystyrene shows remarkably high catalytic activity for the hydrogenation of the methyl ester of soybean fatty acid and the monomeric complex $PdCl_2(PPh_3)_2$.

For example other polymer-metal complexes used for hydrogenation are Rh, Ru or Ni carbonyl complexes of biphenyl phosphonated supported polystyrene.

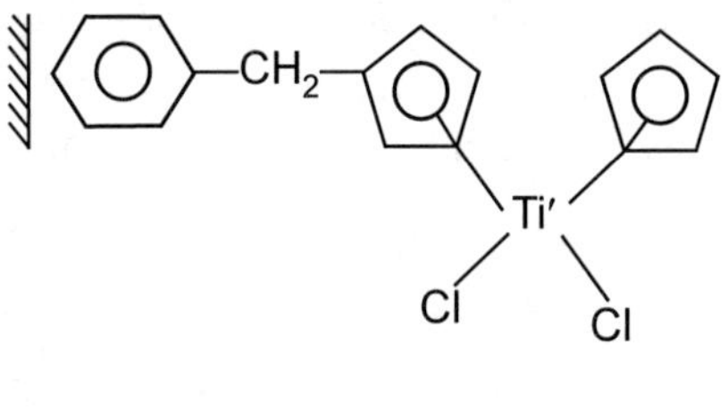

The polymer complex of titanocene dichloride indicated 15-20 times higher chemical activity for the homogeneous hydrogenation of olefins than the monomeric specy.

Selective hydrogenation of olefins than catalysed by polymer-metal complexes was noticed. The diphenyl phosphinated polystyrene-Ru complex or diacetyl iminomethylated polystryrene-Pd complex catalyzed the hydrogenation of octadiene giving selectively Cyclooctene.

This is because the rate of hydrogentiaon of the diene to monoene is much higher than that of the momoene Cyclooctane.

Recently the use of polymer as matrix for the dispassion of metal particles has gained importance. A colloid has been prepared by reduction of Rh(II) chloride with methyl alcohol in presence of PVA(polyvinyl acetate) The colloidal dispassion of Rh in methyl alcohol water solution was very effective for the hydrogenation of olefins at 30°C under an atmospheric hydrogen pressure. The colloided dispassion contains rhodium particles of small size i.e. 8 and 40 Å. The reactions are like this:

$$\text{Rh Cl}_3 \longrightarrow \text{RH(III)—PVA Complex} \xrightarrow[\text{CH}_3\text{OH}]{\text{Reduction}} \text{Rh particles (8 Å)} \longrightarrow \text{Rh particles (40 Å)}$$

Porphyrin containing polymers were found to be remarkably effective as a catalyst in photo redox systems. In such systems various porphyrin groups such a protoporphyrin and chlorine as well as metalloporphyrin groups such as Mg(II) and Cu(II) chlorophyllin and Fe(III) and Co(II) protoprophyrin were incorporated into vinyl polymers.

12.4.3 Hydrolysis

The hydrolyses of sodium pyrophosphate was done by using some metal complexes as catalysts. The catalytic activity of the polymer-complexes declined in the order:

$$\text{Zr(IV)} > \text{U(vi) O}_2 > \text{Cr(III)} \equiv \text{Ce(III)} \equiv \text{Cu(II)}$$

The increase in the catalytic activity was ascribed to electrostatic effect of the polymer complexes as well as to the possible multiplying effect by $ZrOL^+$ and ZrO(OH) along the polymer-ligand chains. The Catalytic hydrolysis of oligo phosphates by poly (L-lysine)-Cu(II) complexes has been reported. Breslow attached the Ni(II) complex to cyclohexa amylase, which formed a hydrophobic cavity as shown and studied the hydrolysis of *p*-introphenyl acetate.

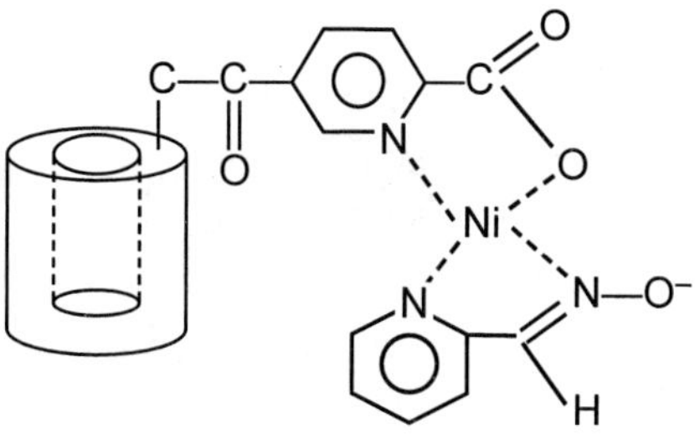

12.4.4 Hydroformylation of Olefins

Hydro formylation is the reaction of olefins, carbon monoxide and hydrogen producing saturated aldehyde or alcohols. Insoluble catalysts of polymer-metal complexes are very useful for the reaction since they are easily separated from the products and used again and again. They sometimes show selectivities for the reaction due to the polymeric structure:

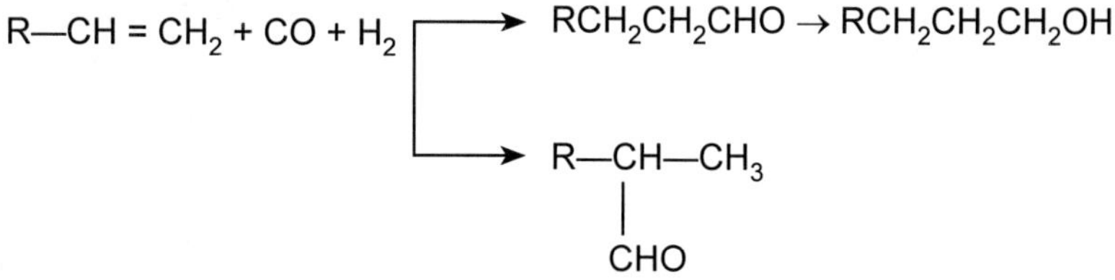

The complex of $[RH(CO)_2Cl_2]$ bound to the cross linked polystyrene containing the —$P(Ph)_2$, —$N(CH_3)_2$ or –CH_2—N—$(CH_2)_2$ group catalysed the hydroformylation of 1-hexane. The polymeric phophine-Rh complex showed selectivity for the aldehyde synthesis and the polymeric amine-Rh complex selectivity for the alcohols as shown above. The complex RhCl(CO) bound to phosphinated silica get coated with polystyrene is a stable hterogenous catalyst for the hydroformylation of ethylene.

The cobalt resin derived from a cross linked poly (chloromethylated styrene) is a good hyroformylation catalyst for 1-pentene and cyclohexene.

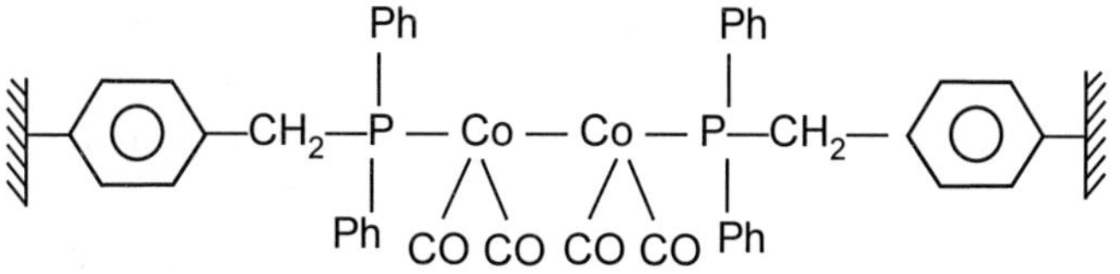

Catalyst recovery be attained by simple filtration and the catalystic activity is retained after prolonged exposure of resin to air.

12.5 INITIATION OF RADICAL POLYMERISATION

Polymer-metal complexes initiate radical polymerisation and often show higher activity than the corresponding monomeric analogue.

Nylon-Cu(II) systems initiate the polymerisation of methyl methacrylate (MMA) in presence of CCl_4 and water.

The study of polymerization by Cu(II) complex of (E amino cronic acid) and other oligo meric nylon showed the increase of the catalytic activity with the increase in the degree of polymesisation of the nylon.

Natural products e.g. cellulose and soluble starch were found to initiate the polymerization of MMA in aqueous solution in presence of Cu^{2+} ion through the complex formation of Cu^{2+} with the natural polymer.

12.6 ASYMMETRIC SYNTHETIC AND OPTICAL RESOLUTION

Steric hindrance of polymer ligand can provide asymmetric structure which bring about asymmetric reaction.

Haleno and Nozawa in 1971 succeeded in carrying out an asymmetrically selective oxidation of 3,4-dihydroxy phenyl alanine DOPA by poly (L-lysine) (PLL) –Cu (II) complex. D-Dopa was found to be oxidized by the PLL-Cu(II) complex. D-DOPA was found to be oxidized by the PLL-Cu(II) complex more rapidly than L-DOPA. Such an asymmetric oxidation was ascribed to this asymmetrically selective interaction of DOPA with PLL-Cu(II) complex. The combination of two or more asymmetric cupric complex units held on the α-helical structure of PLL was considered to cause the asymmetric catalysis.

Rh complex of Disphosphine compounds bound to polystyrene catalysed the asymmetric hydrogenation of ketone e.g. acetophenone, methyl benzyl ketone with the optical yield of 58%.

Asymmetric structure provided by polymer-metal complexes makes optical resolution of the adsorbed organic molecules. The Cu^{2+} complex of the polymer containing N, N-dibenzyl lucine adsorbed D-isomer of amino-acids more than L-isomer. The sterio selectivity was opposite to the complex monomeric analogue.

12.7 THE DECOMPOSITION OF H_2O_2

The decomposition of H_2O_2 is affected by using various metal complexes as catalysts. The catalysts has its origin in redox action of a metal complex. The reaction may proceed either through a transient intermediate formed between the metal catalyst and the substrate or through chain decomposition.

The decomposition of H_2O_2 is employed as a standard reaction to determine the catalytic activity of polymer-metal complex.

Polymer Metal complexes exhibit high efficiency in the catalytic decomposition of H_2O_2 because:

(*i*) Some polymer-metal complex contain incomplete complexes due to steric hindrance and increases catalytic activity.

(*ii*) In some polymer complexes the coordination bound between polymer ligand and metal ion is relatively weak than substrate coordinate with high frequency.

(*iii*) Chain decomposition of substrate proceeds rapidly because concentration of catalytic site is locally enhanced. Polymer-metal complexes showed much higher catalytic activity than the corresponding monomeric analogue.

The catalytic activity of Fe(III) tri-ethylene tetramine (trien) complex was found to be 10^7 times higher than the ferric aqua in the decomposition of H_2O_2. In the mechanism proposed OOH ion coordinates to ferric ion as a bidentate ligand, resulting in the cleavage of the (O—O) bond with the evolution of oxygen.

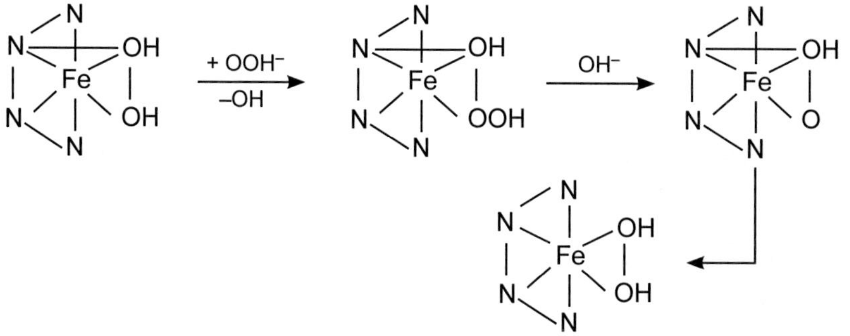

Poly(β-diketone)-Cu complex and poly (hydroxoxamic acid)—Cu complex showed higher catalytic activity than the monomeric catalyst .

High activity of polymer complexes is attributed to weak coordination bonds between polymer ligands and the metal ions. A higher molecular weight of the polymer ligand gave higher activity.

Polymethacrylate acetone-Cu, polyglycine Cu and ferric complexes of polyacrylic acid showed high activity for decomposition of H_2O_2.

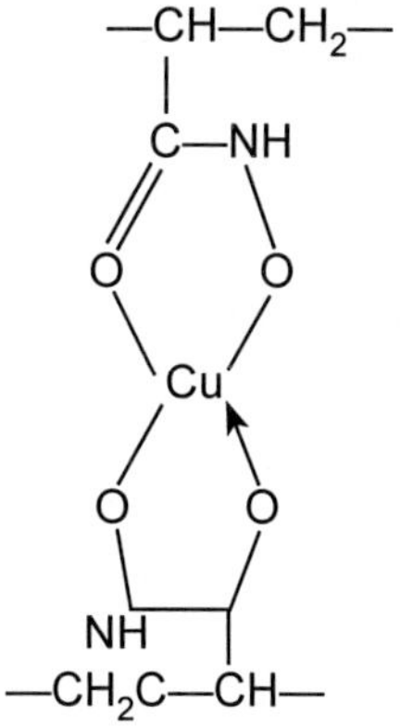

12.8 POLYMER-METAL COMPLEXES AS HETEROGENEOUS BIOLOGICALLY ACTIVE CATALYSTS

Heterogeneous catalyst is prepared because the main purpose of supporting a catalyst is to be able to recover and recycle it easily.

The three distinct stages in the preparation of a supported catalyst are

1. The choice of a suitable support material.
2. The introduction of a suitable functional group on the support which would interact with the catalyst species in the loading process.
3. The choice of such functional groups depends on the catalysts chosen. The incorporation of the catalyst on the support is carried out under mild condition so as to retain the activity of the catalyst.

Metal ions are important in the activity of bioinorganic materials in which the metal ion are bound to proteins, nucleic acids and related ligands. Metallo enzymes are generally formed between a polypeptide and metal ion. The coordination structure of the complex, the conformation of the polypeptide which is dependent on the sequence of amino acids in the polypeptide, stiffness of the backbone and the interaction between the pendent groups cause specificities in metallo enzymes.

In metallic enzymes the large protein ligand decides the character and the activity of the central metal ion by providing a selective binding site for the substrate. The selectivity is generated by the long chain apo proteins. In blue-copper proteins the geometry of the central copper ion is not square planer but- a distorted tetrahedral structure which is caused by the steric strain of the peptide chain.

Macromolecular effects on catalytic activities of polymer-metal complexes

12.8.1 Porosity and Surface Area

The pore-size and surface area of the polymers varied with the nature and extent of cross linking agent. Optimum cross linked polymeric ligands should have a large surface area and average pore-radius. The reaction rate increase with the exposed area of the catalytic species.

To obtain reasonable reaction rates, such area has to be increased to an adequate level per unit volume and this can be achieved by making the catalyst-highly porous and spreading the active species over the entire surface. The development of a catalyst formulation should take into consideration not only the chemical nature and dispersion of the active species but also the design of the catalyst besides the factors such as pore-size and adequate mechanical strength.

The characteristics of the pore-structure also influence the response of the catalysts to poison. Hydrogenation of butadiene to butane, the oxidation of o-xylene to analydride require highly selective catalysts. Rapid diffusion of the gas require macro-micro combination of pore-sizes.

12.8.2 Extent of Cross-Linking

The complexation behavior of ligand function supported on polymer is different from corresponding low molecular weight analogue.

The variation depends upon the polymer-matrix in which the ligand function is attached. Complexing groups in a macro molecular ligand may adopt conformation different from that required by the preferred stereo chemistry of a given metal ion. The factors like matrix structure, matrix configuration, matrix nature, spacer between the cross linked matrix, the ligand function and the extent of cross linking has a significant effect on complexation .

The complexation characteristics of insoluble macromolecules are strongly influenced by the nature and the extent of cross-linking in the polymer ligand . in the case of poly (4-vinyl pyridine)-Cu complexes with increasing divinyl benzene content, conversion of planar coordination centre to tetrahedral ones occur. The change results form an increase in steric hindrance to complex formation with a polymeric ligand.

12.8.3 Spacer Groups

The complexing power of macro molecular ligand depends upon the position of functional groups with regard to the main chain.

The shorter distance between them, the lower the efficiency of complex owing to the steric hindrance. By introducing a spacer to polymer matrix the less reactive functional groups can make more reactive and cause high catalytic activity.

Functional groups directly attached to the backbone chain of high polymers are often less reactive than if these groups were not attached to a polymer. This reduced activity is caused by steric hindrance or by the fact that all or part of the polymer chains interferes with the reactivity of the individual functional groups. Spacer groups could be designed to be flexible or stiff. The commonly used spacer are hexa methylene diamine, poly ethylene glycol and α,w-dibromoalkane. If the functional groups initially present in the polymer are not isolated , the spaces may become double coupled with the polymer and is effectively lost. Such reaction have been used to study site isolation .

12.8.4 Coordination Complex in Analytical Chemistry

Analytical chemistry is divided into two aspects.

(*i*) Qualitative inorganic analysis

(*ii*) Quantitative inorganic analysis

Qualitative in Organic Analysis

It is that method of analytical chemistry which seeks to find elemental composition of inorganic compounds. It is mainly focussed on detecting ions in aqueous solution. Materials in other forms may be first brought into this state by using standard methods.

The solution thus obtained is then treated with various reagents to test for reactions characteristic of certain ions which may cause colour change, precipitate formation and other visible changes. Qualitative inorganic analysis is that branch which seeks to establish elemental composition of inorganic compounds through various reagents.

This may be achieved by

(*i*) Physical appearance of inorganic salts.

S. No.	Salt	Colour
1.	MnO, MnO_2, FeO, CuO, Co_3O_4, Ni_2O_3, Sulphides of Ag^+, Cu^+, Ca^{2+}, Ni^{2+}, Fe^{2+}, Co^{2+}, Pb^{2+}, Hg^{2+}	Black
2.	Hydrated Cu^{2+} Salts	Blue
3.	HgO, HgI_2, Pb_3O_4	Red
4.	Cr^{3+}, Ni^{2+}, hydrated Fe^{2+} salts	Green
5.	Hydrated Mn^{2+} salts	Light Pink
6.	KO_2, $K_2Cr_2O_7$, Sb_2S_3, Ferrocyanide	Orange
7.	Hydrated Co2+ Salts	Reddish Pink
8.	Chromates, AgBr, Ag_2S_3, AgI, PbI_2, CdS	Yellow
9.	Cdo, Fe_2O_3, PbO_2, $CuCrO_4$	Dark Brown

12.8.5 Detecting Cations

Cations are usually classified into six groups. Each group has a common reagent which can be used to separate them from the solution. Separation of ions must be done in the sequence specified in the next step. Some ions from the earlier groups may react with the reagent of a later groups and cause ambiguity. This happens because cation analysis is based an the solubility product of the ions as the cations attain the solubility product concentration it precipitates and hence allowing the ions to be detected. The precise details have been given in each group analysis .

12.8.6 Group I Cations

In this group cations are precipitated as insoluble chlorides. Therefore group reagent to separate them is HCl of 1-2 M concentration. Hence dilute HCl is used (conc. HCl forms soluble $[PbCl_4]^{2-}$ with Pb^{2+}).

This cations in the first groups are Ag^+, Hg_2^{2+} and Pb^{2+}. They all form white solid compounds such as AgCl, Hg_2Cl_2 and $PbCl_2$. Out of these white precipitates $PbCl_2$ is soluble in hot water and thus gets separated.

12.8.7 Confirmatory Test for Pb^{2+} Ions

1. To a part of the solution containing Pb^{2+} ions potassium iodide solution is added.

$$Pb^{2+} + 2KI \rightarrow \underset{\text{Yellow PPT}}{PbI_2} + 2K^+$$

2. Kr_2CrO_4 solution is added to Pb^{2+} ions

$$Pb^{2+} + K_2CrO_4 \rightarrow \underset{\text{Yellow PPT}}{PbCrO_4} + 2K^+$$

The remaining white precipitate contains Ag^+ and Hg_2^{2+} (ous) ions.
On adding ammonia solution to the precipitate AgCl dissolves in ammonia

$$AgCl + 2NH_3 \rightarrow \underset{\text{Soluble complex}}{[Ag(NH_3)_2]Cl}$$

whereas Hg_2Cl_2 gives a black precipitate consisting of a mixture of chloro-mercuric amide and elemental mercury .

12.8.8 Confirmatory Test for Ag^+ Ions

1. AgCl is reduced to Ag under sun light which gives violet colour due to colloidal silver .
2. Potassium iodide solution is added to the Ag^+ ion solution to get a yellow precipitate

$$Ag^+ + KI \rightarrow AgI + K^+$$

3. To another portion of the Ag^+ solution is added K_2CrO_4 solution

$$2Ag^+ + K_2CrO_4 \rightarrow \underset{\text{Red PPT}}{Ag_2CrO_4} + 2K^+$$

12.8.9 Confirmatory test for mercury (1) ion

1. Add KI solution to the Hg_2^{2+} ion solution

$$Hg_2^{2+} + 2KI \rightarrow \underset{\text{Scarlet PPT}}{Hg_2I_2} + 2K^+$$

2. Caustic soda solution is added to Hg_2^{2+} solution

$$2Hg_2^{2+} + 2NaOH \rightarrow \underset{\text{grey}}{2Hg_2O} + 2Na^+ + H_2O$$

12.8.10 Group II Cations

Second group cations include Cd^{2+}, Bi^{3+}, Cu^{2+}, As^{3+}, As^{5+}, Sb^{5+}, Sn^{4+} and Hg^{2+} and Pb^{3+}. Lead is present in sufficient amount to be tested in group I. However, if Pb^{2+} is not detected in group I than it appears in group II sometimes if HCl concentration may not be sufficient to detect Pb^{2+} in group I then it is tested in group II.

12.8.11 Group Reagent

All these group II cations are precipitated as sulfides. The sulfides of these metal ions are formed in the presence of dilute HCl. If dilute HCl is not present then pre mature precipitation of group III cation may occur thus leading to misleading result. Other acids such as sulphuric acid may precipitate group V cations while nitric acid may oxidise sulphide ions to colloidal sulphar .

The precipitates of these cations are indistinguishable except CdS which is a yellow precipitate. Hg S is soluble in aqua regia (3HCl:1 HNO_3) and can be separated from the rest. CuS dissolves in ammonia forming intense blue solution where as CdS dissolves to give a colorless solution. The sulphides soluble in yellow ammonia sulphide where they soluble form poly sulphide complexs e.g., As^{3+}, As^{5+}, Sb^{3+}, Sb^{5+}, Sn^{2+}, Sn^{4+} etc.

This group is determined by adding salt in water and then adding dilute HCl to make the medium acidic followed by H_2S gas.

If it forms reddish brown black precipitate than Bi_{3+}, Cu^{2+}, Hg^{2+} or Pb^{2+} are present. In the case of yellow precipitate Cd^{2+} or Sn^{4+} is present ;brown precipitate then Sn^{2+} is present; red orange precipitate indicates Sb^{3+}.

Confirmatory Test for Pb^{2+}

$$Pb^{2+} + K_2CrO_4 \rightarrow \underset{\text{yellow ppt}}{PbCrO_4} + 2K^+$$

Confirmatory Test for Cu^{2+}

$$2Cu^{2+} + K_4[Fe(CN)_6] + CH_3COOH \rightarrow Cu_2[Fe(CN)_6] + 4K^+$$

$$Cu_2+ + 2NaOH \rightarrow Cu(OH)_2 + 2Na^+$$

$$Cu(OH)_2 \rightarrow CuO + H_2O$$

Confirmatory Test for Bi^{3+}

$$Bi^{3+} + 3KI(excess) \rightarrow BiI_3 + 3K^+$$

$$BiI_3 + KI \rightarrow K[BiI_4]$$

$$Bi^{3+} + H_2O\ (excess) \rightarrow BiO^+ + 2H^+$$

Bismathyl ion (BiOCl) Turbid

Confirmatory test of mercury

$$Hg^{2+} + 2KI\ (Excess) \rightarrow \underset{\text{Red PPT}}{Hg\,I_2\downarrow} + 2\,K^+$$

$$HgI_2 + 2KI \rightarrow K_2[Hg\,I_4]$$

Red PPT dissolves

$$2Hg^{2+} + SnCl_2 \rightarrow 2Hg + SnCl_4$$

white precipitate turns grey

12.9 ROLE OF HCL IN GROUP II ANALYSIS

The solubility of second group sulphides is low, therefore, only low concentration of sulphide ions will precipitate them. If the concentration of sulphide ions is not kept low then the group III cations will also get precipitated but we want II cations. Cations are present in salt. So to add sulphide S^{2-} ion in the water extract we add H_2S, which dissociate more than required and give excess S^{2-} ions which is not favourable for the group II cation precipitate. So, in order to reduce the concentration of sulphide ions, we add a solution which contains some H^+ ions so that by the common ion effect H_2S dissociates less and we get less concentration of sulphide ions to precipitate only the group II sulphides hence we add dilute HCl in the water extract e.g.

$$HCl \rightarrow H^+ + Cl^-$$

$$H_2S \rightarrow 2H^+ + S^{2-}$$

Common ion effect of H^+ ions is required for less concentration of sulphide ions.

12.9.1 Group III Cations

Group III cations have ions that form hydroxides which are insoluble even at low concentration. The precipitation of hydroxides occur at pH 8 to 9. Therefore buffer solution is used to ensure this pH.

Cations present in group III are Fe^{2+}, Fe^{3+}, Al^{3+} and Cr^{3+}

This group may be determined by making a solution of the salt in water and adding ammonium chloride (electrolyte) and then ammonium hydroxide. NH_4Cl is added to ensure low concentration of hydroxide ions.

The formation of reddish brown precipitate indicates Fe^{3+} sometimes adding a drop of concentrated HNO_3 helps as it oxidises the Fe^{2+} present to Fe^{3+} ions.

A gelatinous white precipitate indicates Al^{3+} and a green precipitate indicates Cr^{3+} or Fe^{2+}.

The last two are distinguished by adding NaOH in excess to the green precipitate. If the precipitate dissolves, Cr^{3+} is indicated otherwise, Fe^{2+} is present.

12.9.2 Confirmatory Test for Fe^{3+}, Al^{3+}, Cr^{3+}

1. $Fe^{3+} + 3SCN^- \rightarrow Fe(SCN)_3$
   Blood Red Solution
2. $Fe^{3+} + [Fe(CN)_6]^{2-} \rightarrow Fe_4[Fe(CN)_6]_3$
   Prussian blue Solution
3. $Al(OH)_3 + 3NaOH \rightarrow Na_3\,[Al(OH)_6]$
   Soluble Solution

4. $Al(OH)_3$ + Blue litmus → Blue floating lake
5. $Cr(OH)_3 + 2NaOH$ → Na_2CrO_4 yellow solution
6. $Na_2CrO_4 + H_2O_2$ → $CrO_5 + 2NaOH$
   Blue coloration

12.9.3 Group IV cations

Group IV includes Zn^{2+}, Ni^{2+}, Co^{2+} and Mn^{2+}. Zinc salts are colourless, Manganese salt are faint pink or colourless, Nickel and cobalt salt are highly coloured often blue green.

12.9.4 Group Reagent

This group cations are precipitated as sulphides. Here the concentration of sulphide ions required is high. To acquire high concentration of sulphide ion H_2S is passed through water extract of the salt plus NH_4OH or NaOH.

ZnS forms a white precipitate
NiS forms a blue-green precipitate
Co S forms a purple or move precipitate
Mn S forms a very faint pink precipitate

The precipitates are washed with water and reacted with extremely dilute hydrochloric acid. This precipitates Nickel salts, if present, The supernatant liquid is filtered and reacted with excess NaOH. This precipitates' Manganese's salts, if present.

12.9.5 Confirmatory Tests

Zinc Salt

$2Zn^{2+} + K_4[Fe(CN)_6]$ → $Zn_2[Fe(CN)_6]$
Bluish precipitate

Amphoteric mature of Zinc

$ZnCl + 2NaOH$ → $Na_2ZnO_2 + 2HCl$
Sodium zincate

Nickel Salt

$Ni^{2+} + 2DMG$ → $Ni(DMG)_2$
Red PPT

$Ni^{2+} + (NH_4)_2SO_4$ → $Ni(NH_4)_2(SO_4)_2$
Blue Solid

$Ni^{2+} + 2NaOH$ → $Ni(OH)_2$

$2Ni(OH)_2 + Br_2 + H_2O$ → $2Ni(OH)_3$
Blue PPT

Cation CO^{2+}

$Co^{2+} + 2HCl$ → $CoCl_2.6H_2O$
V. Light pink

$Co^{2+} + KNO_2 + H_3COOH \rightarrow K_3[Co(NO_2)_6]$
Potassium cobalti nitrite test yellow ppt.

$Mn^{2+} + 2NaOH \rightarrow MnO(OH)_2\downarrow$
Manganic hydroxide Brown PPT.

Mn^{2+} + Conc.$5HNO_3 + 5PbO_2 \rightarrow 2HMnO_4$. Pink Manganic acid solution $+ 5Pb(NO_3)_2 + 2H_2O$.

12.9.6 Group V cations

The Most important ions in group V are Ba^{2+} Ca^{2+} and Sr^{2+}. Group V cations form carbonates that are insoluble in water. The reagent used to precipitate group V cations is $(NH_4)_2CO_3$ (around 0.2M) in neutral or slightly basic pH. The cations in the previous groups are separated before hand since many of them also form insoluble carbonates.

The best way to distinguish them after separation is by using flame test

(*a*) Calcium gives brick red

(*b*) Barium gives grassy green

(*c*) Strontim gives crimson color to the flame.

The white precipitate of $BaCO_3$, $CaCo_3$ and $SrCO_3$ is dissolved in dilute HCL and then to a test portion is added K_2CrO_4.

$BaCl_2 + K_2CrO_4 \rightarrow BaCrO_4 + 2KCl$
Yellow ppt.

$CaCl_2$ + COOK–COOK or $COONH_4$–$COONH_4$ $\rightarrow Ca(C_2O_4)_2$
White ppt.

$SrCl_2 + H_2SO_4 \rightarrow SrSO_4 + 2HCl$
White ppt.

12.9.7 Group VI Cations

This group includes cations like Mg^{2+} and K^+. The group reagent used is sodium phosphate. After light warming and rubbing the test tube with a glass rod a dirty while precipitate of $Mg_3(PO_4)_2$ or $Mg_2P_2O_7$ is obtained. Like wise K_3PO_4 is formed. On performing the flame test potassium gives a violet flame.

Potassium is further confirmed by cobaltinitrite test.

$KCl + Co(NO_3)_2 + CH_3COOH \rightarrow K_3[Co(NO_2)_6]$
Yellow ppt

12.9.8 Zero Group cations

Ammonium ion has no place in the main scheme of cation analysis because it is made of non metals nitrogen and hydrogen.

12.9.9 Confirmatory Test for NH_4^+ ion

It is tested by taking the water extract of the salt, and subsequently adding mercuric chloride and potassium iodide solution a brown precipitate develops which is also known as Nesler's reagent.

$$NH_4Cl + KI + HgCl_4 \rightarrow \underset{\substack{\text{Brown ppt.} \\ \text{Nesler's reagent}}}{K_2[HgI_4]}$$

12.10 DETECTING ANIONS

12.10.1 Group I Anions

It consists and CO_3^{2-}, CH_3COO^-, HCO_3, S^{2-}, SO_3^{2-}, $S_2O_3^{2-}$ and NO_2^-. The reagent for this group is dilute HCl or dilute H_2SO_4.

1. Brisk effervescence due to release of CO_2 gas which turns lime water milky due to the formation of $CaCO_3$. Then on passing excess of CO_2 milkiness disappears due to formation of $Ca(HCO_3)_2$.
2. Vinegar smell of CH_3COOH and blood red colouration is produced with $FeCl_3$ solution due to the formation of ferric acetate.
3. Smell of rotten eggs H_2S.
   Confirmatory Tests of S^{2-} ion
   (*a*) Lead acetate paper, turns black due to the formation of PbS.
   (*b*) S^{2-} ion turn red sodium nitro prusside solution purple
   $Na_2S + Na_2\,[Fe(CN)_5NO] \rightarrow Na_4[Fe(CN)_4NOS]$ Purple
4. Smell of SO_2 gas, smells of burning sulphur.
   It turn $K_2Cr_2O_7$ green:
   $K_2Cr_2O_7 + 4SO_2 + H_2O \rightarrow K_2SO_4 + Cr_2\,(SO_4)_3$ Green
5. Produces SO_2 and a deposit of sulphur. Indicates $S_2O_3^{2-}$ ion $S_2O_3^{2-}$ ion turns silver nitrate solution black (brown to black)
   $$Na_2S_2O_3 + 2AgNO_3 \rightarrow \underset{\text{Brown}}{Ag_2S_2O_3} \rightarrow \underset{\text{Black}}{Ag_2S}$$
6. Reddish brown fumes are obtained which turn
   (a) $FeSO_4 + NO \rightarrow \underset{\text{black}}{(NO)\,FeSO_4}$
   (a) $KI + \text{Starch} + NO \rightarrow$ blue

12.10.2 Group of II Anions

The group consists of Cl^-, Br^-, I^-, NO_3^- and $C_2O_4^{2-}$. The group reagent is concentrated sulphuric acid (H_2SO_4) after addition of Conc. H_2SO_4 Cl^-, Br^- and I^- form precipitates with $AgNO_3$. The precipitates being white, pale yellow and yellow respectively. The silver halides formed are completely soluble, partially soluble and insoluble respectively in aqueous ammonia solution.

12.11 CONFIRMATORY TEST FOR Cl^- ION

12.11.1 Chromyl Chloride Test

Solid salt is heated with $K_2Cr_2O_7$ and cone. H_2SO_2 forming red vapours of chromyl chloride (CrO_2Cl_2). On passing the gas through the solution of NaOH yellow solution of Na_2CrO_4 is obtained. This yellow solution forms yellow precipitate with $(CH_3COO)_2$ Pb.

$$2NaCl + K_2Cr_2O_7 + \text{conc. } H_2SO_4 \rightarrow CrO_2Cl_2 + K_2SO_2 + Na_2SO_2$$

$$CrO_2Cl_2 + 2NaOH \rightarrow Na_2CrO_4 + 2HCl$$

$$Na_2CrO_4 + (CH_3COO)_2Pb \rightarrow \underset{\text{Yellow ppt}}{PbCrO_4\downarrow} + 2CH_3COONa$$

12.11.2 Br^- and I^-

Layer Test

A sodium carbonate extract is made from the solution containing Br^- or I^- and $CHCl_3$ or CS_2 is added to the solution which forms two layer; an orange colour in the $CHCl_3$ or CS_2 layer confirms Br^- and a violet colour indicates the presence of I^-.

12.11.3 Test for NO_3^- Ion

Nitrate ion gives brown fumes with concentrated H_2SO_4 due to the formation of NO_2 gas. The NO_2 gas is intensified by adding copper turnings. To this is added aqueous solution of $FeSO_4$ salt and addition of conc. H_2SO_4 along the side of the test tube. This produces brown ring around the walls of the tube at the junction of the two liquids caused by the formation of $|Fe(NO)|^{2+}$ ion.

12.11.4 Test for $C_2O_4^{2-}$ Ion

Upon treatment with conc. H_2SO_4 oxalates yield CO_2 and CO gas which burn with a bluish flame (CO), turn lime water milky and decolorise $KMnO_4$ solution and gives a while precipitate with $CaCl_2$.

12.11.5 Group III Anions

The group III anions such as SO_4^{2-}, PO_4^{3-} and BO_3^{3-} neither react with concentrated or dilute H_2SO_4.

12.11.6 SO_4^{2-} Ion Test

The soda extract solution is neutralized with dilute acetic acid and then $BaCl_2$ solution is added. While precipitate insoluble in any acid or base confirms SO_4^{2-} ion.

$$S.E + BaCl_2 \rightarrow BaSO_4\downarrow$$

in soluble in any
acid or base

12.11.7 PO_4^{3-} Ion Test

To the soda extract solution is added few drops of conc.HNO_3 followed by the ammonium molybdate solution. A crystalline yellow precipitate confirms PO_4^{3-} ion.

12.11.8 BO_3^{3-} Ion Test

Add to the soda extract solution concentrated H_2SO_4 and ethanol and burn in a flame, green edged flame confirms ethyl borate formation and this confirms BO_3^{3-} ion.

$$2Na_3BO_3 + 3C_2H_5OH + \text{conc. } 3H_2SO_4 \rightarrow (C_2H_5)_3BO_3 + 3Na_2SO_4$$

Green-edge

12.12 MODERN TECHNIQUES

Modern techniques, instrumental analysis give quick and accurate results such as atomic absorption, spectroscopy and ICP-MS are more in use. Qualitative analysis is only pedagogical tool in the present day analysis.

12.12.1 Quantitative Analysis

The composition of a sample is very important. There are two types of methods used to find out the composition of a sample (*a*) Volumetric analysis and (*b*) Gravimetric analysis. Out of the two gravimetric yields more accurate results but it is time consuming. The volumetric analysis, on the other hand produce satisfactory result but takes much less time.

Volumetric analysis is based on titration. It may include

(*a*) neutralisaiton reaction
(*b*) precipitation reaction
(*c*) complex forming reaction and
(*d*) redox reaction.

(*i*) The neutralisation reaction is based on acid and a base reaction, hence it is called neutralisation.

(*ii*) In the precipitation reaction the standard solution in most of the cases is silver nitrate which forms highly insoluble precipitate of silver salts. This is simply called argentometry.

(*iii*) In the rest of the two methods, complex formation reaction between the metal ion and the standard solution of ethylene diamine titra acetic acid (EDTA) occurs, and (*iv*) In the redox reaction, is carried out between an oxidizing agent and a reducing agent.

Further, qualitative analysis can be performed by (a) mass spectrometry on biological samples can tell relative abundance ratio of specific proteins, indication of certain diseases, like cancer.

12.12.2 Quantitative Vs. Qualitative

The term qualitative analysis, refers to analysis in which substances are identified on the basis of qualitative properties such as chemical reactivity, solubility, molecular weight, melting point, radioactive properties (emission and absorption) mass spectra, nuclear half life etc.

Quantitative analysis refers to analysis in which concentration of an analyte may be estimated and expressed as a numerical value in appropriate units. Qualitative analysis may take place with quantitative analysis, but the qualitative analysis requires identification of the analyte for which numerical value has to be found *i.e.* Quantitative analysis. For example for a given solid sample qualitative techniques such as IR, NMR spectroscopy are used to identify the compounds present and then quantitatively determine the amount of each compound in the sample. Procedures for estimating the presence of metal ions have been developed but they are largely replaced by the modern instruments which are collectively known as qualitative inorganic analysis. Similarly tests for identifying organic compounds are also known.

Many techniques are used for both qualitative and quantitative measurements. For instance, indicator solution changes colour in the presence of a metal ion can be used as a qualitative test when a drop of sample is added. It could also e used as a quantitative test, by studying the colour of the indicator solution with different concentrations of the metal ion (ultraviolet-visible spectroscopy).

12.13 EXTRACTION OF METAL FROM COMPLEX ORES

The extraction of metals from complex ores is called metallurgy. In complex ores iron is located in oxides in a form of chemical compounds or in solid solutions with oxides of harder reducible metals like magnesium, chromium or titanium. Ore grains in disseminated ores are separated by gangue material layers, mainly by silicates of different composition.

The complex chromite spinal crystal lattice (Mg^{2+} Fe^{2+}) [Fe^{3+}, Al^{3+}, Cr^{3+}] O_4, close packing of anions, double-charged cations in the tetra and triply charged cations in the anions lattice.

The methods of separation of metal from the complex ores include mechanical reparation, thermal decomposition, displacement of one metal by other, high temperature chemical radiation and electrolytic refining.

In the case of mechanical separation metals in free elemental form such as coinage metals, Pt and Au are separated by panning.

In some metals thermal decomposition of unstable compounds

(*a*) $Ag_2O \rightarrow 2Ag + 1/2\ O_2$

(*b*) Marsh Test

As, Sb salt + Zn/ H_2SO_4

$\rightarrow$ As/SbH_3 $\rightarrow$ Silver mirror test of the metal

(*c*) Decomposition of NaN_3

$$2Na\ N_3 \rightarrow 2Na + 3N_2$$

(*d*) Mond's Process

$$Ni(CO)_4 \xrightarrow{180°C} Ni + 4CO$$

and Van Arkel Process

Displacement of one element by other

This is based on negative E° in electrochemical series

$$Cu^{2+} + Fe \rightarrow Fe^{2+} + Cu$$

$$-0.44 \qquad +0.16$$

$$Cl_2 + 2Br^- \rightarrow 2Cl^- + Br_2$$

$$+1.36 \qquad +1.09$$

12.14 HIGH TEMPERATURE CHEMICAL REDUCTION

Metals are mostly available as their oxides. Oxide ores are directly reduced to the metal by reducing agents like C, Al, Si, H_2, carbon are the most widely used reducing agents but it has ability to form carbides.

Sulphide ores are first roasted so as to covert to oxides and then reduced to the metal. In some cases self reduction may also give metals. In other cases other metals are also used as reducing agents

Another method of electrolytic reduction may be followed in highly electropositive metals *e.g.* alkaline earth metals (electrolytic reduction of fused halides)

It is an excellent-method but is a little expensive, H_2 is a poor reducing agent

After the isolation of metals the next step is purification which may be done by following process

1. Electrolytic refining
2. Zone refining
3. Chromatographic methods
4. Ion-exchanges method.

12.14.1 Photography

In the photographic techniques complexes are formed at many stages until the final photograph is obtained.

The light sensitive silver bromide film is used which on exposure to sunlight undergoes the chemical reaction:

$$2Ag\,Br + \xrightarrow{hv} Br_2 + \underset{\text{colloid}}{2Ag}$$

This gives us the negative colloidal effect of sunlight-on the object. The part of the object where more light is imparted appears dull and the part of object where the less light is incident is bright.

To convert the negative into the positive the film is sent for developing. In the developers solution the chemical reactions taking place are:

$$2Ag + \underset{\substack{\text{Developer}\\\text{solution}}}{3Na_2S_2O_3} \rightarrow \underset{\text{Soluble complex}}{2Na_3[Ag(S_2O_3)_2]}$$

12.14.2 Artificial Silk

Combining an ammonia solution with copper carbonate yields a copper carbonate complex with a precipitate. The precipitate is discarded and the solution is transferred to a new beaker. Remarkably, cotton wool completely dissolves is this solution to form viscose. Injecting sulfuric acid into the viscose well cause polymerization and rayon is formed. The rayon is initially blue but on standing it will turn colourless. This viscose solution is turned into the fibers is called artificial silk.

Similarly iron (III) complex and cobalt (III) complex is made easily soluble in chloroform or carbon disulphide to turn into a viscose which is turned into a artificial silk fiber.

12.15 METAL COMPLEXES IN MEDICINE

The application of inorganic chemistry to medicine. Elemental medicine is a rapidly developing field and therapeutic and diagnostic metal complexes are now having an impact on medical practice. Advances into bio coordination chemistry are crucial for improving the design of compounds to reduce toxic side effects and understand their mechanisms of action. Here we shall tackle two key areas:

1. Transformation of metal complexes and
2. Targetting with special reference to platinum anticancer, gold anti arthritic and bismuth anticancer drugs.

Most of the major classes of pharmaceutical agents contain metal compounds. The new areas of their application are rapidly emerging. Some of these are discussed in short with emphasis on the targetting of metal complexes and their bio-transformation. Targetting is important because of the toxicity associated with metal complexes. It they can be delivered only at the tissues, cells and receptors, where they are needed then their toxic nature may be reduced. Metal complexes may undergo ligand substitution and redox reactions is likely to mean that the active species are bio transformation products of the administered complex. Identification of these active species will lead to the more effective use of metal compounds as drugs.

12.15.1 Platinum Anticancer Agents

Platinum (II) complexes are amongst the most widely used treatment of cancer. These injectable diammine complexes have been approved while others are under trial. There is presently emphasis on reducing the toxic effect of platinum anticancer complexes towards normal cells, circumventing acquired resistance to cisplatin and increasing the activity of platinum complexes towards a wide range of types of cancer.

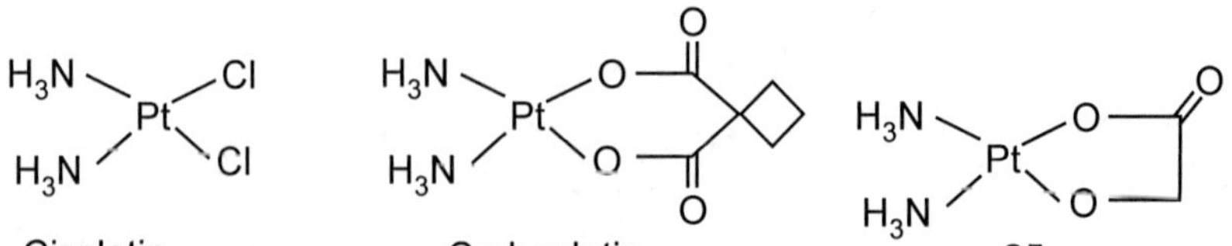

Cisplatin Carboplatin 25

The ultimate target for Pt is DNA and certain platinated DNA adducts trigger DNA degradation and apoptosis (slow programmed cell death). The usefulness of cisplatin in the clinic is limited by (*i*) spectrum of anti cancer activity (not active enough against several types of cancer) (*ii*) the development of resistance after continued treatment and (*iii*) its high toxicity to normal cells.

Three resistance mechanisms are acceptable:

(*i*) Reduced transport across the cell membrane

(*ii*) Strong binding to inactivating thiolate ligands inside the cell

(*iii*) Repair of platinated lesions on DNA by enzymes

Cellular resistance to cisplatin can be overcame by changing the ammine ligands to 1,2 diaminocyclohexane ((DACH) and currently there is much interest in the DACH complex and platinum (iv) analogues which are orally active. This generates interest to investigate how the structures of amine ligands affect the activity of platinum complexes. Pt(II) complexes of some trans Pt (iv) complexes have been reported to be inactive. Photo sensitivity of many Pt(iv) complexes may limit their clinical use.

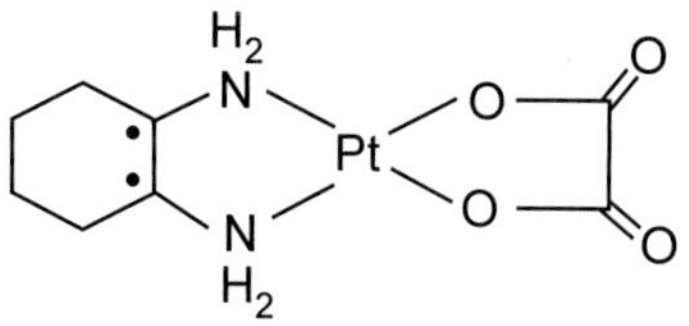

Oxaliplatinum
(4)

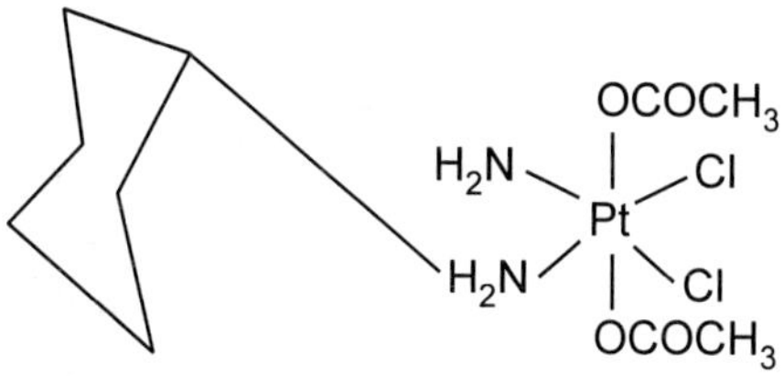

JM 216
(5)

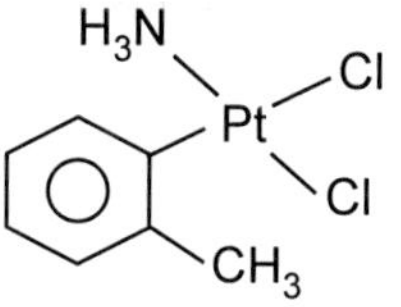

AMD 473
(6)

12.15.2 Hydrolysis

Intra cellular hydrolysis has been thought to be an important process for the activation of chloro-platinum anti cancer diammine ligands, can markedly effect the rate of hydrolysis and the PKa of the resulting aqua complexes . The aqua ligands are very reactive towards the DNA, hydroxo ligands are relatively inert. The pKa value of the aqua ligands are >0.3 units lower than those of cisplatin. A higher proportion of hydroxo species would be expected to be present under intercellular conditions.

12.15.3 Attack on DNA

Use of 15N-labeled complexes and 2 D single quantum coherence ($^{1}H/^{15}N$) NMR spectroscopy allows the detailed path ways of reactions of cisplatin to be followed. Intermediates at concentrations as low as 50μm can be detected, and the ^{15}N chemical shift is diagnostic of the trans ligands which makes it a powerful method for identification of species. These studies clearly demonstrate that equation of cisplatin is the rate limiting step. The formation of mono functional adducts with Pt bound to N^7 is followed by ring closure to form stable G G chelate. The G G chalate is known to be important adduct in cells. Pt is selected for G G sequences because of high electron density at such sites (most easily oxidized sites of DNA).

12.15.4 Monofunctional Adducts

Unexpectedly two monofunctional adducts formed during the reaction of cisplatin with DNA duplex had a long half life 80 hours at 298 K. This suggests that 3 D DNA plays a role in stabilising the long lived adduct either by shielding Cl ligand from hydrolysis or by hindering the approach of the in coming N^7

ligand. Molecular modeling studies demonstrate that H-bonding between the NH_3 ligands and carbonyl groups on DNA play a major role in determining the orientation of the Pt—Cl bonds and their accessibility. Molecular mechanics calculations show that although the chloride ligand in the mono functional adduct faces outward, away from the helix, the aqua ligand which replaces it after hydrolysis faces inwards on the bases of strong H–bonding properties. The mono functional intermediates show that platination and chelation rates for cisplatin diaqua complex $[Pt(NH_3)_2(H_2O)_2]^{2+}$ depend on the DNA base. Long lived mono functional adducts may also promote the formation of DNA—proteins cross-links. Platinated lesions on DNA would normally be repaired by enzymes which may be protected from repair by high mobility proteins.

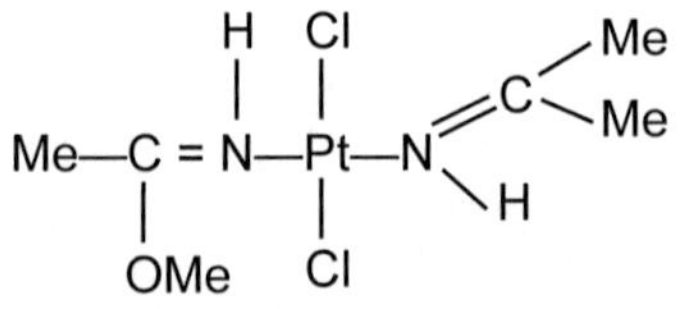

Monounctional adducts

12.15.5 DNA Base Recognition by Pt Complexes

Physiologically cisplatin does not attack the DNA base thymine (T) but on changing the ligands on Pt(II) to amino phosphine this may be achieved. Amino phosphine ligands bind strongly to Pt(II) but in bis chelated cis complexes the Pt—N bond is relatively labile on account of the high trans influence of P and steric interaction between the substituent's on the N-atoms. Therefore, chelate ring opening in these complexes can be controlled by the substituent on N, by the size of the chelate ring, by the PH and concentration of competing ligands such as Cl^-.

In the complex , despite the presence of four phenyl groups in the complex they are salable in water

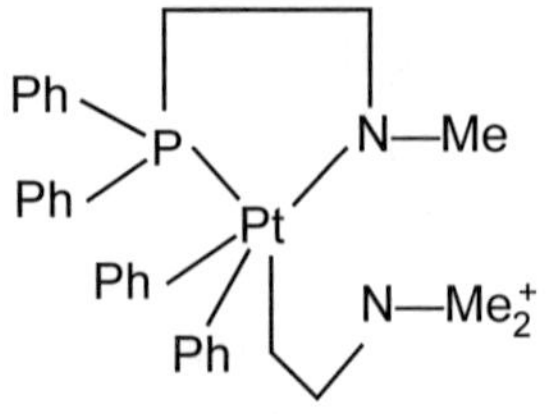

(8)

Complex (8) exhibits activity against cisplatin-resistant tumor cells. two mechanisms are suggested for the cytotoxicity firstly the complex may act as a positively charged lipophilic anti mitondrial agent similar to $[Au(dppe)_2]^+$. Secondly , in the ring opened form, it can bind to DNA bases and form lesions different from cisplatin.

12.15.6 Binding to Meta Thiamine

In the case of cisplatin binding to meta thioamine (Met) would be considered as an activation step. The meta bolites (Pt $(Met)_2$] has been detected in the urine of patients treated with cis platin and it is an uncreative complex, existing in solution as a mixture of 3 diastereomers of each of the 2 geometrical isomers. The cis isomer predominates over trans is over by 87;13. The intercanversion between the cis and trans is slow, cis to trans conversion has a half life of 22.4 H and trans to cis 3.2 H at 310 K). Transfer of Pt onto DNA via met-containing peptides or proteins may therefore be possible. Mono functional adducts of the type [Pt(en)(G)(L-Met-S)] appear to be very stable, therefore, meta thionine may play a big role in trapping these adducts. High trans influence of S as a Pt(II) ligand can lead to facile labilisation of N-ligands and this allows cisplatin to reaction with GMP faster in the presence of L-met than in the absence of L-Met.

12.15.7 Metallo Thionein and Glutathione

The binding of Pt(II) to thiolate S tends to be irreversible in contrast to thioether S. Reaction between cisplatin and intra cellular thiols such as glutothione may therefore inactivate the drug and part of the cellular resistance. There is over—expression of the pump for glutothione conjugates in cisplatin-resistant cells suggesting that Pt-glutothione complexes are pumped out of the cell.

Cisplatin resistance in same cell types involves the low-molecular-mass, system-rich protein metathione in (MT). Cisplatin administration leads to the induction of MT in for example, the liver, and may bind and inactivate Pt, but-MT may be involved in scavenging free radicals. Reactions between MT and Cisplatin lead to displacement of the ammine ligands and give rise to PtS_4 clusters containing 7-10 Pt per mol. Pt binding in Ca is stranger than Zn (II) and Cd(II). Tranplatin reacts with MT faster than Cisplatin.

12.15.8 Targeting Via Transferring Receptor: Ruthenium and Titanium

Transferrin receptors are known to be over-expressed on cancer cells and ^{67}Ga is known to be transported in this way and to localise in tumour tissues. $Ga(NO_3)_3$ has been used as an anti cancer agent. Ru(III) is also taken up by cells as a transferrin complex but in the case of anti cancer trans-$[RuCl_2(indozole)_2]$ (InH) a complex which hydrolyses at about the same rate as ciplatin binds to one of the ligands in the Fe (III) binding site, without the concomitant binding of the usual synergistic anion carbonate. The binding can be reversed by citrate and the Ru-transferrin complex has a high anti proliferative activity. The Ru(III) complex Na[trans-$RuCl_4$ (DMSO) (inidazole)] is more active against secondary than against primary tumours and this anti meta static activity may be of clinical value. It also binds reversible to transferrin. Two Ti(IV) complexes are under clinical trial (*i*) budotitane and (*ii*) titanocene dichloride. There compounds show strong tendency towards hydrolysis. The reactive ligands ac ac and cyclopentadiene may play a major role in the activity of the complexes. It has been demonstrated that Ti can bind strongly to transferrin and so this may provide a further route for Ti Transport and delivery to cells.

12.16 GOLD ANTIPATHETIC DRUGS

Injectable Au(I) thiolate drugs such as aurothiomolate (myocrisin), aura thioglucose (solganol) and aurothio-propanol sulphonate (allochrysin) and the oral drug auranofin (Rid aura) are widely used for the treatment of difficult cases of rheumatoid arthritis. Injectable thiolate complexes are polymers containing linear bridging thiolate sulphurs in chain or cyclic forms. None of the injectable thiolate drugs have been crystallized, but the related cyclic hexa meric 1;1 Au(I) thiolate complex has been reported. Au(1) has a much higher affinity for thiolate S canpared to thioethers and lower affinity for N and O ligands . Au(I) binds , to DNA weakly and is not carcinogenic or mutagenic. Gold (I) has highest affinity for thiols. Consequently, in blood most of the circulating Au (I) is bound to cysteine –34, Gold binding causes a 'flipout' of cysteine-34. Auranofin binds relatively faster but the negatively-charged aurothiomalate polymer binds slowly.

The thiomalate and PEt_3 ligands of aurothiomalate and auronofin are retained on binding of gold to albumin.

Alb-cys34-SH Au STM→ Albumin-cyst. 34-S-Au STM

Albumin can transfer Au(I) into cells. Glutathione is known to bind intercellular Au(I) and the gold is mobile within the cell.

Another important metabolite is $[Au(CN)_2]$. The urine and blood plasma of most patient treated with gold drugs contain $[Au(CN)_2]$. The enhanced uptake of gold into red cells of smokers is due to the formation of this species after inhaling HCN in smoke $[Au(CN)_2]$ may be key metabolite since it can inhibit oxidative bursts in while cells, and itself has anti arythritic, anti cancer and anit-HIV activity.

Under the oxidative conditions that exist in inflamed joints, oxidation of Au(I) to Au (III) is likely and $[Au(CN)_2]^-$ or $[Au(CN)_2Cl_2]$ may be formed. The formation of Au(III) is responsible for some of the side effects. Antigen processing by proteases takes place in lysosomes where Au(III) may be formed. Ultimately lot of gold is deposited in lysosomes (aurosomes) of cells but it is not metallic Au but a protein complex. Here, gold may inhibit lysosomes enzymes which are responsible for destruction of joint tissue.

12.17 BISMUTH ANTICANCER DRUGS

Bismuth (III) compounds such as bismuth sub citrate and sub salicylate are wiedly used for the treatment of diarrhea, dyspepsia and gastric and duodenal ulcers. Bi (III) is active against the bacterium Helicobacier pylari which is associated with mucus layer of ulcers. This bacterium relies on the activity of the Ni enzymes urease to protect it from highly acidic condition via the production of ammonia. The chemical studies show a good similarity between the chemistry of Bi(III) and Fe(III). Metabolism of Fe(III) may partly account for the antibacterial activity. Strains of bacteria which produce siderophores are more resistant to Bi(III).

Bi(III) is a highly acidic metal ion with PKa for hydrolysis is as low as 1.5 and its aqueous chemistry is dominated by hydroxide and oxide complexes. The hydrolysis of Bi(III) and formation of polycatinic species such as $[Bi_6O_4(OH)_4]^{6+}$ even at acidic pH, may allow it to block Ca^{2+} channels. Coordination numbers for Bi(III) vary from 3 to 10 and the geometries of ten are irregular. In some structures there seems to be clear role of 6 S^2 lone pair of electrons. The structures of Bi(III) citrate complexes contain chains sheets and other network. They are generally referred to as colloids. They are based on the dimer $[Bi_2(cit)_2]^{2-}$ as shown

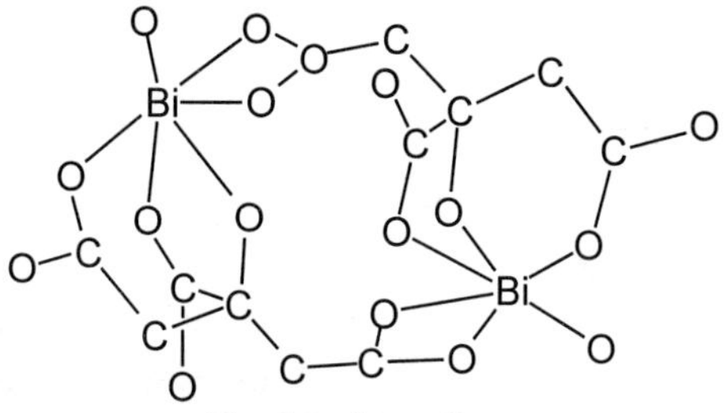

Bismirth citrate dimer

Which contains strong Bi(III) alkoxide bonds and a double corboxylate bridge. The terminal coordination is vacant which may allow the unit to aggregate. The nature of aggregates depends on the pH, ionic strength and nature of counter cation.

Bi(III) bonds strongly to both O and S ligands. The thiol glutathionate is probably involved in the transport of Bi(III) in vivo. Bismuth binds to GSH inside red blood cells and the rate of uptake a of Bi(III) seems to be limited by passage due to the cell membrane.

Despite the large size of ionic radius of Bi(III) (1.03A°) it binds strongly to both the N and C lobe binding the sites of the Fe(III) transport protein transferrin which is present in blood serum. The strength of metal binding in transferrin correlates with metal ion acidity and allows prediction of the strength of metal binding for other metal ions.

12.18 CONCLUSION

To summarises the progress which is being made in understating the mechanisms of action of three metals widely used in therapy are:

Pt, Au and Bi. For Pt (II) there is good evidence that DNA is the target. Pt-N bonds (trans Cl and O) are strong and relatively inert whereas Bi(III) and Au(I) are more labile metal ions. Actually their target seems to be proteins. Despite the widespread use of these metal ions in drugs, their bio coordination chemistries are poorly understood. Further studies are likely to be focussed on the selective metal complexes to a particular type of cell. In this way it will be possible to increase their effectiveness and control side-effects. The study of elemental medicine is likely to make major contributions to understanding of natural biological processes.

12.18.1 Metal Complexes as Drugs and Chemotherapeutic Agents

Platinum complexes as therapeutic agents

(*a*) Clinically used anticancer agents
- (*i*) AMD 473 (ZDO473)
- (*ii*) JM-216 (satrapltin)
- (*iii*) Poly (di and tri)-nuclear platinum complexes
- (*iv*) Transplatinum compounds

(*b*) Non-platinum anticancer agents
- (*i*) Ruthenium complexes
- (*ii*) Arsenic trioxide
- (*iii*) The mitochondrion ion as target gold-platinum complexes
- (*iv*) Manganese-base superoxide dismatase mimics
- (*v*) Titanium compounds
- (*vi*) Gallium Nitrate

(*c*) Antibacterial Agents
- (*i*) Sliver and mercury salts
- (*ii*) Bismuth-containing anti ulcer drugs
- (*iii*) Metal-containing drugs as anti parasitic agents

(*d*) Paramacodymic uses of metal complexes
- (*i*) Lithium carbonate
- (*ii*) Vanadium complexes in diabetes
- (*iii*) Gold compounds as ant arthritic agents
- (*iv*) Nitric oxide in physiology and medicine
- (*v*) Lanthanum carbonate.

■ ■ ■

13

Metal Carbonyls and Nitosyls

13.1 INTRODUCTION

CO, CNR, R_3P and NO are those ligands which have filled donor orbitals as well as empty π-orbitals. These ligands are called π-bonding ligands. They combine with transition metals in zero, –1 and +1 oxidation states to form π-bonding complexes. These complexes exhibit ligand to metal L → M as well as metal to ligand M → L bonds. The compounds formed by CO and metal atom in –1, 0 and +1 oxidation states are called metal carbonyls.

In the lower oxidation states metal has good number of electrons or high electron density. Metals are basically electropositive. During the course of combination of CO and metal the two way bonding between metal and CO takes place as analysed below:

Ligand CO has the bonding between carbon and oxygen as shown:

$$:C \leftleftarrows O:$$

Due to the existence of coordinate bond from donor 'O' and acceptor 'C', a pair of electrons fall vacant on the carbon atom. Therefore, there is a greater possibility of carbon acting as a donor although oxygen is more electronegative. As a result L → M bond is formed.

$$\overset{\delta^-}{M} \leftarrow :\overset{\delta^+}{C} \leftleftarrows O:$$

Accordingly, metal acquires greater electron density and becomes electronegative as the bonding metal is in low oxidation state. According to Pauling's electroneutrality principle metal is comfortable when it acquires a positive charge.

Therefore, metal back donates the pair of electrons to CO in order to maintain its neutrality and metal to ligand bond results

$$\overset{\delta^-}{M} \rightarrow :C \stackrel{\leftarrow}{\equiv} O$$

This gives both ways bonds between metal and ligand. Due to multiple bonding the bond between metal and ligand acquires π-bond character.

Classification of Carbonyls

There are two broad classification of carbonyls as shown in the flow-sheet diagram:

Metal Cabonyls

- Mononuclear
  These carbonyls contain one metal centre e.g., $[Ni(CO)_4]$, $[Fe(CO)_5]$, $[Ru(CO)_5]$ $[V(CO)_6]$, $[Cr(CO)_6]$, $[W(CO)_6]$ etc.
- Polynuclear
  These carbonyls contain more than one metal centre e.g., $[Fe_2(CO)_4]$, $[CO_2(CO)_8]$, $[Mn_2(CO)_{10}]$ $[MnCo(CO)_9]$, $[Mn\ Re(CO)_{10}]$

Polynuclear carbonyls are further classified into non-bridged and bridged carbonyls on the basis of difference in their structures:

Polynuclear carbonyls

- Non-bridged carbonyls These carbonyls contain terminal ligands and metal-metal bond *e.g.,* $[(CO)_4Co—Co(CO)_4]$ $[(CO)_5Mn—Mn(CO)_5]$ These exist in solution
- Bridged carbonyls
  They contain terminal carbonyls and carbonyl groups binding the metal atoms through a bridge *e.g.,* $[Fe_2(CO)_9]$, $[CO_2(CO)_8]$ They exist in solid state

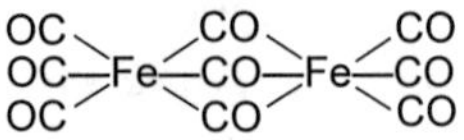

13.2 PREPARATION OF METAL CARBONYLS

General Method

1. **Direct Combination:** Simple carbonyls of Ni and Fe can be prepared **by** the direct combination of CO and finely divided metal (to increase the surface area at a definite temperature).

$$Ni + 4CO \xrightarrow[\text{and 1 atm.}]{\text{Room temp.}} [Ni(CO)_4]$$

This is called Mond's process for preparation of metal carbonyls. For iron carbonyl the temperature and pressure conditions are high.

$$Fe + 5CO \xrightarrow[\text{100 atm.}]{\text{200°C}} Fe(CO)_5$$

2. **Reduction of Metal Carbonyls in Presence of CO:** Carbonyls can also be prepared by treating transition metal halides in solvents like

THF (tetrahydrofuran), ether, diethylene glycol, dimethyl ether etc. with CO at 300°C and 200–300 atm. in the presence of reducing agents like H_2, metal alkyls R_3 Al etc.

$$CrCl_3 + 6CO \xrightarrow[\text{ether}]{(C_2H_5)_3Al} [Cr(CO)_6] + AlCl_3$$

$$MoCl_5 + 6CO \xrightarrow[(CH_3)_2O]{\text{Na in}} [Mo(CO)_6] + 5NaCl$$

3. **From $Fe(CO)_s$:** CO ligand in $[Fe(CO)_5]$ is easily replaceable. Hence certain carbonyls may be prepared from it.

$$WCl_6 + 3[Fe(CO)_5] \xrightarrow[\text{Ether}]{100°C} W(CO)_6 + 3FeCl_2 + 9CO$$

4. OsO_4 may be reduced by CO and subsequently $Os(CO)_5$ is obtained as:

$$OsO_4 + 4CO \xrightarrow[350 \text{ atm}]{250°C} Os(CO)_5 + 4CO_2$$

5. Thermal or photochemical decomposition. This is suitable for obtaining polynuclear carbonyls of Fe, Co, Rh, Os etc.

$$\underset{(M = Fe)}{2M(CO)_5} \xrightarrow[\text{light}]{UV} [M_2(CO)_9] + CO$$

Physical Characteristics

$[Ni(CO)_4]$, $[Fe(CO)_5]$, $[Ru(CO)_5]$ and $[Os(CO)_5]$ are liquids at room temperature while all other carbonyls are solids but they melt or decompose at low temperature. They are all covalent compounds and hence soluble in non-polar solvents. Except $[Fe_2(CO)_9]$ all are soluble in organic solvents. All carbonyls are diamagnetic. $[V(CO)_6]$ is paramagnetic and contains an unpaired electron. Carbonyls have metals in zero oxidation states most of the time.

Chemical Characteristics

The carbonyls very easily undergo substitution reactions. CO group is either completely substituted or substituted partially by other monodentate ligands like CNR, PR_3, PCl_3, py and CH_3OH. For example

(*a*)

$$[Ni(CO)_4] + 4CNR \longrightarrow [Ni(CNR)_4] + 4CO$$

$$[Fe(CO)_5 + 2NO \longrightarrow [Fe(CO)_2(NO)_2]$$

$$[Mn_2(CO)_{10}] + 2PR_3 \longrightarrow [2Mn(CO)_4(PR_3)] + 2CO$$

$$[Ni(CO)_4] + 4PCl_3 \longrightarrow [Ni(PCl_3)_4] + 4CO$$

$$2[Fe_3(CO)_{12}] + 3py \longrightarrow [Fe_3(CO)_9(py)_3] + 3[Fe(CO)_5]$$

Bidentate ligands replace two or more CO groups at a time like diars, 0-phen and NO_2.

$$[Fe(CO)_5] + \text{diars} \longrightarrow [Fe(CO)_3\,\text{diars}] + 2CO$$

$$[Ni(CO)_4] + 2NO_2 \longrightarrow [Ni(NO_2)_2] + 4CO$$

(*b*) Reduction with Na/NH_3 or Na metal and NaOH

$$[Fe(CO)_5] + 3NaOH \longrightarrow Na[H^+Fe(CO)_4^-] + Na_2CO_3 + H_2O$$

Fe atom in $Na[H^+Fe(CO)_4^-]$ is in –2 oxidation state.

In $[Fe_2(CO)_9]$, $[Co_2(CO)_8]$, $[Mn_2(CO)_{10}]$, $[Cr(CO)_6]$, the carbonylate anion formation shows that these carbonyls are reduced.

$$[Fe_2(CO)_9] + 4Na \longrightarrow \overset{+\ \ 2-\ +\ 0}{2Na_2[(Fe(CO)_4]^{2-} + CO}$$

$$CO_2(CO)_8] + 2Na^+ \longrightarrow 2Na^+\,[(Co(CO)_4]$$

$$[Mn_2(CO)_4] + 2Na \longrightarrow 2Na^+\,[Mn(CO)_6]^-$$

(*c*) **Reaction with halogens:** Most of the carbonyls react to form carbonyl halides. For example,

$$[Fe(CO)_5] + Cl_2 \longrightarrow Fe(CO)_4Cl_2 + CO$$

$$[Co_2(CO)_8] + 2Br_2 \longrightarrow 2CoBr_2 + 8CO$$

$$[Mn_2(CO)_{10}]\ 2Br_2 \longrightarrow 2Mn(CO)_5Br$$

$$[Ni(CO)_4] + Br_2 \longrightarrow NiBr_2 + 4CO$$

(*d*) **Combination with NO**

$$[Fe(CO)_5] + 2NO \xrightarrow{95°C} [Fe(CO)_2(NO)_2] + 3CO$$

Electronically NO is a 3-electron donor therefore 3CO are exchanged for 2NO.

$$[Co_2(CO)_8] + 2NO \xrightarrow{40°C} 2[Co(CO)_3NO] + 2CO$$

$$3[Fe_2(CO)_9] + 4NO \longrightarrow 2[Fe(CO)_2(NO)_2] + [Fe(CO)_5] + [Fe_3(CO)_{12}] + 6CO$$

$$2[Ni(CO)_4] + 2NO + 2H_2O \longrightarrow 2[Ni(NO)(OH)] + 8CO + H_2$$

(*e*) **Action of H_2:** Formation of carbonyl hydrides is foreseen as in the following cases:

$$2[Fe(CO)_5] + H_2 \xrightarrow{200°C} 2[Fe(CO)_5H]$$

$$[Co_2(CO)_8] + H_2 \xrightarrow[200°C]{165°C} 2[Co(CO)_4H^+]$$

(*f*) **Heat:** On heating different carbonyls give different products *e.g.*

$$[Fe(CO)_s] \xrightarrow{250°C} Fe + 5CO$$

$$2[Co_2(CO)_8] \xrightarrow[\text{inert atm.}]{50°C} [Co_4(CO)_{12}] + 4CO$$

$$[Ni(CO)_4] \xrightarrow{180°C} Ni + 4CO$$

13.3 STRUCTURE AND BONDING IN CARBONYLS

Mononuclear Metal Carbonyls

These metal carbonyls are formed with coordination number 4, 5 and 6. They are all covalent in nature and give the desired hybridisation on the basis of valence bond theory. Mononuclear carbonyls have linear M—CO bond. The metal atom is bonded to CO through a metal carbon bond although oxygen is more electronegative.

Coordinate Mononuclear Metal Carbonyls

$Ni(CO)_4$ analysed on the basis of VBT acquires sp^3 hybridisation and a tetrahedral structure.

$$Ni^0_{28} \rightarrow 3d^8\, 4s^2\, 4p^0\, 4d^0\, 4f^0$$

In $[Ni(CO)_4]$, nickel is zero valent

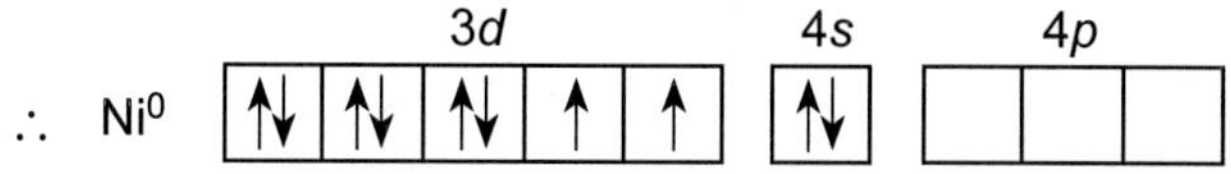

CO group is a very strong field ligand and provides enough energy to pair all the electrons. It gets paired electrons in 4*s* get unpaired, two electrons in 3*d* paired making available one 4*s* and three 4*p* orbitals to donor pair from CO molecule.

∴ Ni^0 electron distribution in the valence shell becomes

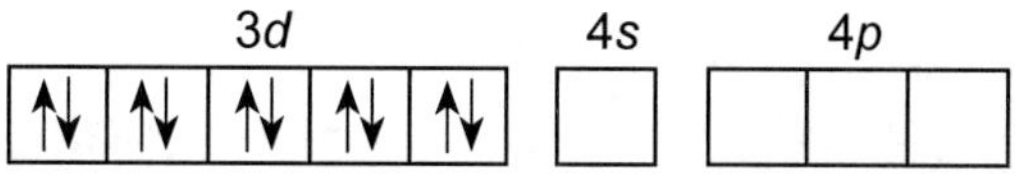

Ligand 4CO donate 4 pairs of electrons to the vacant 4s and 4*p* orbitals so finally:

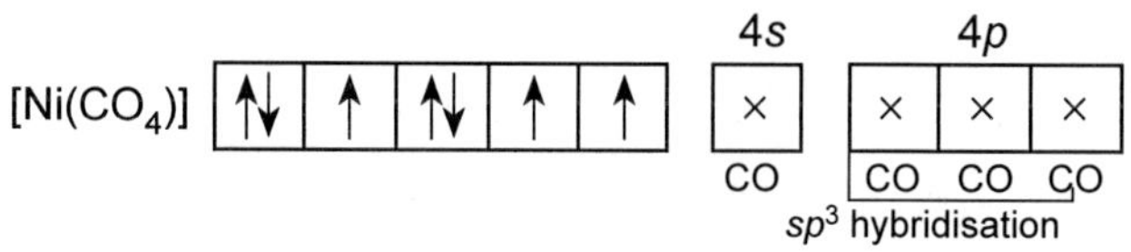

This leads to sp^3 hybridisation opting tetrahedral structure.

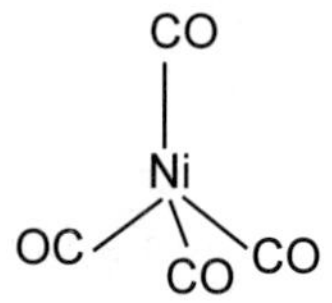

Since the M ←:CO bond is a α bond formation and M: → CO is a π bond the overall π bond character develops. When :CO donates a pair of electrons to zero valent Ni, it acquires negative charge. Since it is improbable for the metals to have negative charge, therefore, it back donates or back coordinates and the metal is left uncharged. This concept of multiple bonding strengthens the M—CO bond and stability of carbonyl complexes is increased. Hence, both metal and CO ligand act as donors and acceptors.

Applying EAN Rule to $[Ni(CO)_4]$ Complex

Tetracarbonyls obey EAN rule without any exception. Accordingly the effective atomic number of zero-valent metal plus the pair of electrons donated by CO groups must be equal to the atomic number of the nearest noble gas *e.g.*

Atomic number of Ni metal = 28

4 pairs of electrons are donated by the four CO ligands = 8

$[Ni(CO)_4] = 36$

This is equal to the configuration of noble gas krypton hence the complex is quite stable. Therefore, Nickel is expected to form $[Ni(CO)_4]$ and not $[Ni(CO)_6]$.

Physical Characteristics

$[Ni(CO)_4]$ is a colourless liquid which boils at 43°C and solidifies at –25°C. It is miscible in benzene. It does not react with acids or alkalis in cold and is highly poisonous.

Chemical Characteristics

It undergoes partial or full substitution of carbonyl ligands, by monodentate ligands like alkyl or aryl isocyanide (CNR), PR_3, PCl_3, py, CH_3OH etc.

For example:

$$[Ni(CO)_4] + 4CNR \longrightarrow [Ni(CNR)_4 + 4CO$$

$$[Ni(CO)_4] + 4PCl_3 \longrightarrow [Ni(PCl_3)_4] + 4CO$$

Bidentate ligands like diars, o-phen, NO_2 etc. replace two or more CO groups at a time:

$$[Ni(CO)_4] + \text{o-phen} \longrightarrow [Ni(CO)_2(\text{o-phen})] + 2CO$$

$$[Ni(CO)_4] + \text{diars} \longrightarrow [Ni(CO)_2(\text{diars})_2] + 2CO$$

$$[Ni(CO)_4] + 2NO_2 \longrightarrow [Ni(NO_2)_2] + 4CO$$

Action with Halogens: Nickel carbonyl reacts with halogens to form metallic halides and CO

$$[Ni(CO)_4] + Br_2 \longrightarrow NiBr_2 + 4CO$$

The other carbonyls are known to form carbonyl halides.

Action with NO: Nickel carbonyl reacts with moist NO to form blue coloured Ni(NO)(OH) while dry NO gives blue solution of the composition $Ni(NO)(NQ_2)$ *e.g.*

$$[Ni(CO)_4] + 2NO + 2H_2O \longrightarrow 2[Ni(NO)OH] + 8CO + H_2$$

$$Ni(CO)_4 + 4NO \longrightarrow [Ni(NO)(NO_2) + 4CO + N_2O$$

Action of Heat: Different carbonyls give different products but $[Ni(CO)_4]$ decomposes to give pure nickel metal according to Mond's.

$$[Ni(CO)_4] \longrightarrow Ni + 4CO$$

This reaction is used to isolate pure nickel metal.

Coordinate mononuclear metal carbonyls

This includes $[Fe(CO)_5]$ and $[Os(CO)_5]$ but $[Fe(CO)_5]$ is more important.

13.4 IRON PENTACARBONYL $[Fe(CO)_5]$

(*i*) It is prepared by the action of CO at 200 atm. pressure on heated iron at 200°C

$$Fe + 5CO \xrightarrow[200°C]{200 \text{ atm.}} [Fe(CO)_5]$$

(*ii*) It can also be prepared by the action of CO at 200 atm. pressure at 200°C on FeI_2 in the presence of Cu. In actual, iron carbonyl iodide is formed first which further reacts with CO in the presence of Cu to form $[Fe(CO)_5]$.

$$FeI_2 + 4CO \longrightarrow [Fe(CO)_4I_2]$$

$$CO + [Fe(CO)_4I_2] + 2Cu \longrightarrow Fe(CO)_5 + Cu_2I_2$$

Physical Characteristics

$[Fe(CO)_5]$ is a yellow liquid immiscible in H_20 but miscible in methanol, ether, acetone and benzene.

Chemical Characteristics

(*i*) Decomposes at 250°C to yield pure Fe.

$$[Fe(CO)_5] \longrightarrow Fe + 5CO$$

(*ii*) It dimerises on irradiation by near UV light

$$2[Fe(CO)_5 \xrightarrow[\text{light}]{\text{UV}} [Fe_2(CO)_9] + CO$$

(*iii*) In the presence of UV light $[Fe(CO)_5]$ is activated

$$[Fe(CO)_5] + hv \longrightarrow [Fe(CO)_5^*]$$

to recombine with $[Fe(CO)_5]$ to form $[Fe_2(CO)_9]$

$$[Fe(CO)_5^*] + [Fe(CO)_5] \longrightarrow [Fe_2(CO)_9] + CO$$

(*iv*) On hydrolysis by H_2O and acid

$$[Fe(CO)_s] + H_2SO_4 \longrightarrow FeSO_4 + 5CO + H_2$$

(*v*) **Action of alkali:** When heated with NaOH, carbonylate anion $[Fe(CO)_4]$ is formed.

$$[Fe(CO)_5] + 3NaOH \longrightarrow Na^+[Fe^{-2}(CO)_4H+]^- + Na_2CO_3 + H_2O$$

(*vi*) **Action with NH_3**

$$[Fe(CO)_5] + H_2O + NH_3 \longrightarrow [Fe(CO)_4H_2] + \underset{\text{Carbonic acid}}{NH_2COOH}$$

(*vii*) **Halogens:** $[Fe(CO)_5]$ reacts with halogens in non-aqueous solvents to form tetracarbonyl halides

$$[Fe(CO)_5] + X_2 \longrightarrow [Fe(CO)_4\,X_2] + CO$$

(where X = Cl, Br or I)

(*viii*) Reaction with **ethylene diamine** gives an addition product of the type $[Fe(CO)_5en]$

(*ix*) **Action with NO:** $[Fe(CO)_5]$ reacts with NO under pressure below 45°Cto give the end product $[Fe(NO)_4]$ *via* intermediate $[Fe(CO)_2(NO)_2]$

$$[Fe(CO)_5] + 2NO \longrightarrow [Fe(CO)_2(NO)_2] + 3CO$$

$$[Fe(CO)_2(NO)_2] + 2NO \longrightarrow \underset{\text{Black solid}}{[Fe(NO)_4]} + 2CO$$

It should be noted that NO molecule is a three electron donor while CO is two electron donor, therefore, 3CO molecules are replaced by 2NO molecules electronically.

(*x*) **Reaction with non-nietal halides:** Here, $[Fe(CO)_5]$ acts as a dehalogenating and carboxylating agent

$$[Fe(CO)_5] + 2CCl_4 \longrightarrow FeCl_2 + C_2Cl_6 + 5CO$$

$$[Fe(CO)_5] + SO_2Cl_2 \longrightarrow SO_2 + FeCl_2 + 5CO$$

(*xi*) **Action with metal halides:** $SnCl_4$ and $SbCl_3$ are reduced to lower halides like $SnCl_2$ and $SbCl_3$

$$[Fe(CO)_5] + SnCl_4 \longrightarrow SnCl_2 + [Fe(CO)_4Cl_2] + CO$$

$$[Fe(CO)_5] + SbCl_5 \longrightarrow SbCl_3 + [Fe(CO)_4Cl_2] + CO$$

$[Fe(CO)_4Cl_2]$ then forms adducts with $SnCl_2$ and $SbCl_3$

$$SnCl_2 + [Fe(CO)_4Cl_2] \longrightarrow SnCl_2.[Fe(CO)_4Cl_2]$$

$$SbCl_3 + [Fe(CO)_4Cl_2] \longrightarrow SbCl_3.[Fe(CO)_4Cl_2]$$

Structure of Adducts

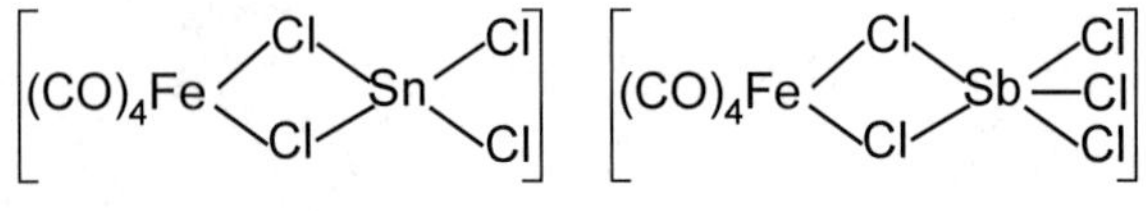

(*xii*) When burnt in air $[Fe(CO)_5]$ forms finely divided Fe_2O_3 which is used as polishing agent.

Application of EAN Rule

EAN rule as applied to [FelCO)$_5$] and [Os(CO)$_5$] is fully satisfied

e.g. Fe° = 26 atomic number

5:CO = 10 electrons [Fe(CO)$_5$] = 36

which is the atomic number of noble gas Krypton. Similarly for [Os(CO)$_5$]

Osmium Atomic number = 76

5:CO electrons = 10

[Os(CO)$_5$] = 86

Atomic number 86 stands for noble gas Radon hence EAN rule is obeyed in this carbonyl.

Structure and Bonding in [M(CO)$_5$]

Pentacarbonyl Mononuclear Carbonyls

Pentacarbonyls like [Fe(CO)$_5$] and [Os(CO)$_5$] can opt for two geometries *viz.* (*a*) Trigonal bipyramidal, (*b*) Square pyramid

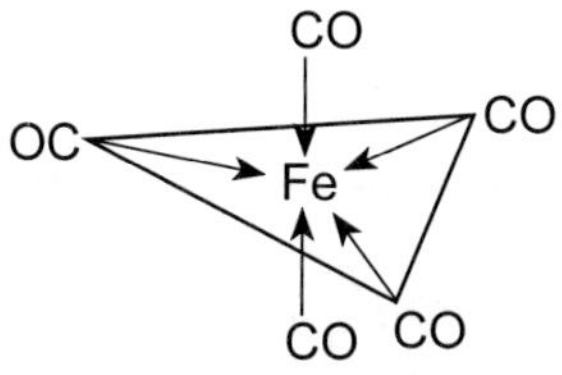

Trigonal bipyramidal

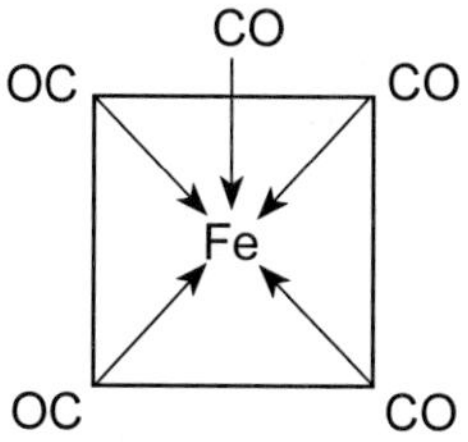

Square pyramid

In order to decide which one of the two geometries is the real structure of [Fe(CO)$_5$], proceed as follows:

(*i*) Run the 1R and Raman spectra and compare the number of bands actually observed and the number of bands calculated for trigonal bipyramidal geometry. If they are in agreement then the trigonal bipyramidal geometry is accepted.

(*ii*) X-ray studies contain that ^{13}C-NM R gives only one type of signal for axial and equatorial CO groups. This is because of flexibility in the structure. It means that axial and equatorial bonds exchange their positions very fast so that on an average ^{13}C resonance signal is observed for all the five carbonyl groups. Electron diffraction study has revealed that the molecule in the vapour state has shown Fe—C axial bond lengths equal to 1.79Å and Fe—C equatorial bonds lengths equal to 1.84Å.

Trigonal bipyramidal structure states that Fe atom is in zero oxidation state and *dsp*^{3} hybridisation is represented. As [Fe(CO)$_5$] is diamagnetic all the

8 electrons in the valence shell of Fe atom get paired owing to the strong field CO ligand which provides energy more than the pairing energy. Therefore, all the unpaired electrons in the valency shell of Fe atom get paired up.

(a) Fe° atom in free state

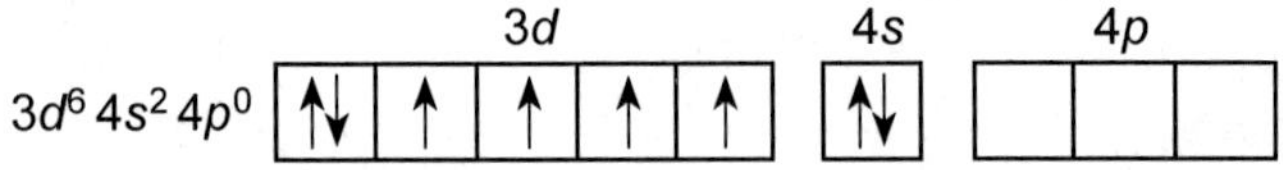

(*b*) Fe° atom in $[Fe(CO)_5]$: Effect of strong field: C—O ligand

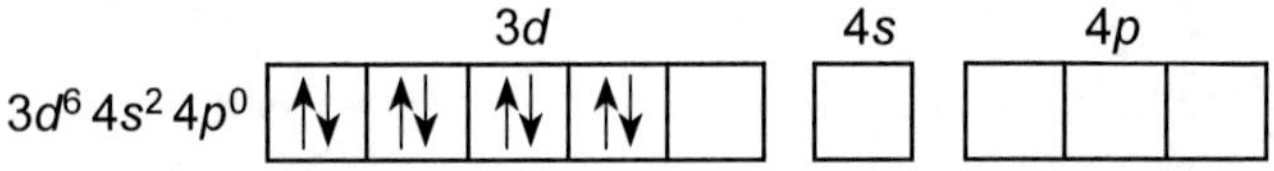

(*c*) $[Fe(CO)_5]$

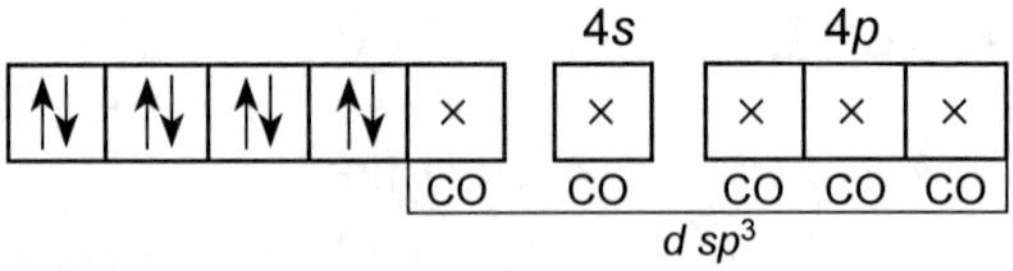

Trigonal bipyramidal structure 'X' indicates the electron pair donated by ←:C≦O ligand.

Coordinate Mononuclear Metal Carbonyls

The important metal carbonyls in this category are $[Cr(CO)_6]$, $[V(CO)_6]$, $[Mo(CO)_6]$ and $[W(CO)_6$.

They can be prepared by the following methods of preparation:

(*i*) **Job's Reaction:** In this CO at 50 atm. pressure at room temperature is passed through a suspension of $CrCl_3$ in ether. CO has already been treated with C_6H_5MgBr at –70°C.

$$CrCl_3 + 6CO \xrightarrow[\text{Room temp.}]{50^\circ\text{ atm.}} [Cr(CO)_6] + C_6H_5MgBr$$

(*ii*) Cr(III) salt dissolved in ether is treated with CO at high temperature in the presence of reducing agent $LiAlH_4$.

$$CrCl_3 + 6CO + LiAlH_4 \xrightarrow[\text{70 atm. press..}]{175^\circ C} [Cr(CO)_6] + LiCl + AlCl_3$$

(*iii*) $CrCl_3$ in the presence of anhydrous $AlCl_3$, powdered Al and benzene is treated with CO.

$$CrCl_3 + Al + 6CO \xrightarrow{\text{anhyd. } AlCl_3} [Cr(CO)_6] + AlCl_3$$

(*iv*) Most commonly used method for the preparation of $[V(CO)_6]$ is when metal salt is reduced in the presence of CO.

$$VCl_3 + 4Na \xrightarrow[\text{diglyme.}]{CO} [Na(diglyme)]^+[V(CO)_6]^- \xrightarrow{H_3PO_4} H_2 + [V(CO)_6]$$

(*v*) $WCl_6 + [Fe(CO)_5]^- \xrightarrow{100^\circ C} [W(CO)_6]$

Physical Characteristics

$[Cr(CO)_6]$, $[Mo(CO)_6]$ and $[W(CO)_6]$ are white solids with melting points 130°C, 150°C and 180°C respectively. $[V(CO)_6]$ is a black green solid melting at 70°C. They are mostly soluble in ether, $CHCl_3$, CCl_4 and benzene. They do not react with air, alkalis and acids (except cone. HNO_3).

They are decomposed by F_2 at –75°C forming CrF_5. Cl_2 and conc. HNO_3 also promote decomposition.

Chemical Characteristics

Reduction:

(*i*) Liquid ammonia and alkali metal borohydrides ($NaBH_4$, $LiBH_4$ etc.)

$$[Cr(CO)_6] \xrightarrow{Na/NH_3} Na_2[Cr(CO)_5] + CO$$

$$2[Cr(CO)_6] \xrightarrow[THF]{NaBH_4} Na[HCr_2(CO)_{10}] + 2CO$$

$$2[Cr(CO)_6] \xrightarrow[NH_3]{NaBH_4} Na_2[Cr_2(CO)_{10}] + 2CO$$

(*ii*) **Action with** amines, pyridine and ethylene diamine replaces CO group in $[Cr(CO)_6]$ molecule. Pyridine reacts at 140°C to give the following products:

$$[Cr(CO)_6] \longrightarrow [Cr(CO)_4(py)_2] \longrightarrow [Cr_2(CO)_7(py)_5] \longrightarrow [Cr(CO)_3(py)_3]$$

$$[Mo(CO)_6] + C_5H_5Na \xrightarrow[\text{Cyclopentadienide}]{\text{Sodium}} Na[\eta^5\text{–}C_5H_5Mo(CO)_3]$$

EAN rule Compliance: They comply with EAN rule as calculated.

Cr° atomic number = 24

6:C = O Electrons = 12

$[Cr(CO)_6]$ = 36

which is the atomic number of Krypton.

Structure

X-ray diffraction study of $[Cr(CO)_6]$ molecule in vapour phase has shown that the molecule is octahedral in shape with Cr—C and C—O bond lengths 1.92Å and 1.16Å respectively. Theoretically valence bond a approach shows d^2sp^3 hybridisation confirming octahedral geometry and is diamagnetic.

$$Cr^0_{24} \longrightarrow (Ar)\ 3d^5\ 4s^1\ 4p^0\ 4d^0$$

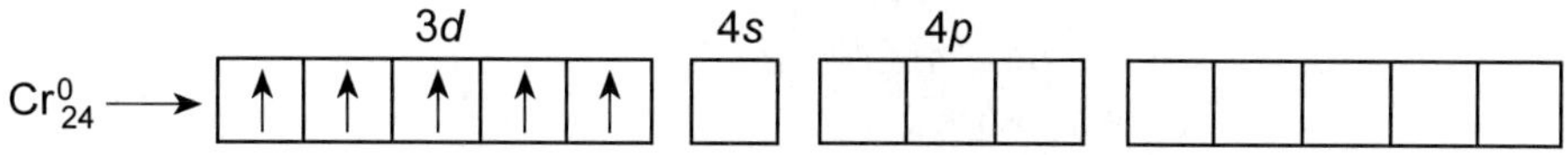

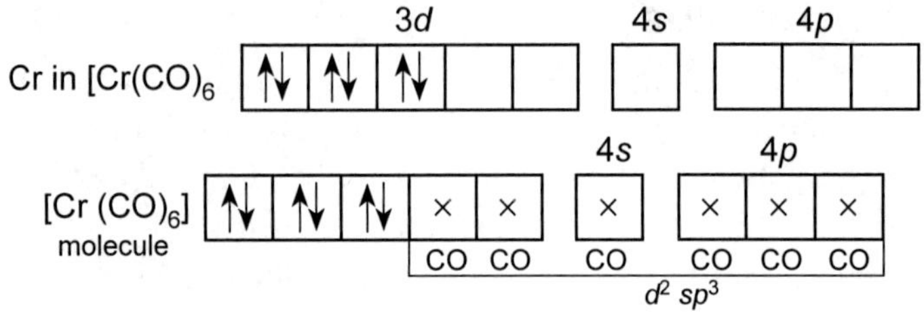

Co being a strong field ligand pairs all the electrons in Cr metal atom to make two 3d, one 4s and three 4*p* orbitals vacant to accept a pair of electrons from CO molecule and show d^2sp^3 hybridisation which calls for octahedral geometry.

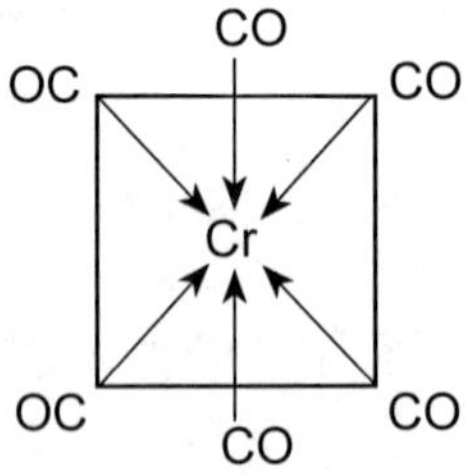

Octahedral structure of $[Cr(CO)_6]$

13.5 POLYNUCLEAR METAL CARBONYLS

Its domain covers binuclear and trinuclear metal carbonyls. Tetranuclear metal carbonyls are quite rare.

Common binuclear metal carbonyls include examples of $[Mn_2(CO)_{10}]$, $[Co_2(CO)_8]$, $[Te_2(CO)_{10}]$, $[Fe_2((CO)_9]$ and $[Re_2(CO)_{10}]$.

Dimanganese Decacarbonyl $[Mn_2(CO)_{10}]$

Preparation

(*i*) MnI_2 is reduced at a high pressure of CO by Mg in diethyl ether.

$$Mn + 2CuI \longrightarrow MnI_2 + 2Cu$$

$$(MnI_2 + 2Cu) + Mg \xrightarrow[25°C/16\ hrs]{CO\ 210\ atm.} [Mn_2(CO)_{10}] + MgI_2 + CuI_2$$

(*ii*) Better yield of $[Mn_2(CO)_{10}]$ can be obtained by the reaction of $Mn(CH_3COO)_2$ with $(C_2H_5)_3$ Al under CO at 20 atm. at 100°C for five hours.

$$[Mn(CH_3COO)_2] + (C_2H_5)_3Al \xrightarrow[25°C/16\ hrs]{CO\ 210\ atm.} [Mn_2(CO)_{10}]$$

(*iii*) Carbonylation of $MnCl_2$ in the presence of benzophenone $(C_6H_5)_2CONa$ at 165°C

$$MnCl_2 + 2(C_6H_5)_2CONa \longrightarrow 2NaCl + Mn[CO(C_6H_5)_2]_2$$

$$Mn[CO(C_6H_5)_2]_2 \xrightarrow[165°C]{CO\ 140\ atm.} [Mn_2(CO)_{10}] + (C_6H_5)_2CO$$

(*iv*) Good yield is obtained by carbonylation of methyl cyclopentadienyl manganese dicarbonyl in presence of Na at 50 atm. in presence of CO at 125°C for 8 hours.

$$(H_3C—C_5H_4)Mn(CO)_3 \xrightarrow[125°C/8\ hrs]{CO,\ 50\ atm.} [Mn_2(CO)_{10}]$$

Physical Characteristics

It is golden yellow crystalline solid melting at 155°C, sublimes in vacuum, soluble in organic solvents. Its solution is slowly oxidised in air.

Chemical Characteristics

(*i*) **Reaction with halogens**

$$[Mn_2(CO)_{10}] + Br_2(l) \xrightarrow{40°C} 2[Mn(CO)_5Br]$$

$$[Mn_2(CO)_{10}] + I_2 \longrightarrow 2[Mn(CO)_5I]$$

order of reactivity is: I > Br > Cl

(*ii*) **Action with Li and Na metals:** Mn—Mn bond is broken by lithium in the presence of THF and Na in liquor NH_3 to form $Li^+[Mn(CO)_5]^-$

$$[Mn_2(CO)_{10}] + 2Li \xrightarrow{THF} 2Li^+[Mn(CO)_5]^-$$

$$[Mn_2(CO)_{10}] + 2Na \xrightarrow[NH_3]{Liquor} 2Na^+[Mn(CO)_5]^-$$

(*iii*) **Reduction by Hydrogen**

$$[Mn_2(CO)_{10}] + H_2 \xrightarrow[200°C]{200\ atm.} 2[HMn(CO)_5]$$

(*iv*) **Action of phosphines, arsines and stilbenes:** With the above compounds it forms paramagnetic monomeric compound of the formula $[Mn(CO)_4(PR_3)]$.

(*v*) Diamagnetic $[Mn_2(CO)_{10}]$ confirms that all electrons in the molecule are Paired with the form of Mn—Mn bond due to strong interactions.

Diironnonacarbonyl $[Fe_2(CO)_9]$

Preparation

$[Fe(CO)_5]$ is first dissolved in glacial acetic acid, cooled and irradiated for 6 hours with near ultraviolet light golden crystals of $[Fe_2(CO)_9]$ are precipitated and isolated by filtration.

Physical Characteristics

It forms non-volatile golden triclinic crystals which are diamagnetic. It is insoluble in water, ether and C_6H_6 but soluble in alcohol, acetone, toluene and pyridine.

Chemical Characteristics

(*i*) **Action of heat:** On warming at 50°C, it gives $[Fe_2(CO)_{12}]$. At 100°C it decomposes to give Fe, CO, $[Fe(CO)_5]$ and $[Fe_3(CO)_{12}]$

(*ii*) **With NO (nitric oxide):** It forms $[Fe(CO)_2(NO)_2]$ alongwith $[Fe(CO)_5]$ and $[Fe_3(CO)_{12}]$.

(*iii*) **Action with phen:** In acetone or benzene solution at 80°C it gives $[Fe(phen)_3]$ $[Fe_2(CO)_g]$ but in pyridine it gives $[Fe(phen)_3]$ $[Fe_4(CO)_{13}]$.

(*iv*) **With Alkali metal in NH_3.** It forms carbonylate anion.

$$[Fe_2(CO)_9] + 4Na \xrightarrow{NH_3} 2Na_2[Fe(CO)_4] + CO$$

EAN rule is satisfied in the case of $[Fe_2(CO)_9]$ as explained:

Fe° Atomic number–26–$3d^6 4s^2$

$$2Fe° = 52$$
$$9CO = 18$$
$$Fe{:}Fe = 2$$
$$\text{Total} = 72$$

Half for one Fe metal as nucleus = 36 which is equal to Krypton noble gas.

Structure

IR and X-ray studies indicate that each Fe atom is directly linked to the other Fe atom by a δ-bond through three bridging carbonyl groups (>, C = O) by a σ bond and in turn to three terminal carbonyl groups. Fe—Fe bond length 2.46Å. The terminal CO bond length are smaller than the bridging CO bond distance. From the diagram it appears that the coordination number of each Fe-atom is 7 and not 6.

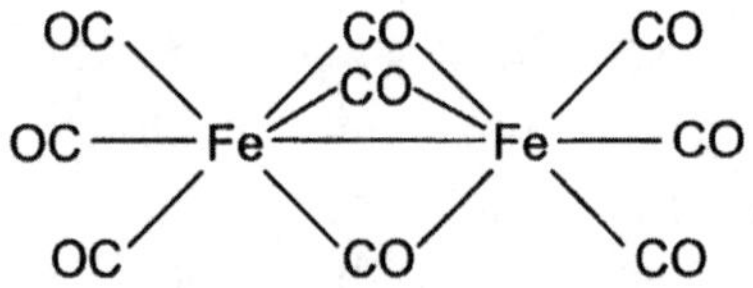

13.6 DICOBALT OCTACARBONYL $[Co_2(CO)_8]$

Preparation: It can be prepared:

(*i*) By the action of CO and reduced metallic cobalt at 220°C and 250 atm.

$$2Co + 8CO \xrightarrow[250\text{ atm.}]{220°C} [Co_2(CO)_8]$$

(*ii*) By the action of dry CO at 200°C and 200 atmosphere in the presence of Cu on binary compounds of Co such as CoS and CoX_2. Copper in turn forms Cu_2S and CuX.

$$2CoS + 8CO + 4Cu \longrightarrow [Co_2(CO)_8] + 2Cu_2S$$

$$2CoX_2 + 8CO + 4Cu \longrightarrow [Co_2(CO)_8] + 4CuI$$

(*iii*) By the action of CO and H_2 on $CoCO_3$ under high pressure and temperature

$$2CoCO_3 + 2H_2 + 8CO \xrightarrow[220°C]{300\text{ atm.}} [Co_2(CO)_8] + 2CO_2 + 2H_2O$$

(*iv*) By the action of acid on solution of $[Co(CO)_4]$. H_2 is evolved and Co_2CO_8 is left behind.

$$2Co(CO)_4H \longrightarrow H_2 + [Co_2(CO)_8]$$

(*v*) By decomposition of $Co(CO)_4H$.

Physical Characteristics

$[Co_2(CO)_8]$ is an orange brown solid melting at 51°C. It is soluble in alcohol, ether and CCl_4. The carbonyl is air sensitive in solid and solution phase.

Chemical Characteristics $[Co(CO)_8]$

(*i*) On heating at 50°C it decomposes in inert atmosphere to form tetra cobalt dodecacarbonyl, *i.e.,* $[Co_4(CO)_{12}]$

$$2Co_2(CO)_8 \xrightarrow{50°C} Co_4(CO)_{12} + 4CO$$

(*ii*) On exposure to air, $[Co_2(CO)_8]$ changes into deep violet basic carbonate of cobalt.

(*iii*) By action of H_2 formation of cobalt carbonyl hydride is observed at 165°C and 200 atm. pressure.

$$[Co_2(CO)_8] + H_2 \xrightarrow[200\text{ atm.}]{165°C} 2[Co(CO)_4H]$$

(*iv*) On reaction with Na-metal in liquid NH_3 below –7.5°C it gets reduced to carbonylate anion.

$$[Co_2(CO)_8] + 2Na \xrightarrow[NH_3]{\text{Liq.}} 2Na^+[Co(CO)_4]^-$$

(*v*) On disproportionation two types of reactions occur

(*a*) strong bases which have nitrogen or oxygen as donor atoms give Co^{+}' and Co^{-1} compounds

$$2[Co_2(CO)_8] + 12NH_3 \longrightarrow 2[Co^{+2}(NH_3)_6][Co(CO)_4)_2 + 8CO$$

(*b*) Isocyanides, phosphines, arsines and stilbines give pentacoordinate cobalt (1) cation

$$[Co_2(CO)_8] + 5PR_3 \longrightarrow [Co^{+1}(PR_3)_5][Co^{-1}(CO)_4]^- + 4CO$$

(*vi*) **Action of NO:** It reacts at 40°C to form cobalt carbonyl nitrosyl, the oxidation state of cobalt is reduced from 0 to –1.

$$[Co_2(CO)_8] + 2NO \longrightarrow [Co\sim^1(CO)_3](NO^{+1})]° + 2CO$$

(*vii*) On action with halogens, $[Co_2(CO)_8]$ decomposes into CoX_2 and CO

$$[Co_2(CO)_8] + 2X_2 \longrightarrow 2CoX_2 + 8CO$$

EAN Rule

$[Co_2(CO)_8]$ is in agreement with the EAN rule as given by the calculation.

$[Co_2(CO)_8]$

In case it is reduced to half then the molecule becomes $Co(CO)_4$

CO° Atomic number	= 27
4 :C$\leftrightharpoons$O	= 8
Co:Co bond	
Share of one Co atom	= 1
Total	= 36

The atomic number 36 happens to be that of krypton, hence the stability.

Structure

Infra-Red studies of $[Co_2(CO)_8]$ in solution show that the compound exists in two isomeric forms. One of the form has a bridged structure and the other non-bridged structure. At very low temperature the bridged structure predominates and as the temperature is raised non-bridged structure appears.

The bridged structure has two bridging carbonyl groups and six terminal carbonyl groups with a Co—Co bond. Each Co-atom has three terminal carbonyl groups as shown in the diagram:

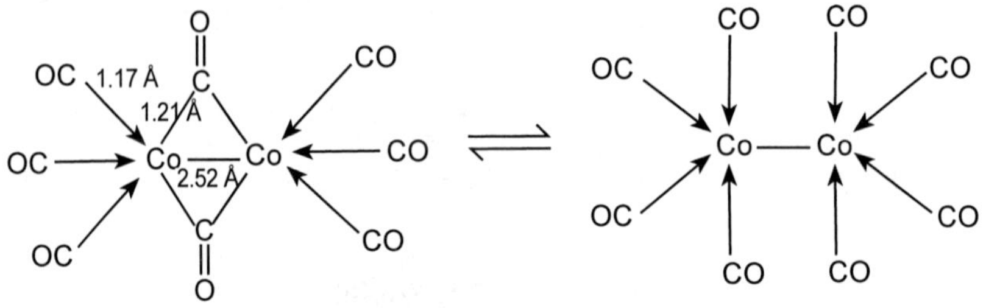

The coordination number of each cobalt atom is six hence each cobalt atom is d^2sp^2 hybridised.

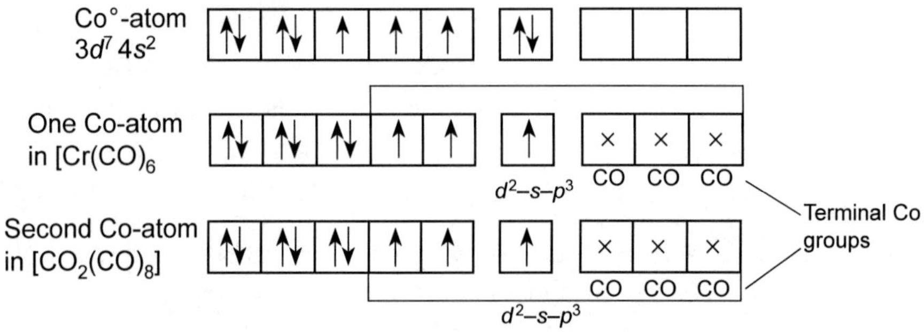

There are three vacant orbitals (4p orbitals) on each Co atom which accept lone pair donated by three terminal CO groups forming 3 coordinate bonds (Co $\leftarrow$:CO). The singly filled orbitals on both Co-atoms overlap with each other and form Co—Co bond. Thus, all the six electrons in $[Co_2(CO)_8]$ become paired to give diamagnetic character to the molecule and three coordinate bonds in each Co-atom are in resonance equilibrium.

The non-bridged has one Co—Co bond and eight terminal carbonyl groups. Each Co-atom is directly joined to the other Co-atom and with four terminal CO groups. In bridged structure two bridging carbonyl groups form Co—Co bonds. The two isomeric forms exist in equilibrium in solution form. The 'X' mark in the vacant orbitals indicate the coordinate bonds in each Co-atom and singly filled form 3 a bonds due to overlap of one Co-atom and the other. The *a* bonds and the coordinate bonds resonate and exist in equilibrium.

Trinuclear Metal Carbonyls

$[Fe_3(CO)_{12}]$ and $[Os_3(CO)]_{12}]$ come in its ambit.

Preparation:

Triiron dodecacarbonyl $[Fe_3(CO)_{12}]$ is prepared as:

(*i*) By heating toluene solution of $[Fe_2(CO)_9]$ to 70°C and then cooling the solution

$$3[Fe_2(CO)_9] \xrightarrow[\text{70°C cooling}]{\text{Toluene}} 3[Fe(CO)_5] + [Fe_3(CO)_{12}]$$

(*ii*) By oxidation of $Fe(CO)_4H_2$ with H_2O_2 or MnO_2

$$3[Fe(CO)_4H_2] + 3H_2O_2 \longrightarrow [Fe_3(CO)_{12}] + 6H_2O$$

$$3[Fe(CO)_4H_2] + 3MnO_2 + 3H_2SO_4] \longrightarrow [Fe_3(CO)_{12}] + 3MnSO_4 + 6H_2O$$

Physical Characteristics

$[Fe_3(CO)_{12}]$ are deep green crystals soluble in organic solvents like toluene, alcohol, ether and pyridine.

Chemical Characteristics

(*i*) **Action of heat:** On heating to 140°C $[Fe_3(CO)_{12}]$ decomposes to give metallic Fe and CO.

$$[Fe_3(CO)_{12}] \longrightarrow 3Fe + 12CO$$

(*ii*) **Action with Na Metal:** It reacts with Na metal in liquid NH_3 to give carbonylate anion

$$[Fe_3(CO)_{12}] + 6Na \longrightarrow 3Na_2^+[Fe(CO)_4]^{2-}$$

(*iii*) **Substitution Reaction:** These reactions occur in pyridine and methanol solutions:

$$2[Fe_3(CO)_{12}] + 3py \longrightarrow [Fe_3(CO)_9(py)_3] + 3[Fe(CO)_5]$$

$$2[Fe_3(CO)_{12}] + 3CH_3OH \longrightarrow [Fe_3(CO)_9(CH_3OH)_3] + 3[Fe(CO)_5]$$

(*iv*) **Action with NO:** It reacts with NO to form iron dicarbonyl dinitrosyl $[Fe(CO)_2(NO)_2]$

$$[Fe_3(CO)_{12}] + 6NO \longrightarrow 3[Fe(CO)_2(NO)_2] + 6CO$$

EAN rule is satisfied in this trinuclear metal carbonyl aptly. According to the structure there arc 3 Fe—Fe bonds and the effective atomic number is calculated as indicate

3Fe—Fe bonds = 6 electrons 3

Fe-atoms = 78

12:C = O donors = 24

Total = 108

For on Fe-atom = 36 which is the atomic number of noble gas krypton.

Structure

Molecular weight determination corresponds to the molecular formula $[Fe_3(CO)_{12}]$. X-ray studies indicate that each of two Fe-atoms is attached to 3 terminal CO groups, two bridging carbonyl groups and the remaining Fe atom is linked to four terminal carbonyl groups and each of the two Fe-atoms three Fe—Fe bonds are present with Fe—Fe bond length equal to 2.8Å.

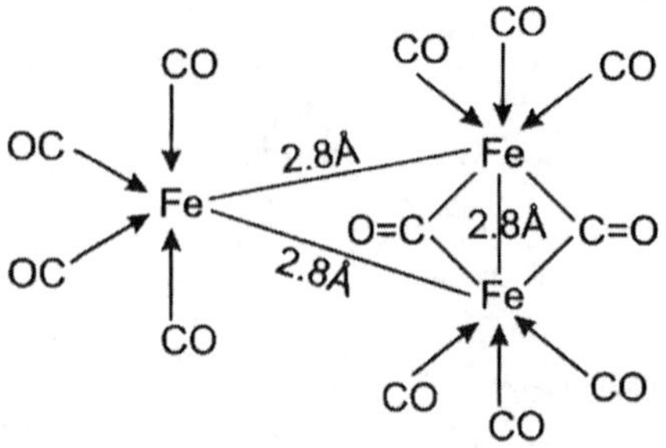

Structure of $[Fe_3(CO)_{12}]$ molecule

On comparing $[Fe_2(CO)_9]$ with $[Fe_3(CO)_{12}]$ we find $[Fe_3(CO)_{12}]$ can be obtained by replacing one bridging CO group in $[Fe_2(CO)_9]$ by $Fe(CO)_4$ group.

$$[Fe_2(CO)_9] \xrightarrow{\ominus > C=0} Fe_3(CO)_8 \xrightarrow{\oplus Fe(CO)_4} [Fe_3(CO)_{12}]$$

Tetranuclear Metal Carbonyls $[M_4(CO)_{12}]$

The important carbonyls in this category are black $[Co_4(CO)_{12}]$ melting at 60°C, red $[Ru_4(CO)_{12}]$ with a melting point of 150°C and yellow $[Ir_4(CO)_{12}]$ which melts at 210°C having a tetrahedron of $Ir(CO)_3$. All the above tetranuclear metal carbonyls have similar tetrahedral arrangement of four metal atoms. In case of Co and Ru, three triangular faces consist of terminal CO groups with one bridging CO group along each edge. The fourth triangular face is occupied by an $M(CO)_3$ group. In case of $[Ir_4(CO)_{12}]$ the structure is symmetrical with four $Ir(CO)_3$ terminal units in a regular tetrahedron. Heavier IR shows preference for non-bridged structure.

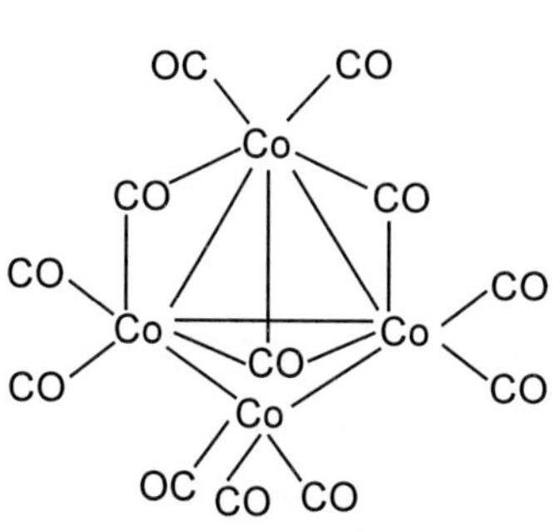

Co or Rh $[M_4(CO_{12})]$

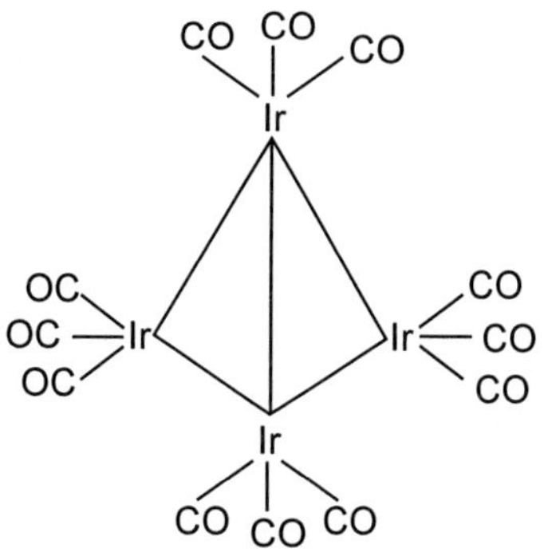

$[Ir_4(CO)_{12}]$

Metal Nitrosyls

These complexes contain NO molecule which is represented electronically by the structure $(:\dot{N}{=}\ddot{O}:)$. On losing an unpaired electron on N-atom it acquires a Positive charge $(:\overset{+}{N} \leftleftarrows\!\!= O:)$ known as nitrosonium ion. On accepting an electron NO molecule becomes NO^- ion represented by the structure $(:\ddot{N}{=}\ddot{O}:)$. In both cases N-atom has paired electron on it. When NO molecule gets attached to the metal atom to form a complex the bonding is represented by $\overset{-}{M}{\leftarrow}\overset{+}{N} \leftleftarrows\!\!= O:$ and $\overset{+}{M}{\leftarrow}\overset{-}{N} \leftleftarrows\!\!= \ddot{O}:$.

These complex compounds are termed as metallic nitrosyls.

In the common metallic nitrosy Is that contain NO^+ ions *e.g.*, $[Fe(NO)_2(CO)_2]$, $[Fe(CN)_sNO]$, $[Fe(NO)(H_2O)_5]$, $[Mn(NO)_3(CO)]$ and $[Co(NO)(CO)_3]$, NO^+ ion presence is confirmed by IR spectra. The NO molecule $:\dot{N}{=}\ddot{O}:$ on getting attached to the metal atom loses its unpaired electron to the metal so that NO acquires a positive charge $:\overset{+}{N} = \ddot{O}:$ and consequently the metal gets a negative charge. Side by side the lone pair on N-atom is donated to M^- and form a coordinate bond. Hence, NO molecule is three electron donor while CO is two electron donor. Nitrosyls are generally prepared from carbonyls or aqua complexes by the substitution reactions. Electronically for every 3CO molecules two NO molecules are substituted.

$$:\dot{N}{=}\ddot{O}: \xrightarrow{-e} :\overset{+}{N} = \ddot{O}:$$

$$M + e \longrightarrow M^-$$

$$M^- + (:)\overset{+}{N} \leftleftarrows\!\!= O: \longrightarrow \overset{\ominus}{M}{\leftarrow}\overset{+}{N} \leftleftarrows\!\!= O$$

NO^+ happens to be isoelectronic with $:C \leftleftarrows\!\!= O$ molecule (both contain 14 electrons). The M—N—O bond angle is 180° therefore it is a linear structure.

Metal Nitrosyls with NO^- anion

The common examples are $[Co(CN)_5NO]^{3-}$ and $[Ca(NH_3)_5NO]^{2+}$ ions. Due to all strong field ligands Co(III) is in the low spin state *i.e.* $t_2g^6e_g^{\circ}$. NO^- ion presence is confirmed by the IR spectra (1100–1200 cm^{-1}). Metal nitrosyls with NO^- ion, $:\overset{\ominus}{\mathrm{N}}=\ddot{\mathrm{O}}:$, out of two lone pairs on N^- one lone pair is donated to the vacant metal orbital to form a bonded coordinate bond.

$$\mathrm{M}\leftarrow(:)\overset{\ominus}{\dot{\mathrm{N}}}=\ddot{\mathrm{O}}:$$

The M—N—O bond angle is 120 to 140°. Hence, it is bent molecule.

Some metal nitrosyls contain both NO^+ and NO^- ions *e.g.,* $Fe(NO)_4$. The complex has $[(ON^+)_3Fe(NO)]$ which is confirmed by IR spectra *e.g.,* (1600 to 1900 cm^{-1}) stands for NO^+(1 100–1200 cm^{-1}) stands for NO^-.

The EAN rule is effectively applicable to metal nitrosyls as N-atom donated a pair of electrons both in NO^+, and NO^- ions *e.g.*

$$\leftarrow(:)\overset{+}{\mathrm{N}}\equiv\mathrm{O}:\ \text{and}\ \leftarrow(:)\overset{\ominus}{\dot{\mathrm{N}}}=\ddot{\mathrm{O}}:$$

Most of the metal nitrosyls are in agreement with EAN rule $_{\mathrm{h}}\mathrm{i}_{\mathrm{ch}}$ jo not obey the rule are listed.

Non compliance to EAN rule

(*i*) $[Cr^{1+}(NO^{1+})(CN^{5-})_5]^{3-}$

$Cr^{+1} = 23$
$NO^+ = 2$
$5CN^- = 10$
Total = 35

(*ii*) $[Mn^{2+}(NO^+)(CN^{5-})_5]^{2-}$

$Mn^{2+} = 23$
$NO^+ = 2$
$5CN = 10$
Total = 35

(*iii*) $[Fe^{1+}(NO^{1+})(H_2O)_5)^{2+}$

$Fe^{+1} = 25$
$NO^+ = 2$
$5H_2O = 10$
Total = 37

Besides mononuclear metal nitrosyls, binuclear metal nitrosyls having bridging groups follow:

EAN rule in monomeric form *i.e.,* $[Fe(NO)_2I]_2$ and $[Fe(NO)_2SCH_3)]_2$

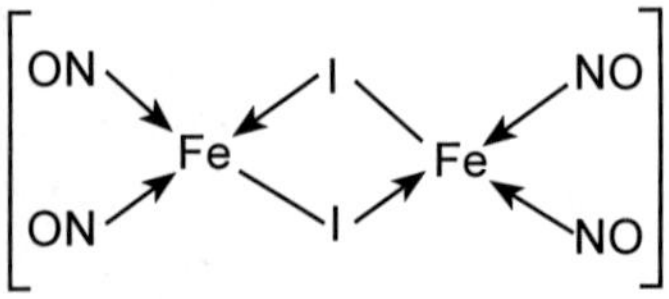

Atomic Number of Fe = 26
Electrons donated by 2NO molecules = 6
Electrons donated by one I-atoms = 2
Electrons donated by another = 1

Electrons due to Fe—Fe = 1

Total = 36 (Kr)

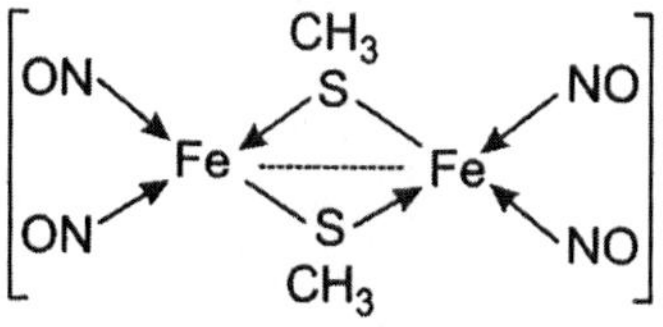

EAN rule in monomeric form in $[Fe(NO_2)(SCH_3)]_2$

Atomic number Fe = 26

Electrons donated by 2NO = 6

Electrons donated by one SCH3 = 2

Electrons donated by othe SCH_3 = 1

Fe—Fe bond = 1

Total = 36(Kr)

Important Metallic Nitrosyls

(*a*) **$[Fe^{+}(NO^{+1})(H_2O)_5]SO_4$ Nitrosoferrous sulphate**

Iron in this complex is in +1 oxidation state

Brown Ring Test

Concentrated H_2SO_4 is added to the freshly prepared $FeSO_4$ solution along the sides of the test tube, a brown ring is formed at the junction of the two layers. The formation of the complex takes place in a number of steps:

(*i*) $NaNO_3$ reacts with concentrated H_2SO_4 to produce HNO_3

$$NaNO_3 + H_2SO_4 \longrightarrow 2HNO_3 + NaHSO_4$$

$$\text{or} \quad NO_3^- + H^+ \longrightarrow HNO_3$$

(*ii*) HNO_3 so produced oxidises Fe^{2+} to Fe^{3+} ions in the presence of H_2SO_4 and gets reduced to NO.

$$2HNO_3 \longrightarrow H_2 + 2NO + \frac{3}{2} O_2$$

$$2FeSO_4 + H_2SO_4 + \frac{1}{2}O_2 \longrightarrow \underset{\text{Redish brown}}{Fe_2(SO_4)_3} + H_2O \times 3$$

$$6FeSO_4 + H_2SO_4 + 2HNO_3 \longrightarrow 3Fe_2(SO_4)_3 + 2NO$$

$$\text{ionically } 3Fe^{2+} + 4H^+ + NO_3^- \longrightarrow 3Fe^{3+} + NO^+ + 2H_2O$$

(*iii*) NO so produced is absorbed by $FeSO_4$ and $[Fe(NO)(H_2O)_5]SO_4$ is obtained

$$NO + FeSO_4 + 5H_2O \longrightarrow \underset{\text{Brown black}}{[FeNO(H_2O)_5]SO_4}$$

$$NO + Fe^{2+} + 5H_2O \longrightarrow [Fe(NO)(H_2O)_5]^{2+}$$

Brown black ring is due to Fe^+—NO^+ charge transfer band which is in visible region.

Physical Characteristics

$[Fe(NO)(H_2O)_5]SO_4$ is stable at low temperature. It decomposes with evolution of NO at high temperature

$$[Fe(NO)(H_2O)_5]^{2+} \xrightarrow{\Delta} Fe^{2+} + NO + 5H_2O$$

This property is used for the purification of NO from other gases.

Application: The formation of $[Fe(H_2O)_5NO]^{2+}$ is used for the test of NO_2^- and NO_3^- ions.

Structure: The magnetic moment value is 3.9BM which suggests three unpaired electrons. In this complex Fe is present as Fe^{+1}. It shows sp^3d^2 hybridisation in $[Fe(NO)(H_2O)_5]^{2+}$. sp^1d^2 hybridisation suggests three unpaired electrons.

Crystal field theory suggests valence shell configuration of $t_{2g}{}^5e_g{}^2$ showing three unpaired electrons:

(*b*) **Sodium pentacyanonitrosyl ferrate (II)**

$Na_2[Fe(CN)_5NO]$ or sodium nitroprusside.

Preparation: A concentrated solution of $K_4[Fe(CN)_6]$ is heated with 50% HNO, in the presence of KNO_3, $K_3[Fe(CN)_6]$ alongwith NO is obtained. This is neutralised with Na_2CO_3 and evaporated to get ruby red crystals of sodium nitroprusside

(*i*) $$3K_4[Fe(CN)_6] + 4HNO_3 \longrightarrow 3K_3[Fe(CN)_6] + NO + 3KNO_3 + 2H_2O$$

or $$3[Fe(CN)_6]^{4-} + H^+ + NO_3^- \longrightarrow 3[Fe(CN)_6]^{3-} + NO + 2H_2O$$

$$2K_3[Fe(CN)_6] + 2NO + 3Na_2CO_3 \longrightarrow \underset{\text{Ruby Red}}{2Na_2[Fe(NO)(CN)_5]} + 2NaCN + 3K_2CO_3$$

(*ii*) Alternatively, it can be obtained by the action of $NaNO_2$ on $Na_4[Fe(CN)_6]$

$$Na_4[Fe(CN)_6] + NaNO_2 + H_2O \longrightarrow \underset{\text{Ruby Red}}{Na_2[Fe(CN)_5(NO)]} + NaOH$$

Physical Characteristics

Sodium nitroprusside exists as ruby red crystals readily soluble in water. On long standing its aqueous solution gets decomposed therefore, whenever required, only freshly prepared solution should be used.

Chemical Characteristics

Action on Sulphides

When freshly prepared $Na_2[Fe(CN)_5NO]$ is added to an alkaline solution of Na_2S (H_2S excluded), red or violet streaks are observed due to the formation of a complex:

$$Na_2S + Na_2[Fe(CN)_5(NO)] \longrightarrow \underset{\text{purple or violet coloured solution}}{Na_4[Fe(CN)_5(NO^+)(S^{2-})]}$$

The test is used to confirm the presence of S^{2-} in inorganic salts. In organic compounds S present as element is converted to Na_2S during Lassaigne's test.

(*ii*) **Action on alkali sulphites or thiosulphates:** Alkali sulphites give **rose** red colour due to the formation of complex:

$$Na_2SO_3 + Na_2[Fe(CN)_5(NO)] \longrightarrow \underset{\text{Rose red}}{Na_4[Fe(CN)_5(NO)(SO_3)]}$$

Thiosulphates do not give this reaction. Hence, it can be used as a mark of distinction between sulphites and thiosulphates.

(*iii*) **Action of $AgNO_3$:** A flesh coloured complex is obtained

$$2AgNO_3 + Na_2[Fe(CN)_5NO] \longrightarrow \underset{\text{Flesh coloured}}{Ag_2[Fe(CN)_5(NO)]} + 2NaNO_3$$

(*iv*) **Action of aldehydes and ketones:** Aldehydes and ketones give red colour with alkaline solution of sodium nitroprusside

(*v*) **Action with alkalis:** With NaOH sodium nitroprusside gives sodium hexacyanoferrate (II)

$$6Na_2[Fe(CN)_5(NO)] + 14NaOH \longrightarrow$$
$$5Na_4[Fe(CN)_6] + Fe(OH)_2 + 6NaNO_3 + 6H_2O$$

Uses: Sodium nitroprusside is a useful reagent to test the presence of S^{2-} ions, aldehydes and ketones and distinguish between sulphites and thiosulphates.

Structure: On the basis of valence bond theory it shows d^2sp^3 hybridisation flaunting octahedral geometry. It has no unpaired electrons as Fe is present as Fe^{2+} and the configuration $3d^64s^0$ (low spin) where all the $3d$ electrons are paired. According to CFT the distribution of electron in Fe^{2+} ion is $t_{2g}{}^6e_g{}^0$.NO molecule appears in the structure as NO^+ ion (nitrosyl ion).

SUMMARY

In this chapter, the select topic of metal corbonyls and nitrosyls has been discussed. Carbonyls are carbon monoxide compounds with metals (transition metals). In CO, carbon atoms act as electron pair donor likewise in NO ligand N acts as donor resulting witc a a bond. Due to high electron density on the metals the electron pair is back donated to fonn rc-bonds. Metallic carbonyls are classified into (*i*) Mononuclear or monomeric carbonyls (*ii*) Polynuclear carbonyls, Polymeric carbonyls may be homonuclea$_r$ *e.g.,* $[Co_2(CO)_8]$, $Fe_3(CO)_{12}$ or hetero nuclear $MnCo(CO)_9$ etc. Metal cabohyls can be prepared by direct combination of metal and ligand, from halids of metals and metal oxides. They are generally crystalline solids except $[Ni(CO)_4]$, $[Fe(CO)_5]$ and $Os(CO)_5$ which are liquids. They can be easily substituted by other legands lake PR, and NO.

They can easily form carbonylate anions with NaOH in a solvent. They get reduced by Na/Hg or borohydrides. Carbonyl chlorides and hydrides are well known.

Chemistry of some important monocarbonyls like [Ni(CO_4)], [Fe$(CO)_5$] and [Cr$(CO)_6$] has been discussed in detail with an emphasis on their bonding, hybridisation, structure and EAN rule. Binuclear, trinuclear and tetranuclear carbonyls have also been discussed on the some lines. The physical and chemical characteristics of these carbonyls have been elaborated. Some examples of bridged carbnyls [$Mn_2(CO)_{10}$], [$Fe_2(CO)_9$], [$Co_2(CO)_8$], [$Fe_3(CO)_{12}$], [$Ru_3(CO)_{12}$] and [$Co_4(CO)_{12}$] and [$Ir_4(CO)_{12}$] have been taken up in this chapter some important metal nitrosyls have been included and the bonding between metal atom and NO ligand have been represented. Their preparation invariable results from replacement of CO. NO is three electron donor. Therefore electronically for every 3CO groups two NO groups are inserted. The theory and reactions of (*i*) brown ring test for nitrate ions and (*ii*) sodium nitroprusside test for S^{2-} ions, SO_3^{2-} ions, $S_2O_3^{2-}$ ions, $AgNO_3$ and aldhydes and ketones has been incorporated.

EXERCISE

1. Explain with reasons:

(*i*) [Ni$(CO)_4$] has tetrahedral structure.

(*ii*) [Fe$(CO)_5$] is known not $[Fe(CO)_6]^{3+}$

(*iii*) Co metal does not form mononuclear carbonyls.

(*iv*) In metal carbonyls M—C bond is shorter than the single bond length.

(*v*) Metal in carbonyls is in lower oxidation state.

(*vi*) Cr, Fe and Ni form 6, 5 and 4-coordinate complexes with carbon monoxide.

(*vii*) Ni forms [Ni$(CO)_4$] whereas [Pd$(CO)_4$] and [Pt$(CO)_4$] do not exist.

(*viii*) [Ni$(CO)_2Cl_2$] does not exist while [Pd$(CO)_2Cl_2$] does exist.

(*ix*) CO is a π-aceptor ligand.

(*x*) CO coordinates to the metal atom through C-atom and not O-atom.

(*xi*) There is variation of charge in [Ni$(CO)_4$], [Co$(CO)_4$] and $[Fe(CO)_4]^{2-}$.

2. How do we prepare metal carbonyls?

3. What is Mond's Process for the manufacture of [Ni$(CO)_4$]?

4. Apply VBT to the following:

(*i*) Fe$(CO)_5$ (*ii*) [$Fe_3(CO)_{12}$] (*iii*) $Co_2(CO)_g$ and (*iv*) [$Fe_2(CO)_9$],
Do they obey EAN rule.

5. Give preparation, structure and bonding of simple metal carbonyl [Fe$(CO)_5$].

6. What are the two types of modes of binding of CO in polymeric carbonyls?

7. What do we understand by π-acid ligands? Explain.
8. Give preparation, properties and structure of metal carbonyls.
9. Which two metals react directly with CO under conditions suitable for practical syntheses.
10. How do you prepare:
    (*i*) $Fe(CO)_5$ from iron powder
    (*ii*) $Cr(CO)_6$ from hydrated $CrCl_3$.
    (*iii*) $CO_2(CO)_8$ from hydrated $CoSO_4$
    (*iv*) $Fe_3(CO)_{12}$ from $Fe(CO)_5$.
11. How is N_2 related to CO? Are N_2 complexes more or less stable than CO complexes? Give example.
12. Give examples of carbonylate and carbonyl hydride ions. Do they obey EAN rule?
13. Describe M—N—O bond. Why is it linear in contrast to M—C—O bond. How many electrons are involved in those bondings?
14. Which PX_3 ligand is similar in bonding to CO?
15. Discuss similarities and difference in CN^- and CO ligands.
16. Which metal carbonyls are isoelectronic with $Cr(NO)_4$, $Mn(CO)(NO)_3$, $Fe(CO)_2(NO)_2$?
17. Write balanced equations:
    (*i*) $Mn_2(CO)_{10}$ heated with I_2
    (*ii*) $Mo(CO)_6$ refluxed with KI in THF
    (*iii*) $Fe(CO)_5$ is agitated with aq. KOH
    (*iv*) $Ni(CO)_4$ is treated with PCl_5.
    (*v*) $Co_2(CO)_8$ + NO in hexane.
18. Which one obeys EAN rule.
    (*i*) $[H_2Fe(CO)_4]$ (*ii*) $[V(CO)_6]$ (*iii*) $[V(CO)_6]^-$
    (*iv*) $[Cr(CO)_5]^-$ (*v*) $[Mn(CO)_4(NO)]$
    (*vi*) $[HMn(CO)_4]$ (*vii*) $[V(CO)_5H]^{2-}$
    (*viii*) $[Mn_2(I)_2(CO)_8]$ (*ix*) $[W(CO)_5Cl]^-$
    (*x*) $[CoH(N_2)(PPh_3)_3]$
19. Why $[V(CO)_6]$ is readily reduced to $[V(CO)_6]^-$?
20. Prepare the following:
    (*i*) $[Na[HCr(CO)_5]$ (*ii*) $K[Mo(CO)_5I]$ (*iii*) $[HCo(CO)_4]$
    (*iv*) $K[Co(CO)_4]$ (*iv*) $[Mn(CO)_3I(Py)_2]$
21. Write balanced equations for the preparation of $[Ru(NH_3)_5(N_2)]$.
22. Give three routes for the preparation of $[Co(CO)_4]^-$. What geometry should it acquire?

■ ■ ■

ORGANOMETALLICS

14

Organometallics

14.1 INTRODUCTION

Organometallics or metallo-organic compounds are those in which carbon atom of the organic group (hydrocarbon) is attached to the metal atoms. The chapter deals with organoderivatives of transition elements in their lower valencies.

To classify further, $(C_3H_7O)_4Ti$ ceases to be an organometallic compound because O atom is directly bound to Ti whereas $C_2H_5Ti(OC_3H_7)$ is organometallic as it contains direct metal to carbon bond.

14.2 DEFINITION

Organometallic compounds are substances which contain at least one direct netal-to-carbon bond.

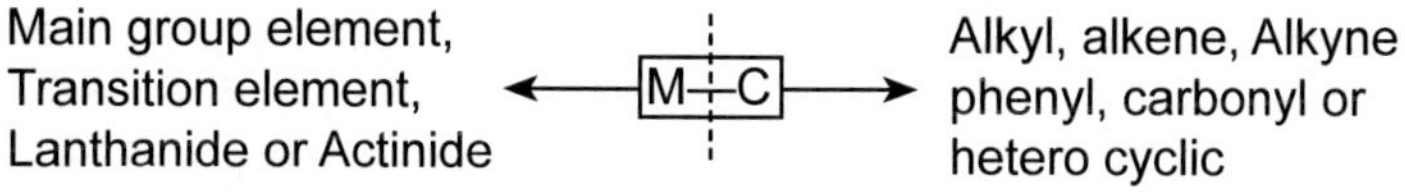

Compounds of the type CaC_2, $CaCO_3$ and $[Fe(CN)_6]^{4-}$ do not represent organometallic compounds although they have direct M—C bond. In the wider term organic groups bond with elements like Si, S, Se and Te fall into the ambit of organometallics. Some common organometallic compounds are RMgX grignard reagent), $(C_2H_5)_4Pb$, $(CH_3)_2Hg$ and $(C_5H_5)_2Fe$ (ferrocene). Even metal carbonyls by definition come in the domain of organometallics but due to difference in their bonding they are dealt as a select topic in chapter XIII. Similarly Grignard reagent is more appropriately discussed as a starting material for syntheses of important organic compounds in organic chemistry.

14.3 CLASSIFICATION

The classification of organometallic compounds is based on the nature of metal atoms and the organo groups:

(*i*) **Ionic organometallic compounds:** They are mostly formed with highly electropositive metals such as sodium, potassium, rubedium and cesium. The common examples are CH_3Na, $Mg(C_4H_9)_2$, $Mg(C_5H_5)_2$ etc. These compounds are reactive towards air and water.

(*ii*) **σ-bonded compounds:** These are formed in which the organic group is bonded to the m«etal atom by two electrons shared between them. They are mostly formed by metals of lower electropositivity or non-metallic elements. The binding is predominantly covalent. They are soluble in org:anic solvents and are volatile, *e.g.*, $Pb(C_2H_5)_4$, $(CH_3)_2Hg$, $(C_6H_5)_2$ Zn, $(CH_3)_3SnCl$ and $(CH_3)SnCl_3$.

(*iii*) **Non-conventionally bonded compounds:** In many organometallic compounds M—C bond cannot be explained by ionic or σ bonds. Transition metals have the capability to bond to more than one C atom of the same organic entity. For example, alkenes act as two-carbon-bonded two electron donors and CH_2=CH—CH=CH_2 may act as four-carbon-bonded 4 electron donors (ligands). Benzene may act as a six-carbon-bonded 6 electron donor in $(C_6H_6)_2Cr$. This characteristic of transition metal arises out of symmetry properties of d-orbitals presemt in transition elements.

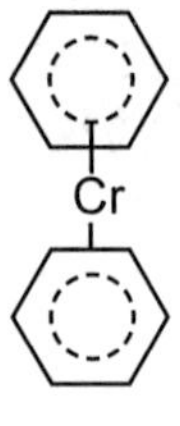

Fig. Dibenzene chromium, (sandwich compound)

There are several examples of one, two, three, four, five, six, seven and eight carbon ligands.

14.4 NOMENCLATURE

The organometallic compounds are named on the basis of IUPAC nomenclature rules pertaining to coordination compounds with the oxidation state of the metal atom given in () in Roman numerals. Alkyl and aryl radicals are considered as anions.

In organo metallic comipounds notation of α and π-bonded are used to distinguish between one carbon bonded ligand and multiple carbon bonded ligands *e.g.*, LiC_5H_5 is written as $(\alpha\text{-}C_5H_5)Li$. $Fe(C_5H_5)_2$ is referred to as

(π-$C_5H_5)_2$ Fe(5 C-bonded ligand). F. A Colton in 1968 introduced the prefixes hapto (fasten) *e.g.*

Monohapto (η^1)(—R) Alkyl, (—Ar) aryl

Dihapto (η^2) (>C=C<) alkene or (—C=C—) Alkyne

Trihapto (η^3) π-allyl (>C=C=C<)

Tetrahapto (η^4) CH_2=CF1—CH=CH_2 butadiene

Pentahapto (η^5)  (Cyclopentadienyl)

Hexahapto (η^5) Arene (benzene)

and so on.

Some of the common examples have been illustrated for the benefit of students.

Formulae	IUPAC Nomenclature
1. π-C_5H_5 $Mn(CO)_3$	Tricarbonyl (η^5-cyclopentadienyl) manganese or Tricarbonyl (pentahapto cyclopentadienyl manganese
2. $(C_6H_6)Cr(CO)_3$	(η^6-benzene) tricarbonyl chromium (written in alphabetical order)
3. $Fe(CO)_2(\sigma$-$C_6H_5)(\pi$-$C_5H_5)$	Dicarbonyl (η^1-cyclopentadienyl) (η^5-cyclopentadienyl) iron
4. $Cr(C_6H_6)_2$	Bis (η^6-dibenzene) chromium
5. Fe(p-$C_5H_5)_2$	Bis (η^5-cyclopentadienyl) iron
6. ReH(p-$C_5H_5)_2$	Bis (η^5-cyclopentadienyl) hydridorhenuim
7. $(CeHg)_3$Si—Si$(CgH_5)_3$	Hexaphenyl disilane
8. $(CH_3)_3SnO_2C(CH_3)$	Trimethyl triacetatetin
9. H_3C—Zn—C_2H_5	Ethyl methyl zinc
10. $Fe(CO)_3(C_4H_9)$	(η^4-butadiene) tricarbonyl iron
11. $\{(H_3C)_3Si\}W(CO)_3$	Tricarbonyl (η^5-cyclopentadienyl) trimethyl silyl tungsten.

Notation Cp is used for cyclopentadienyl.

14.5 SYNTHESIS OF ORGANOMETALLIC COMPOUNDS

1. Direct Action of Metals

$$Mg + C_2H_5Br \xrightarrow{\text{Ether}} C_2H_5\,Mg\,Br$$

This synthesis was first performed by french chemist V. Grignard and the product was called Grignard reagent. This can also be obtained by C_6H_5Br. Direct interaction of alkyl and aryl halides with electropositive metals like Li, > Na, K, Mg, Ca, Cd and Zn will give similar product.

2. Alkylating Agents

The compounds obtained in the reaction (l) can be used to prepare other organometallic compounds, *e.g.* Al, Mg and sodium alkyls can be used **as** alkylating agents in ether solvemt.

$$C_5H_6 + Na \xrightarrow{\text{Ether}} C_5H_5Na + \frac{1}{2}H_2$$

Given below are some examples of metal and non-metal halides alkylated in ether or benzene (toluene)

$$PCl_5 + 3C_6H_5M\,gCl \xrightarrow{\text{Ether}} (C_6H_5)_3P + 3MgCl_2$$

$$2C_5H_5Na + TiCl_4 \xrightarrow{CHCl_3} (\pi\text{-}C_5H_5)_2\,TiCl_2 + 2NaCl$$

3. Non-metal and Metal Hydrides with Alkenes and Alkynes

Hydroboration of alkenes may serve as the best example:

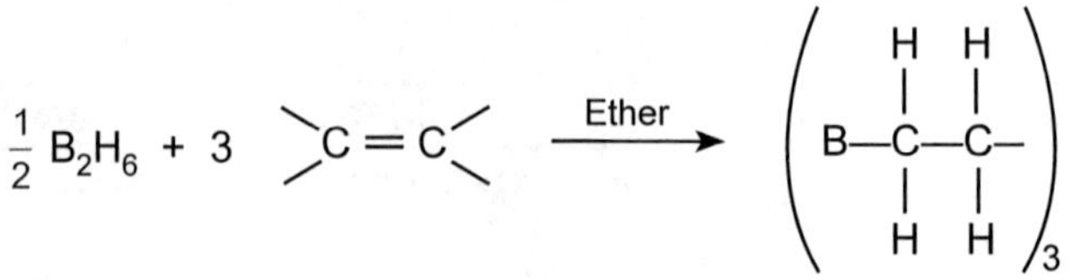

This is addition across the double bond of alkene. The trialkyl borane obtained may be further protonated in the reaction:

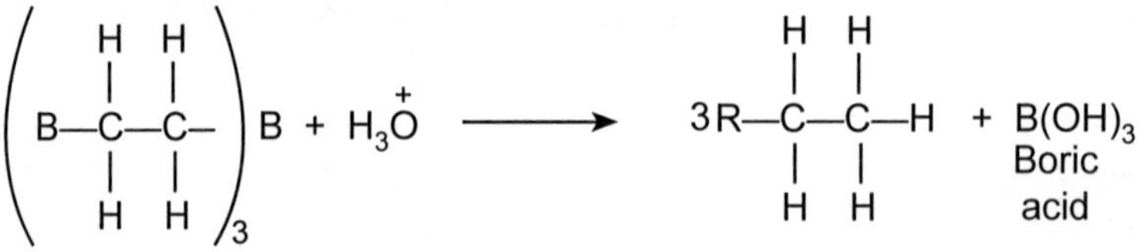

Oxidation of trialkyl borane may yield alcohol

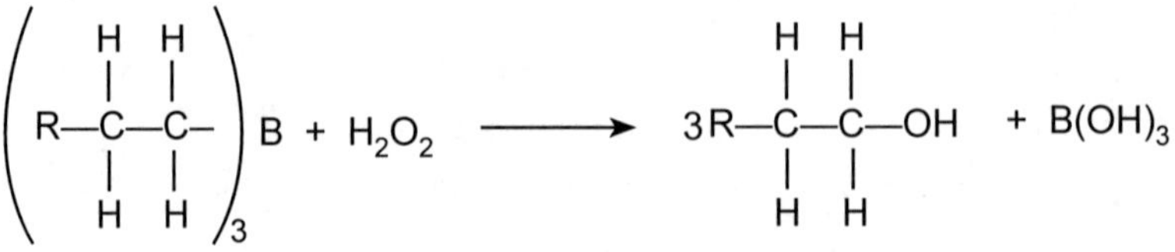

For transition metal hydride complexes these reactions are very important *e.g.*

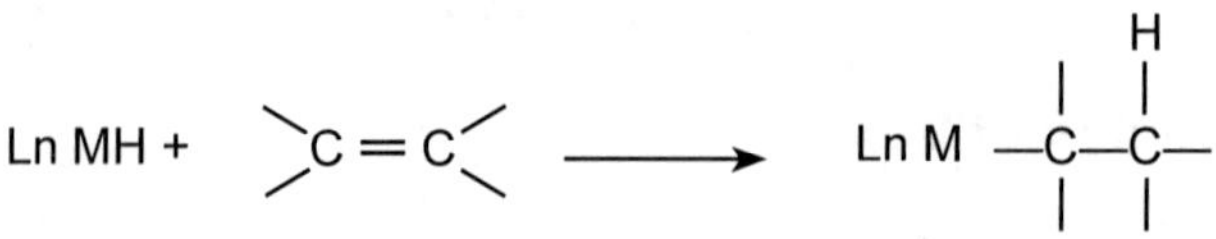

4. Oxidative Addition

In reactions where alkyl and aryl halides are added to unsaturated transition metal compounds, a new metal-carbon bond is realised.

$$RhCl(PPh_3)_3 + MeI \longrightarrow RhCl\,I\,Me(PPh_3)_2 + PPh_3$$

5. Insertion Reactions may also generate metal to carbon bond.

(*a*) $$\left[(CN)_5{-}Co{-}Co(CN)_5\right]^{4-} + HC{\equiv}CH \longrightarrow \left[(CN)_5Co{-}\underset{H}{\underset{|}{C}}{=}\underset{H}{\underset{|}{C}}{-}Co(CN)_4\right]^{4-}$$

(*b*) $$SbCl_5 + 2HC{\equiv}CH \longrightarrow Cl_3Sb(CH{=}CHCl)_2$$

Synthesis of Sandwich Compounds

This is done in the following steps:

(*i*) The liquid cyclopentadiene is taken in $CHCl_3$ solvent and sodium or potassium metal is added to it. Immediate effervescence due to the liberation of H_2 gas takes place.

$$C_5H_6 + K \xrightarrow{CHCl_3} \underset{\text{Potassium cyclopentadienide}}{C_5H_5\,K} + \underset{\text{Effervescence}}{\frac{1}{2}H_2\uparrow}$$

(*ii*) Potassium cyclopentadienide is now taken in THF (tetrahydro furan) solvent and the appropriate metal salt is added in a long jacketed flask. A moderate to violent reaction takes place with flashes of fire (keeping the flask dipped in ice mixture). After cooling the shining crystals of sandwich compound separate out.

$$2C_5H_5K + FeCl_2 \xrightarrow{THF} \underset{\text{Ferrocene}}{(C_5H_5)_2Fe\downarrow} + 2KCl$$

$$2C_5H_5Na + TiCl_4 \xrightarrow{THF} \underset{\text{Biscyclopentadienide titanium (IV) dichloride}}{(\pi\text{-}C_5H_5)_2TiCl_2} + 2NaCl$$

This compound is commonly represented by the formula Cp_2TiCl_2 (orange red shining crystals)

$$2C_5H_5Na + CrCl_2 \longrightarrow 2NaCl + \underset{\text{Chromocene}}{(\pi\text{-}C_5H_5)_2Cr}$$

$$2C_5H_5Na + MnCl_2 \longrightarrow 2NaCl + \underset{\text{Manganocene}}{(\pi\text{-}C_5H_5)_2M}$$

(*iii*) The organometallic compound is dried by evaporation of solvent with the help of vacuum.

(*iv*) The compound isolated is further crystallised and purified by a suitable solvent.

Synthesis of Alkene Complexes

Alkene complexes are synthesised by the interaction with metal carbonyls or metal chlorides with $>C{=}C<$ alkenes.

$$Mn(CO)_6 + C_7H_6 \xrightarrow{\text{Reflux}} Mn(CO)3\ C_7H_8 + 3CO$$

$$RhCl_3(aq) + C_2H_4 \xrightarrow{25°C} [(C_2H_4)_2RhCl]$$

Preparation of Zeise's Salt

The monoalkene complexes like Zeise's salt was first synthesised by W.C. Zeise, a Danish chemist in 1890. The Zeise's salt was synthesised by taking PtClo in benzene and then adding >C=C< alkene.

$$KCl + PtCl_2 + C_2H_4 \xrightarrow{\text{Benzene}} \underset{\text{Zeise's salt}}{K^+[PtCl_3(\eta^2\text{-}C_2H_4)]\cdot H_2O}$$

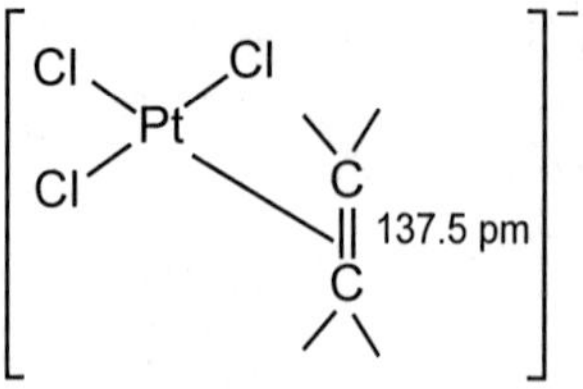

Synthesis of Alkyne Complexes

Due to 2π-bonds in alkynes (—C≡C—) at right angle to each other, alkynes are rich in energy and therefore more reactive than alkenes. Some of the reasons given are:

(*i*) Alkynes are better π electron acceptors than alkenes.

(*ii*) Alkynes have two orthogonal π and π* orbitals. Therefore, they can donate 2 or 4 electrons.

(*iii*) Alkynes can easily cyclo-trimerise to give arenes.

The simplest alkyne complex is coordinated to only one metal atom similar to that of olefin complexes. The —C≡C— is perpendicular to the plane of coordination. Out of 2π electron pairs only one can be donated or take part in the reaction.

$$M(CO)_e + RC = CR' \xrightarrow[\text{Benzene}]{\Delta} M(CO)_5(RC = CR') + CO$$

M may be Mo, W.

Both the π-electron can be donated to a single metal atom (four electrons donor)

$$V(CO)\ (\eta^6\text{–}C_5H_5)_2 + RC{=}CR \longrightarrow V(CO)_2\ (\eta^2\text{–}RC{\equiv}CR)\ (\eta^5\text{–}C_5H_5)_2 + 2CO$$

In case the two pairs or π-electrons are donated to two metal atoms then the synthesis involved may be:

$$Ni(CO)_2(\eta^5\text{-}C_5H_5) + PhC{\equiv}CPh \xrightarrow{\text{Benzene}} PhC{\equiv}C\ PH$$

$$(\eta^5\text{—}C_5H_5)Ni\text{---}Ni(\eta^5\text{—}C_5H_5) + 2CO$$

Allyl Complexes

Allyl is a three carbon alicyclic group and it can bind to the metal atom in two ways: η^3-allyl

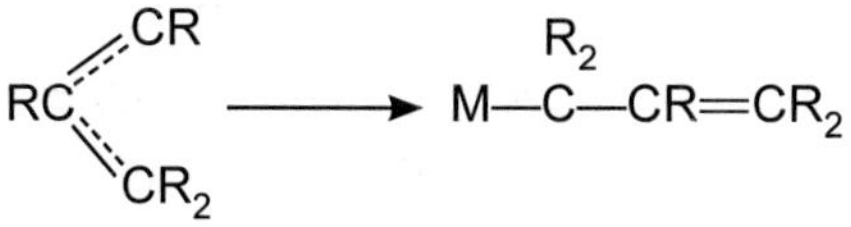

Bent

They are prepared from Grignard reagent or allyl chloride:

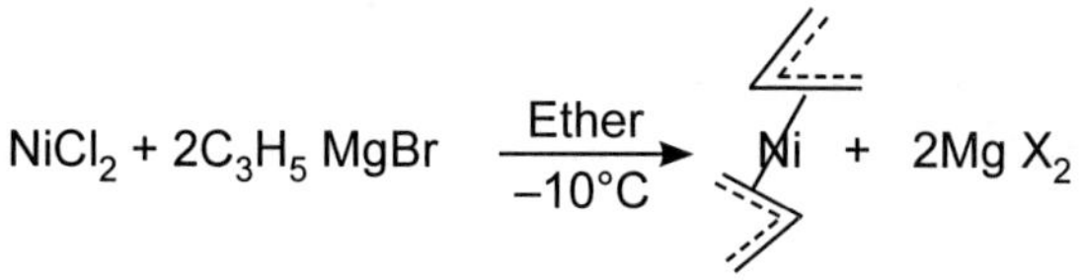

From Allyl Chloride

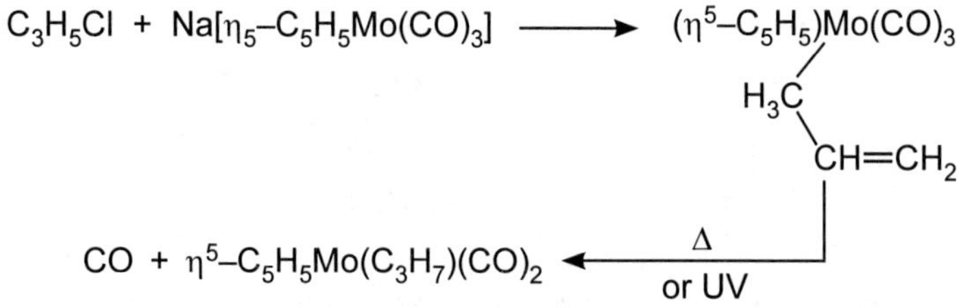

They can also be prepared from butadiene complexes *e.g.,*

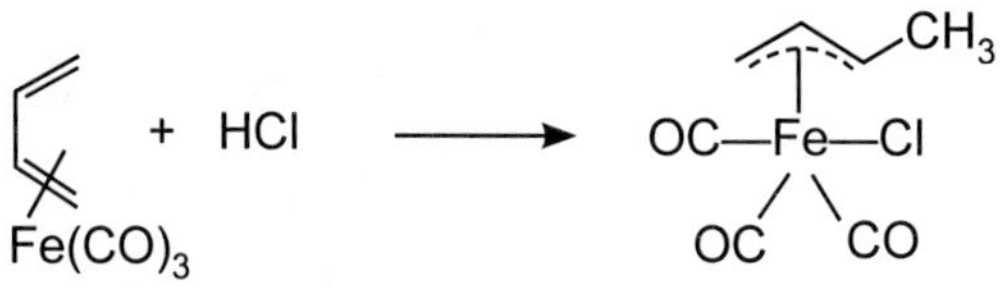

Metal-Benzenoid Complexes

These organometalic complexes were obtained from the reaction of $CrCl_3$ with C_6H_5MgBr (Grignard reagent)

$$CrCl_3 + nC_6H_5\,MgBr \longrightarrow (C_6H_5)_n\,Cr$$

where $n = 2, 3$ or 4

These are also sandwich compounds containing benzene ring.

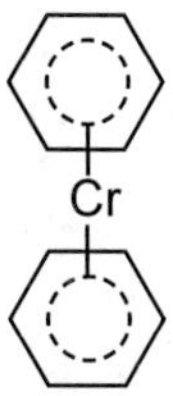

Dibenzene Chromium

Dibenzene chromium are dark brown crystals more sensitive to air than ferrocene.

Carbenes and Alkylidene Complexes

Carbenes: CR_2 are unstable and get destroyed in the free state. They can be easily stabilised by bonding to the transition metals. The lone pair present can be very easily donated. The metal-to-carbon bond is akin to—C=C— double bond or a bond.

$$M \longleftarrow :CR_2 \text{ and } M{=}CR_2$$

Carbone complexes are of two types (*i*) one in which metal has lower oxidation state and (*ii*) in which metal is in higher oxidation state.

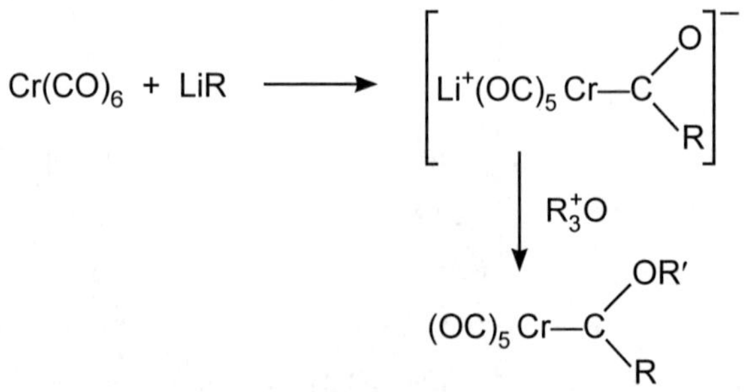

Complexes where metal is in high oxidation state are called **alkylidene complexes** they are obtained by deprotonation reaction of alkyl groups.

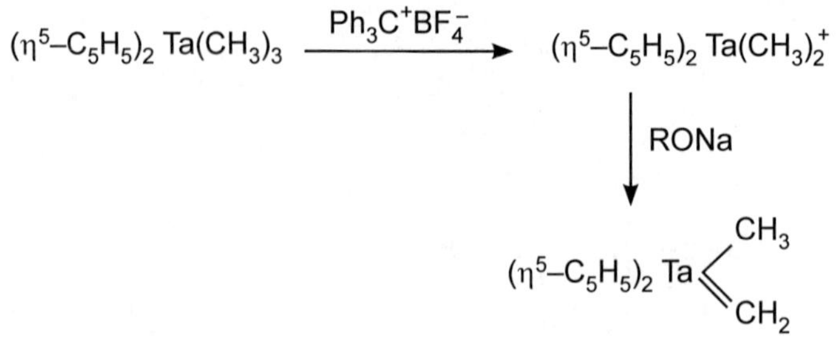

Above mentioned reaction of Ta(V) gives a good comparison of metal-to-carbon single bond and metal-to-carbon double bond.

Carbene and Alkalidyne Complexes

These complexes have M—CR bond. With metal in the lower oxidation state carbene compounds are obtained and metal in the higher oxidation state yields alkylidyne compounds.

Carbene Compounds

They are obtained from carbene compounds as given:

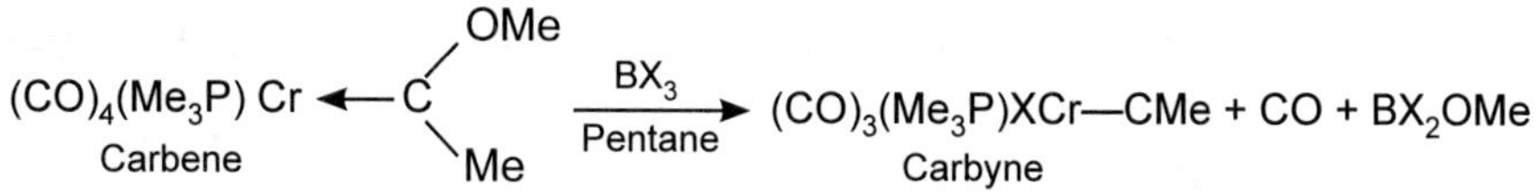

Alkalidyne compounds in which metal has higher oxidation states are formulised as M≡C. They can be obtained by deprotonation of alkylidenes as given:

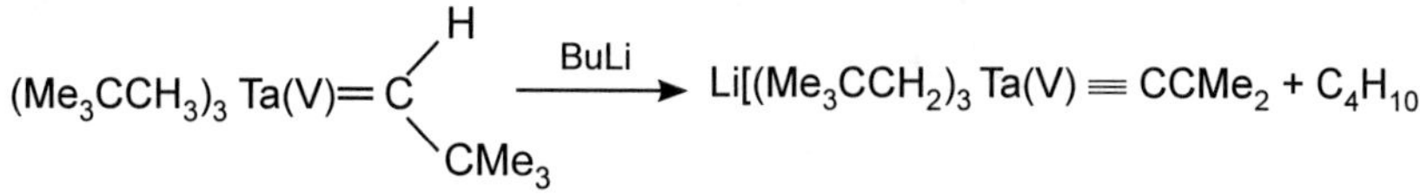

Physical Characteristics

The organometallic compounds are largely covalent in character as shown by their properties. They are low melting solids, liquids or gases at room temperature. They are soluble in non-polar or weakly polar organic solvents such as benzene, toluene, ether or alcohols.

Chemical Characteristics

1. **Decomposition:** Some organometallic compounds like $(CH_3)_4Si$ can remain unchanged for days while $(CH_3)_4Ti$ decomposes even at room temperature.
2. **Oxidation:** Organometallic compounds are generally prone to oxidation in air at room temperature but $(H_2C)_2Hg$ is almost inert. Group IV_B (C, Si, Ge, Sn and Pb) elements are easily oxidised while group IV_A elements (Ti, Zr and Hf) are immune to oxidation. Susceptibility to oxidation is due to empty low lying orbitals in Me_3Ga or lone pair of electrons in Me_3Bi. Group IV **A** alkyls do not possess these features.
3. **Hydrolysis:** Organometallic compounds of group IV_A and V_A are not attached by water whereas those of group I_A, II_B and of Zn, Cd, Al, Ga, In are readily hydrolysed. This is due to nucleophilic attack by water which is facilitated by empty low lying orbital on the metal atom *e.g.,* Al. In tetra-alkyls of groups IV_A and trialkyl of group V_A low lying orbitals are filled hence hydrolysis is not favoured. The rate of hydrolysis depends upon the polarity of metal-carbon bond.
4. **Action on Carbon dioxide:** Grignard reagent reacts with CO_2 and yields acids on subsequent hydrolysis.

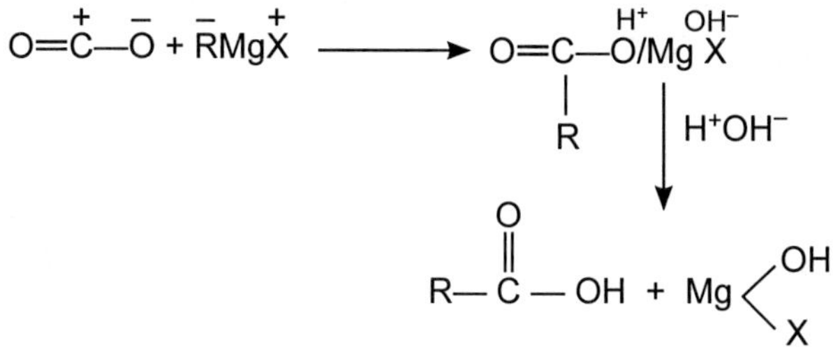

5. **Reaction with Ethers:** Alkyls of alkali metals produce **hydrocarbon** and an alkoxide.

$$RNa + R'\text{—}O\text{—}R'' \longrightarrow R'ONa + R'\text{—}R''$$

$$R'ONa + HOH \longrightarrow R'OH + NaOH$$

6. **Action on Non-metallic Element (Oxygen Excluded):** Halogens react with organometallic compounds to form hydrocarbon halides.

$$C_2H_5Li + Br_2 \longrightarrow C_2H_5Br + LiBr$$

$$Cl_2 + H_3C\text{—}\overset{|}{\underset{|}{C}} \longrightarrow ClH_2C\text{—}\overset{|}{\underset{|}{Si}}\text{—} + HCl$$

Important Organometallic Compounds

1. **Organometallic Compounds of Alkali Metals:** Lithium forms very stable organometallic compounds

Preparation: Li-compounds are prepared by:

$$Li(n\text{-}C_4H_9) + C_6H_5I \xrightarrow{\text{Ether}} Li(C_6H_5) + n\text{-}C_4H_9I$$

$$2Li + C_2H_5I \xrightarrow{\text{Ether}} Li\,C_2H_5 + LiI$$

Organometallic compounds of Na and K are prepared from mercury compounds.

$$2K + HgR_2 \longrightarrow 2KR + Hg$$

Physical Characteristics

Organometallic compounds of alkali metals are not very stable thermodynamically but polymeric compounds are relatively stable. Polymeric lithium alkyls do not decompose at 100°C.

Chemical Characteristics

Higher ionic strength of Li—C bond is responsible for greater reactivity of lithium reagents than Grignard reagents.

Action of Heat

$$LiC_2H_5 \xrightarrow{\Delta} LiH + \underset{\text{alkene}}{H_2C = CH_2}$$

Reaction with CO_2 followed by Hydrolysis

$$CO_2 + LiAr \longrightarrow ArCOOH$$

Reaction with C_2H_5I

$$C_2H_5I + UR \longrightarrow LiI + R\text{—}C_2H_5$$

Action of BCl_3

$$BCl_3 + 3R\text{—}Li \longrightarrow LiCl + BR_3$$

With Alkanes

$$R'H + LiR \longrightarrow LiR' + R{-}H$$

With Ketones

$$LiR + Ar_2CO \xrightarrow{H_2O} Ar_2(R)C{-}OH$$

With R′CON(CH$_3$)$_2$

$$LiR + R'CON(CH_3)_2 \longrightarrow LiN(CH_3)_2 + R'COR$$

Structure

Alkyl and aryl lithium compounds are polymeric and therefore Li—C bond has a covalent character. Organometallic compounds of other alkali metals have largely ionic character and can be represented as M^+R^- *e.g.* alkyls and aryls of sodium and potassium. These compounds catch fire in air violently as regard to lithium compounds.

Organometallic Compounds of Beryllium

BeR_2($R = CH_3$, C_2H_5 etc), $Be(C_5H_5)X$, where X = Cl, Br, I are examples of beryllium compounds.

Preparation

Dialkyls of Be can be obtained by:

$$2LiR + BeCl_2 \xrightarrow{Ether} BeR_2 + 2LiCl$$

$$Be + HgR_2 \xrightarrow{\Delta} BeR_2 + Hg$$

Properties

BeR_2 are colourless solids or viscous liquids. Dimethyl Be is polymeric and forms Be—CH_3 bridges in its structure. Polymerisation decreases with the bulk of alkyl group so that di-t-butyl beryllium is monomeric. Be forms complexes with cyclopentadiene. The cyclopentadienyl group is linked to Be through a σ or π bond *i.e.*, [Be η^5-(C_5H_5)X], where X = Cl, Br or I.

Organometalliic Compounds of Mg

They include compounds of the type MgR_2 (R = alkyl or aryl group) or Grignard reagents RMgX or R—Mg—OR′ having one alkyl and one alkoxy group.

Preparation

RMgX or ArMgX can be prepared by the reaction:

$$LiR + RMtgX \xrightarrow{Ether} R_2Mg + LiX$$

$$HgR_2 + Mg \xrightarrow[solvent]{Ether} R_2Mg + Hg$$

Dimethyl magnesium is polymeric and has Mg—CH_3—Mg bridges. Diphenyl magnesium dissolves in ether to form $Ph_2Mg(OC_2H_5)$ as under:

$$Ph_2Mg + C_2H_5OC_2H_5 \longrightarrow PhMg(OC_2H_5) + Ph.C_2H_5$$

$$LiPh + Ph_2Mg \longrightarrow Li[MgPh_3]$$

Grignard reagent can be synthesised by refluxing alkyl or aryl halide with Mg metal in dry ether:

$$RBr + Mg \xrightarrow[\text{Ether}]{\text{Ether}} RMgBr$$

Grignard reagents are used for the synthesis of important organic compounds.

R—Mg—OR′ can be prepared by the reaction:

$$MgR_2 + R'OH \longrightarrow R\text{—}Mg\text{—}OR' + ROH$$

These are polymeric compounds containing bridges of alkyl and alkoxy groups between atoms CH_3—Mg—OR exists in two polymeric forms shown on next pages:

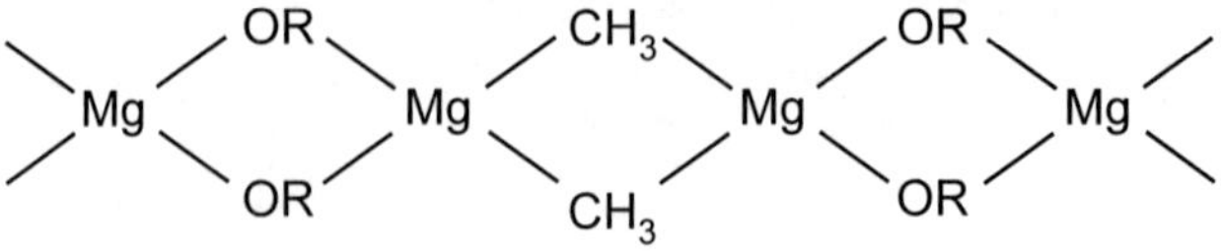

(R = Me, Et) Chain Polymer

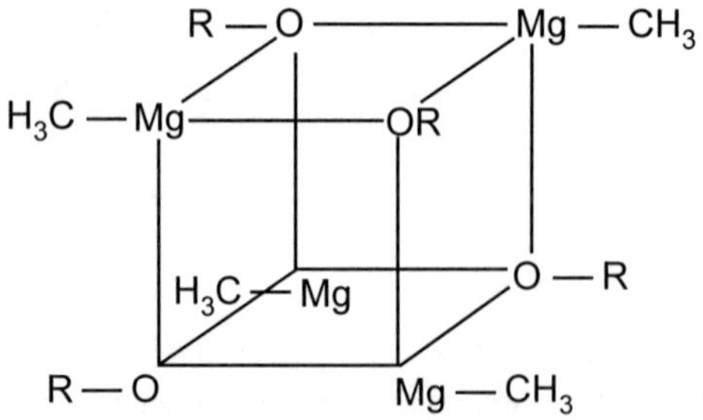

Three dimensional polymer

A brief summary of some typical reactions of Grignard reagents

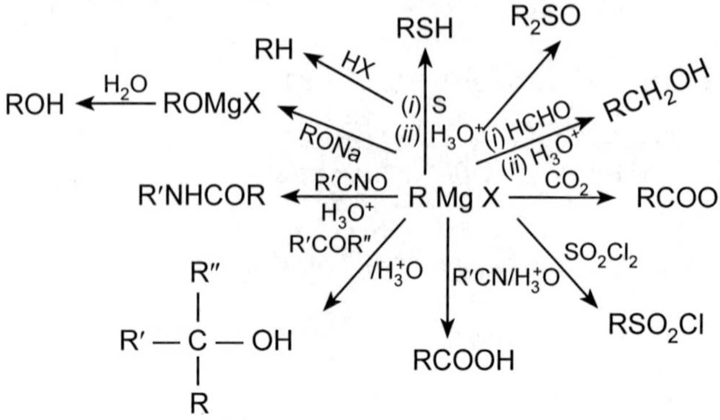

Besides, RMgX is used in a number of synthesis of other organic compounds. The alkaline earth metals use *ns* and *np* orbitals in bonding. They can utilize *sp*, sp^2 and sp^3 hybridisation giving linear, trigonal planar and tetrahedral coordination geometries.

Group 3 or Group 13 Elements

Out of B, Al, Ga, In and Tl, organometallic compounds of B and At are important. Organometallic compounds of these elements have electron deficient character particulary in case of trialkyls of B and Al. They are sensitive to oxygen, water, alcohols, amines and hydrogen chlorides.

R_3Al is prepared by the following methods:

$$2Al + 3R_2Hg \longrightarrow 2AlR_3 + 3Hg$$

$$AlCl_3 + RMgX \longrightarrow [AlR_3, RAlCl_2, R_2AlCl] \text{ Mixture}$$

$$AlH_3 + 3C_2H_4 \longrightarrow Al(C_2H_5)_3$$

Trialkyl aluminium compounds are dimeric because of Al—R—Al bridge formation and terminal alkyl groups

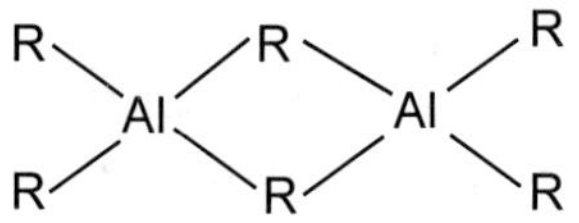

Both the alkyl group undergo very fast exchange even at room temperature. Trialkyl aluminium acts as Lewis acid and has a tendency to form adduct with Lewis base.

Some typical reactions of organo-aluminium are listed below:

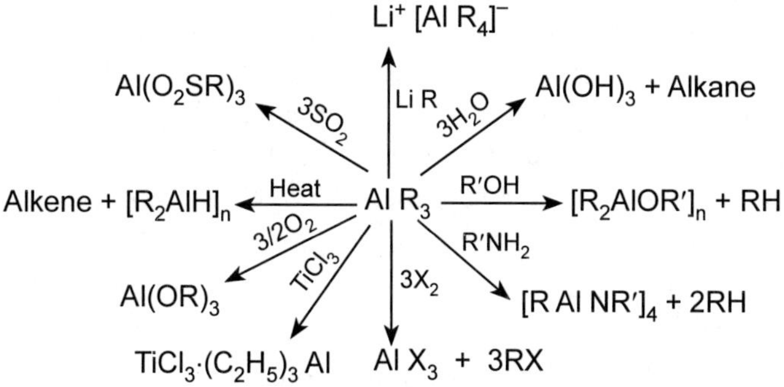

Zeigler Natta catalyst

The alkyl bromides and iodides of aluminium have Al-halogen-Al bridge similar to Al-alkyl-Al bridge. Alkyl fluorides of Al have cyclic tetrameric structure having Al—F—Al bridges.

The structure of $(CH_3)_2AlF$ is shown below:

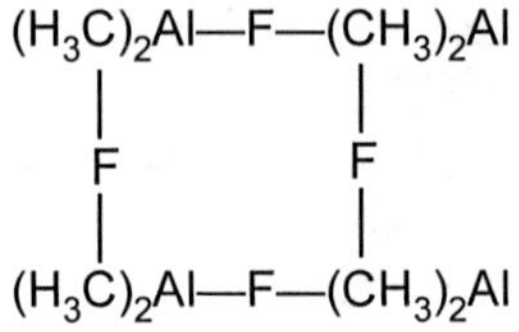

Cyclic $(CH_3)_2$ AlF

Group 4 or Group 14 Elements

The reactivity of M—C bond in tetra-alkyls or aryls of MR_4 increases from Silicon to lead *e.g.,* $Si(C_2H_5)_4$, $Ge(C_2H_5)_4$, $Sn(C_2H_5)_4$, $Pb(C_2H_5)_4$ *i.e.,* M—C bond energy decreases and so does thermal stability. Tetraethyllead is used as an anti-knock in petrol engines.

The electronegativity value of Pauling's scale hardly show any change from Si to Pb (1.8). Therefore, there occurs similarity in these tetra-alkyls.

The coordination geometries of group IV organometallics shows:

In the tetra-alkyls of $Si(CH_3)_4$, $Ge(CH_3)_4$, $Sn(CH_3)_4$ and $Pb(CH_3)_4$ the coordination number is 4. These have sp^3 hybridisation acquiring tetrahedral geometry. Out of the elements Si, Ge, Sn and Pb, Si and Sn have been receiving more attention owing to their abundance. Organogermanium are scarce and organolead are avoided due to environmental factors although they were earlier used in gasoline. ,

14.6 METAL-ALKENE COMPLEXES OF π-BONDED ORGANOMETALLICS

π-bonded organometallics of transition elements occur in a variety of fascinating structures. The interaction of the orbitals of unsaturated hydrocarbons with metals are responsible for these novel structures. They play an important role in the catalytic reactions of alkenes, alkynes and their derivatives. π-bonded organometallics involve more than one carbon atom within the bonding distance of the metal. Ethylene, propylene and a number of olefins form complexes with Pd(II), Pt(II), Hg(II), Cu(I) and Ag(I) salts.

Preparation

(*i*) (*a*) Olefins complexes are mostly prepared by the displacement of loosely held ligand *e.g.,* solvent molecules

$$2RhCl_3 + 2H_20 + 6C_2H_4 \longrightarrow 2[RhCl(C_2H_4)_2] + 2CH_3CHO + 4HCl$$

(*b*) Ionic ligand

$$K_2PtCl_4 + C_2H_4 \longrightarrow \underset{\text{Zeise's salt}}{K[PtCl_3(C_2H_4)]} \xrightarrow{C_2H_4} [PtCl_2(C_2H_4)_2]$$

(*c*) Neutral ligand

$$[Rh(CO)_2Cl]_2 + 2C_8H_{12} \longrightarrow [RhCl(C_8H_{12})_2]_2 + 4CO$$

(*ii*) Direct action of salt with olefin

$$2PdCl_2 + 2C_2H_4 \longrightarrow [PdCl_2(C_2H_4)]_2$$

Physical Characteristics

Stabilities of metal-olefins vary widely. They arc usually labile encouraging displacement reaction by ligand phosphine. Olefins in complexes are susceptible to nucleophilic attack.

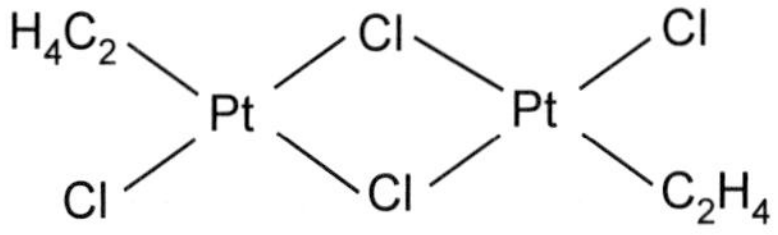

Dimer $[PtCl_2 \cdot C_2H_4]_2$

Bonding and Structure

The $[PtCl_2.C_2H_4]_2$ is dimeric as shown by molecular weight determinatiion by freezing point depression method. It shows a bridged structure with C_2H_4 molecules at trans-position. X-ray examination of Zeise's salt shows Pt at the centre of a square plane with Cl atoms at three corners. Ethylene is at the fourth corner and is perpendicular to the plane with the C = C bond of ethylene. It is perpendicular to the metal-ligand bond axis. The two carbon atoms are equidistant from. the metal.

Bonding in $[PtCl_3(C_2H_4)]^-$

(*a*) 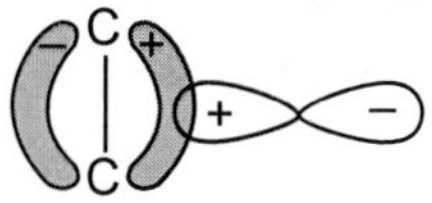 σ-type bond

(*b*) 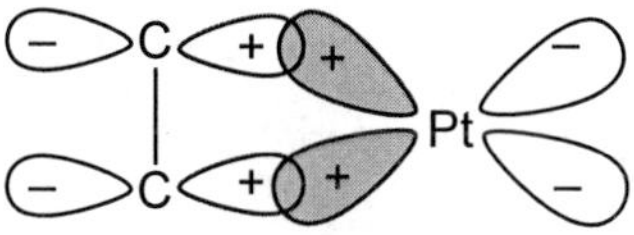 π-type bond

(*c*) 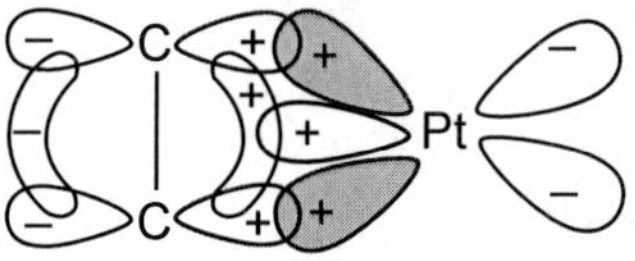 cooperative σ and π-bonding

Ethylene molecule has $2p_z$ orbitals on two carbon atoms >C = C<. It has π-bonding and π*–antibonding molecular orbitals. π-bonding orbital is filled while π*-antibonding molecular orbital is vacant. The filled π-bonding orbital of C_2H_4 molecule overlaps with empty hybrid orbital (σ type orbital) of Pt^{2+} ion. This forms *d*π-*p*π back bonding.

The two steps, *i.e.,* donation of π-bonding electrons of C_2H_4 to the metal σ orbital and back donation of electrons from the metal to the anti-bonding orbitals of the olefin are synergically related (step *c* in the figure) and weaken the π-bonding in the olefin except in anion of Zeise's salt in which there is marked increase in C—C bond length.

Due to the lengthening of C—C bond, V(C = C) 1525 cm^{-1} decreases in metal olefin complexes with respect to free ligand by 60-150 cm^{-1}.

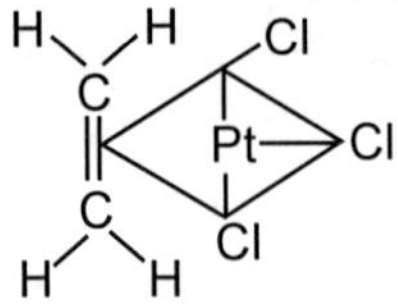

Structure of anion $[PtCl_3.C_2H_4]^-$

14.7 SANDWICH ORGANOMETALLICS OR METALLOCENES

Complexes formed by cyclopentadienyl radical with transition elements are called metallocenes with general formula $(C_5H_5)_2M$. The metal M may be Cr^{2+}, Mn^{2+}, Fe^{2+}, V^{2+}, Ti^{2+}, CO^{2+}, Ni^{2+} but $(C_5H_5)_2Fe$ ferrocene happens to be very important. Cyclopentadienyl group is a five electrons ligand. They form useful derivatives like $[(C_5H_5)Mn(CO)_3]$, $[(C_5H_5)M(NO)]$ etc.

Preparation of Metallocenes

(*i*) **By action of Alkali metal on cyclopentadiene and cyclopentadienides with transition metal chloride**

$$2C_5H_6 + 2Na \longrightarrow 2C_5H_5Na + MCl_2$$

$$MCl_2 + 2C_5H_5Na \xrightarrow[\text{reflux}]{\text{THF}} (\eta^5\text{-}C_5H_5)_2M + 2NaCl$$

C_5H_5 group is also denoted by Cp

$$2NaC_5H_5 + [Co(NH_3)_6](SCN)_2 \longrightarrow [Co(NH_3)_6]\{C_5H_5\}_2 + 2NaSCN$$

$$\underset{\text{unstable}}{[Co(NH_3)_6]\{C_5H_5\}_2} \longrightarrow 6NH_3 + (\eta^5\text{-}C_5H_5)_2Co$$

Complexes of Fe, Co and Ni are obtained by this method.

In liquid NH_3 as solvent the reaction proceeds:

$$Cd(SCN)_2 \xrightarrow{4NH_3} Cd(NH_3)_4(SCN)_2$$

$$[Cd(NH_3)_4(SCN)_2] + 2C_5H_5Li \longrightarrow Cd(NH_3)_4(C_5H_5)_2 + 2LiSCN$$

$$Cd(NH_3)_4(C_5H_5)_2 \xrightarrow[\text{Pressure}]{\text{Reduced}} 4NH_3 + (\eta^5\text{-}C_5H_5)_2Cd$$

(*ii*) **By reaction of halides with cyclopentadiene in the presence of a base such as diethylamine, piperidine or KOH.**

$$MCl_2 + 2C_5H_6 + 2Et_2NH \longrightarrow M(\eta^5\text{-}C_5H_5)_2 + 2Et_2NH_2^+Cl^-$$
(M may be Fe, Co or Ni)

$$FeCl_2.4H_2O + 2C_5H_6 + 3KOH \longrightarrow Fe(\eta^5\text{-}C_5H_5)_2 + 2KCl$$

(*iii*) **Reaction of cyclopentadienes with the metal or metal carbonyls**

$$Mg(s) + 2C_5H_6(g) \longrightarrow [Mg(\eta^5\text{-}C_5H_5)_2] + H_2$$

$$Fe + 2C_5H \longrightarrow [Fe(\eta^5\text{-}C_5H_5)_2 + H_2$$

$$[Fe(CO)_5] + 2C_5H_6 \longrightarrow [Fe(\eta^5\text{-}C_5H_5)_2(CO)_2]$$

$$FeO + 2C_5H_6 \xrightarrow{250°C} [(\eta^5\text{-}C_5H_5)_2Fe] + H_2O$$

$$2Cr(CO)_6 + 2C_5H_6 \longrightarrow [(\eta^5\text{-}C_6H_5)Fe(CO)_3] + 6CO + H_2$$

(*iv*) **In the laboratory it is prepared by the reaction of Grignard reagent with metal halides.**

$$2C_5H_5MgBr + FeCl \longrightarrow (\eta^5\text{-}C_5H_5)Fe + 2MgClBr$$

Cr and V complexes can also be prepared by this method.

Physical Characteristics

Metallocenes of 3*d* transition metals are formed in various oxidation states of metals from 0 to +5. They are mostly paramagnetic except in cases of $Fe(\eta^5\text{-}C_5H_5)_2$, $Cr(\eta^5(C_5H_5)_2$ and $Ti(\eta^5\text{-}C_5H_5)_2$. They are soluble in organic hydrocarbon solvents and are sublimed at 100°C. They have remarkably close melting points from 172° to 174°C. Metallocene with heavier metals melt at higher temperatures. Ferrocene is very stable in air while others are labile. The order of stability is Ni > Co > V ≫ Cr. In metallocenes all the —C—C— bond lengths are similar and the two rings are parallel. They have staggered or eclipsed structures with covalent characteristics. The manganocene and titanocene are anomalous. They have sandwich structures but behave like ionic compounds. Metallocenes are stable towards hydrolysis. They are easily oxidised in acid solution to give cations

$$(\eta^5\text{-}C_5H_5)_2Fe \underset{\text{solution}}{\overset{\text{acid}}{\rightleftharpoons}} (\eta^5\text{-}C_5H_5)_2Fe + \text{Ferrocenium ion}$$
Blue green

$$(\eta^5\text{-}C_5H_5)_2C \rightleftharpoons (\eta^5\text{-}C_5H_5)_2Cr + \text{Chromocenium ion}$$

$$(\eta^5\text{-}C_5H_5)_2 \rightleftharpoons (\eta^5\text{-}C_5H_5)_2V + \text{Vanadocenium ion}$$
Purple

$$(\eta^5\text{-}C_5H_5)_2C \rightleftharpoons (\eta^5\text{-}C_5H_5)_2Co + \text{Cobaltocenium ion}$$
Yellow green

Hydrogenation

Ferrocene is resistant to catalytic hydrogenation than benzene because of delocalisation of π-electron density of the ring.

Ferrocene

It is an orange yellow solid melting at 178°C. It is sublime and is soluble in organic solvents. It is stable in air and towards hydrolysis.

Chemical Characteristics

(*i*) **Oxidation:** Ferrocene is oxidised by acid solution as well as bromine.

$$Fe(\eta^5\text{-}C_5H_5) \xrightleftharpoons{\text{acid}} Fe^{+}(\eta^5\text{-}C_5H_5)_2$$

$$C_{10}H_{10}Fe\ \frac{1}{2}\ Br \longrightarrow C_{10}H_{10}FeBr \xrightarrow{6Br_2} FeBr_3 + 2C_5Fl_5Br_5$$

Pentabromo cyclopentane

(*ii*) Cyclopentadienyl rings in the molecule of ferrocene are aromatic in nature and therefore undergo normal aromatic substitution reactions:

(*a*) **Friedel Craft's Acylation:** Ferrocene and acetyl chloride are taken in equimolar quantities to give monosubstituted derivative and in double proportions they form disubstituted product

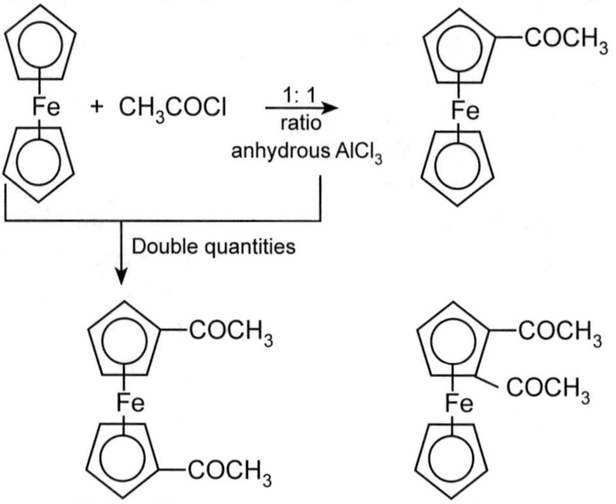

(Yield 60%) (Yield 1%)

(*b*) **Friedel Craft's Alkylation**

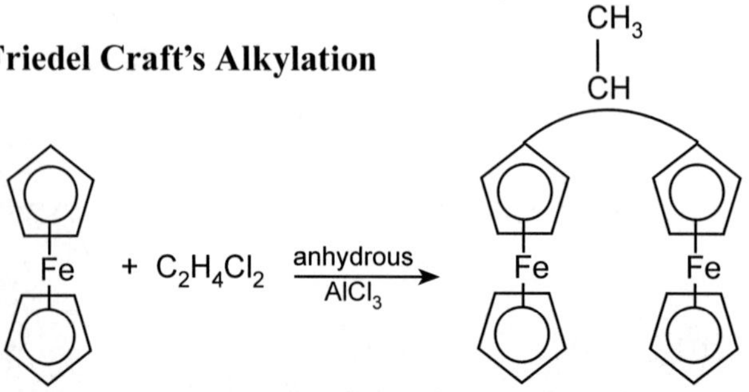

Diferrocenyl ethane

(c) **Aminomethvlation**

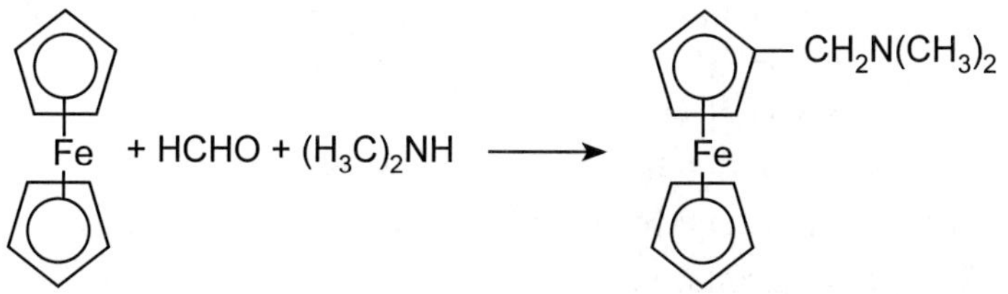

(*d*) **Sulphonation:** It takes place in presence of concentrated H_2SO_4 or chlorosulphonic acid alongwith acetic anhydride to give mono substituted or heterosubstituted compounds.

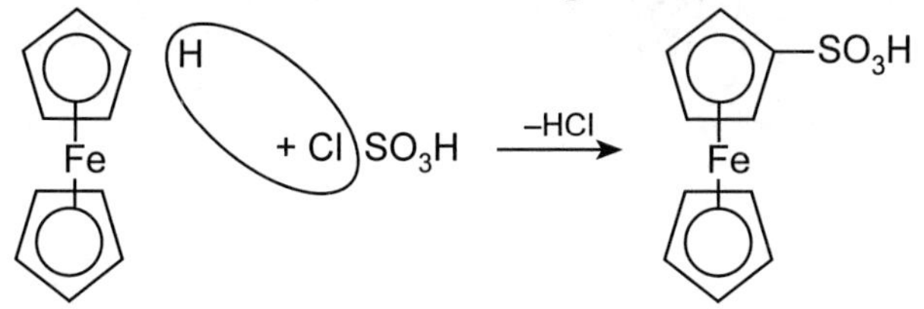

(*e*) Formylaion can take place with N, N-dimethyl formaldehyde in presentee of $POCl_3$.

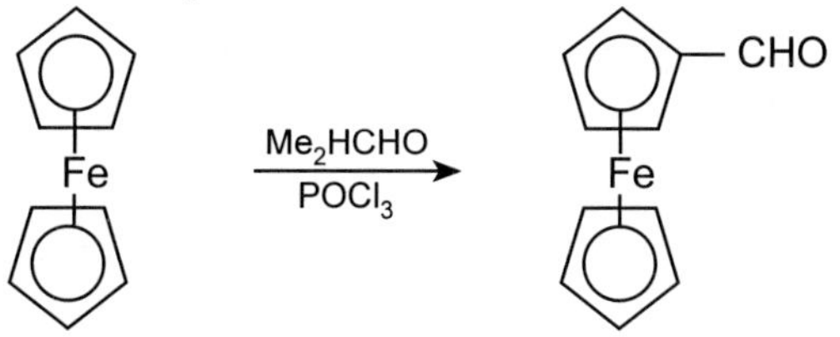

(*f*) Condensation with HCHO in presence of concentrated H_2SO_4 or liquid HF and subsequent reduction gives 1, 2 differrocenyl ethane.

(*g*) **Arylatiom:** It is affected by ArN_2Cl, *i.e.* benzene diazonium chloride

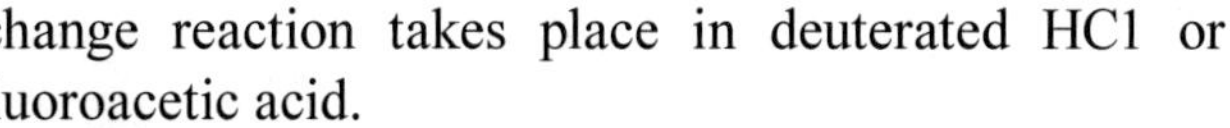

(*h*) Exchange reaction takes place in deuterated HCl or deuterated trifluoroacetic acid.

(*i*) **Metallation:** Reaction with mercuric acetate in alcohol at room termperature gives mono and disubstituted products

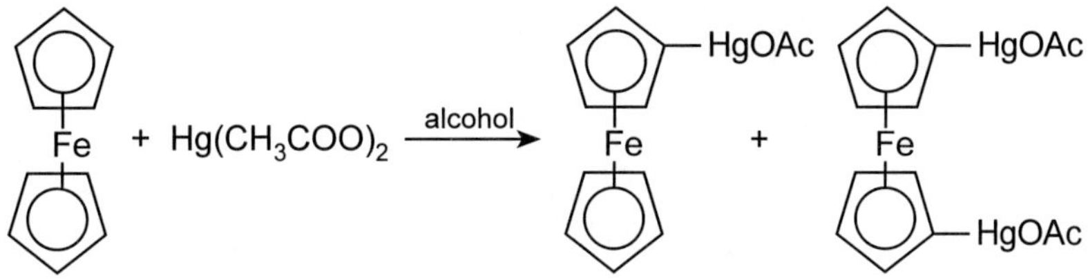

Other cyclopentadienyl sandwich compounds which also undergo this substitution are $Os(\eta^5\text{-}C_3H_5)_2$, $Mn(\eta^5\text{-}C_5H_5)_2$, $Ru(\eta^5\text{-}C_5H_5)_2$ and $[(\eta^S\text{-}C_5H_s)V(CO)_4]$

Addition and reduction reactions are well known in metallocene; other than ferrocene such as nickelocene and cobaltocene.

(*j*) Addition of Tetrafluroethene

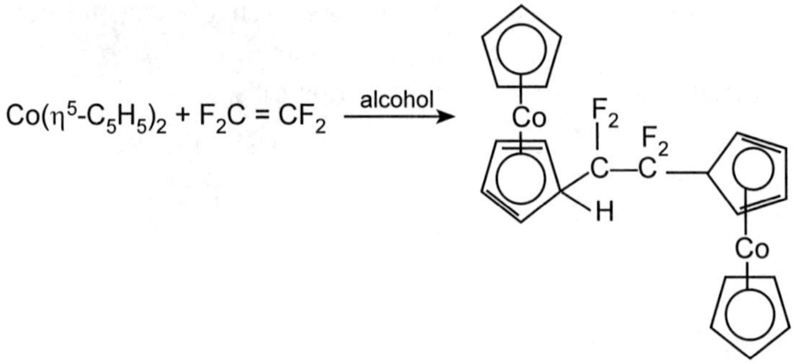

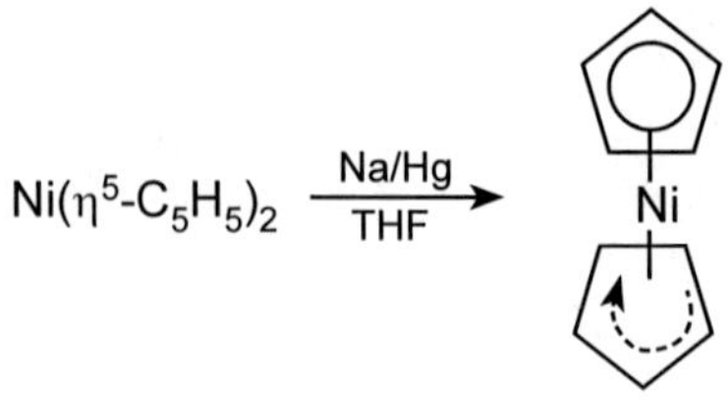

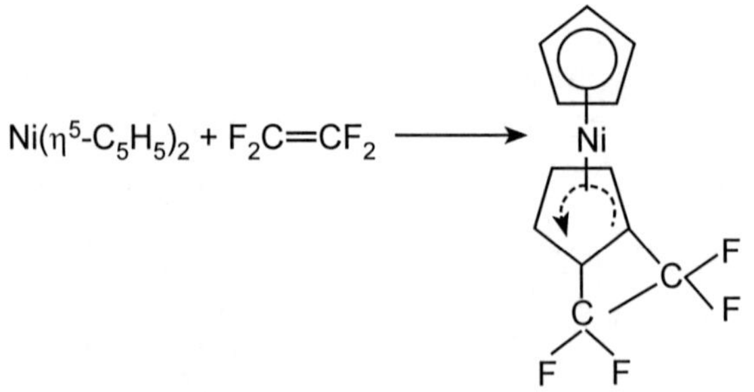

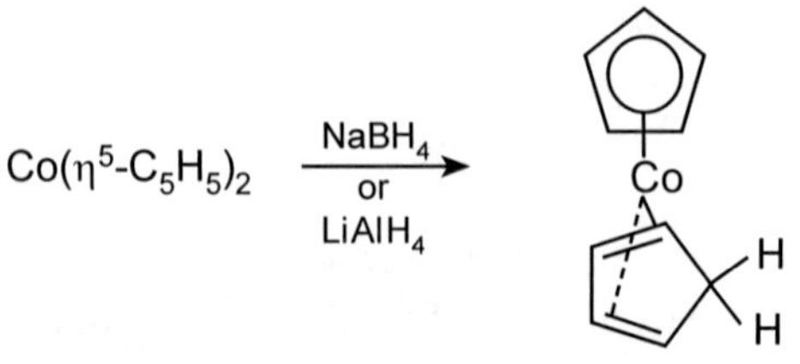

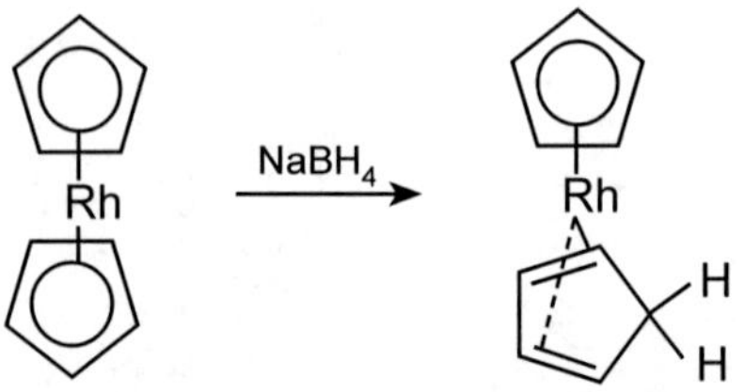

Structure and Bonding in $(\eta^{S}\text{-}C_5H_5)_2$ Fe: Structure and bonding has been discussed in the light of (*i*) Valence Bond Theory and (*ii*) molecular orbital theory.

Valence Bond Approach: According to this theory, cyclopentadienyl anion is considered $C_5H_5^-$ as a resonance hybrid of five degenerate structures (same energy).

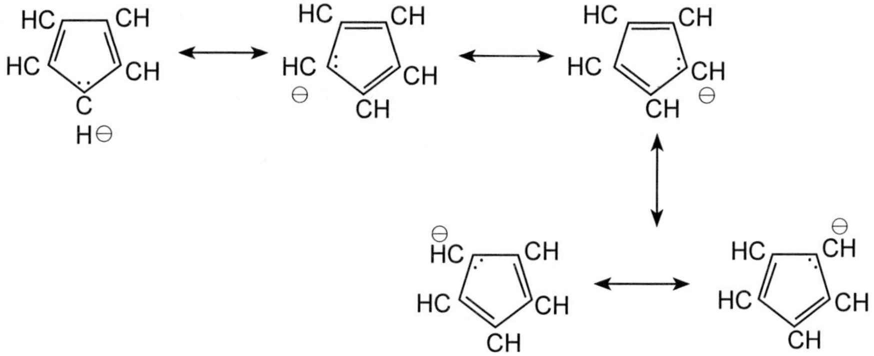

With the help of valence bond theory we are able to predict the molecular geometry of the metal complex with respect to the number and types of metal orbitals available for bonding. Two types of forms of bonding were given by (*a*) Fisher and (*b*) Dunitz and Orgel.

Fisher's Bonding

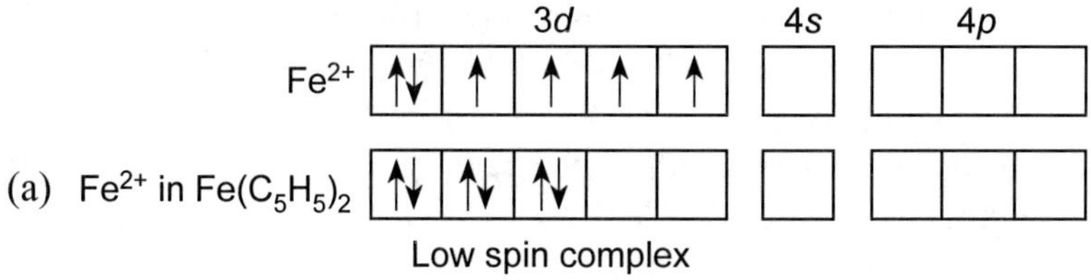

Each $C_5H_5^-$ donates its 3 pairs of π-electrons to the vacant d^2sp^3 hybrid orbitals.

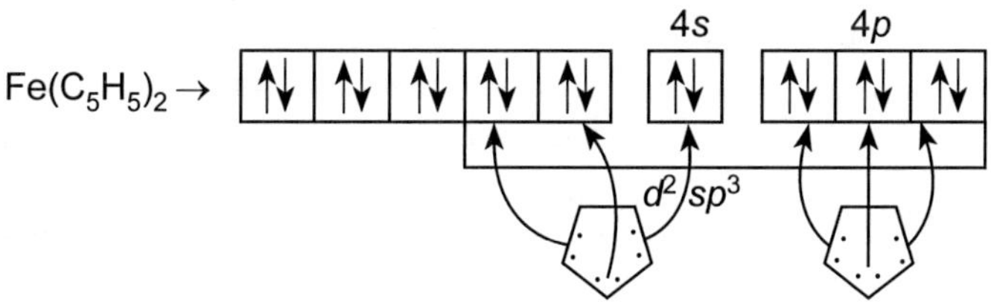

The aromatic nature of ferrocene makes available π-electrons in each ring but it does not agree with the involvement of six electrons in bonding with the metal.

(*b*) Dunitz and Orgel stated that iron is in zero oxidation state and two covalent σ bonds of the two cyclopentadienyl rings bind the iron metal.

> The π-bonded structure C_5H_5—Fe—C_5H_5 is in consonance with the properties of ferrocene. The metal atom is not attached to any individual carbon atom, it is equidistant from all the five carbon atoms and lies on the axis passing at right angles to the centre of the ring.

Valence bond is thus useful in predicting the geometry of η^5-complexes. The geometry was further confirmed by X-ray and electron diffraction studies.

Crystallographic studies authenticated the sandwich structure of ferrocene which possesses staggered conformation with two planar cyclopentadienyl rings. Gas phase electron diffraction data recommend eclipsed structure as shown

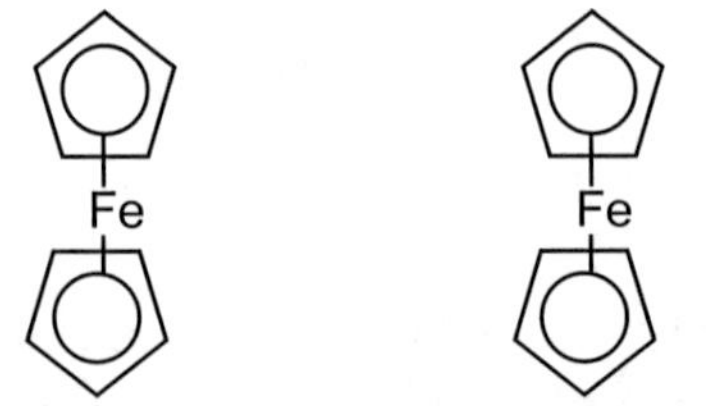

14.8 MOLECULAR ORBITAL THEORY (MOT)

For better understanding of the nature of bonding one has to develop molecular orbital picture for cyclopentadienyl ligand and metal orbitals. The ligand and metal orbitals should have correct orientation and energies for better overlap for effective bonding.

Cyclopentadienyl anion contains delocalised *n* orbitals. All the Five carbon and five hydrogen atoms are coplanar with —C—C— bond length (equal to 1.39Å) which is in between that of ethene and benzene.

On the molecular orbital basis five *p* orbitals (one on each C-atom) are coplanar. Due to linear combination they form three bonding and two anti-bonding molecular orbitals. The most strongly bonded is the one in which all the five orbitals overlap (delocalize) and are designated 'X'. At a little higher energy there are two doubly degenerate set of orbitals E_1 each of which has one nodal plane. After this another set of double degenerate orbitals are found at a little higher energy with two nodal planes with energy increasing from bottom to top.

There are ten ligand group orbitals of two cyclopentadienyl anions, five group molecular orbitals (Ψ_1 – Ψ_5) from one cyclopentadienyl anion are combined in pairs with similar set from the second $C_5H_5^-$ in symmetric and unsymmetric fashion. The combination of ligand group and iron group orbitals are shown in diagram on next page:

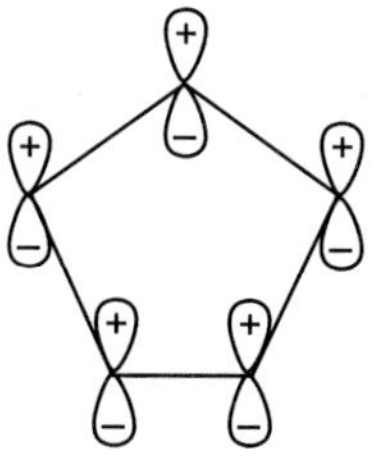

Fig. I: Orbitals in cyclopentadienyl ring

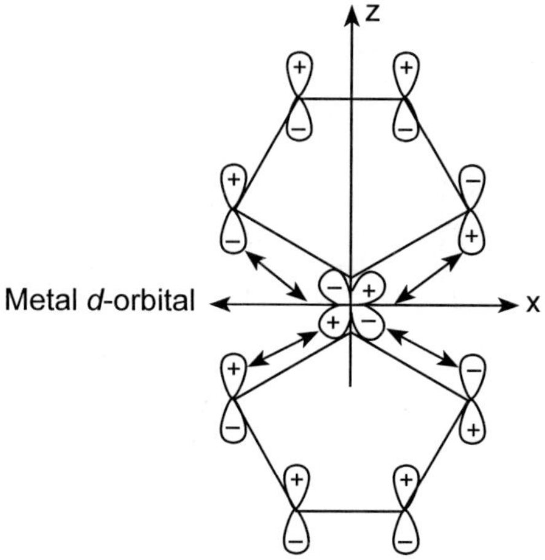

Fig. II: π-bond resulting from overlap of ***dxz*** and ligand orbitals

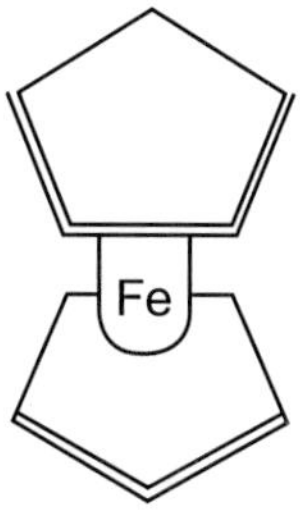

Fig. III: Ferrocene

In the LCAO molecular orbital treatment of bonding in metallocenes, the atomic orbitals of metal ion are grouped by symmetry. The iron metal orbitals are grouped as s, $dz^2(a_1g)$, $a_{2u}(4Pz)$; d_{xy}, d_{xy}, d_{zx} (e_{ig}); px, py (e_{iu}); d_x, $d_{x^2-y^2}$ (e_{2g}) and '*a*' and '*e*' stand for non-degenerate and degenerate whereas 'g' and 'u' represent symmetry and antisymmetry of the orbitals.

In the metal ligand bond, *d*-orbitals of metal will overlap with the ring orbital when positive lobe of the metal is nearer to positive lobe of the ring (s, dx^2). But when negative lobe of the metal and ring orbitals are close then they overlap (pz orbital has the correct orientation).

In case the energies of the ligand and metal orbitals is not very different then bonding molecular orbitals will result otherwise it is antibonding. No metal orbital of proper orientation is available which can overlap with degenerate orbitals of ligand such as e_{2u}.

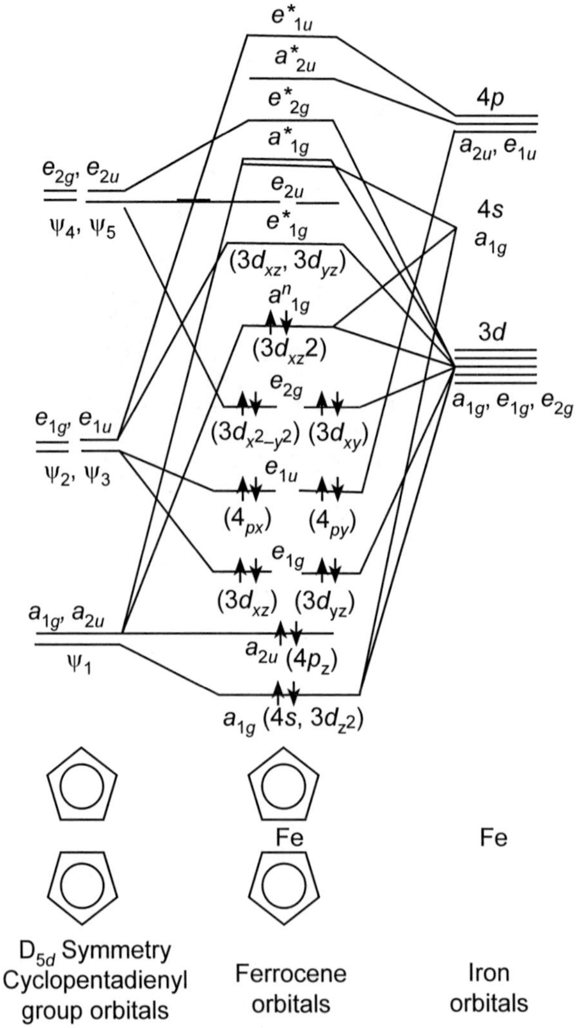

A qualitative molecular orbital diagram for ferrocene.

Intermixing of metal and ligand orbitals requires that their orbitals must have appropriate energy, *i.e.,* the energy difference between the two orbitals must not be too different and their orientation must be the same. According to the MO diagram the effective overlap can result into σ-bonds ψ_1, π-bonds in the form of ψ_2 and ψ_3 and δ-bonds ψ_4 and ψ_5.

In the case of ferrocene molecule the ligand orbitals are classified in three pairs; (*i*) filled pair a_{1g} and a_{2u} symmetry, (*ii*) higher energy filled e_{1g} and e_{1u} orientation and still higher unfilled e_{2g} and e_{2u} symmetry. The metal *dxz* and *dyz* orbitals with e_{1g} orientation interact with cyclopentadienyl ring orbitals than the metal orbitals a_{1g}, *px*, *py*(e_{1u}) and *pz*(a_{2u}). These interactive bonds give the required stability to ferrocene molecule so that they can hold together due to bulkiness.

Antibonding e_{2g} orbitals of ligands interact with $d_{x^2-y^2}$ and d_{xy} orbitals giving weakly bonded molecular orbitals.

Ferrocene is an 18 electron system (8 electrons in the valency shell of iron plus 10 electrons from the 2 cyclopentadienyl rings) require nine molecular orbitals of lowest energy to form stable metallocene. MO diagram also indicates the magnetic and chemical properties of metallocenes *e.g.,* cobaltocene (19e) and nickelocene (20e) have electrons in antibonding degenerate e_{1g} orbitals (Paramagnetic). Similarly chromocene (16e) and vanadocene (15e) have unfilled bonding orbitals and therefore, are highly reactive.

SUMMARY

In this chapter, emphasis has been laid on M—C bond, definition of organometallic compounds. Classification of organometallic compounds has been discussed in the light of M—C bonding. Some examples of IUPAC nomenclature have been explained. The different methods of synthesis of organometallic compounds have been elaborated with stress on its isolation. The synthesis of sandwich compounds like ferrocene, chromocene, manganocene aqd nickelocene have been highlighted. Besides, synthesis of Zeise's salt, alkene alkyne, alkyl, carbene and carbyne complexes have been taken up. Physical and chemical characteristics of important organometallic compounds have been incorporated. Organometallic compounds of Be, Mg, B, Al, group IV elements have been associated. It also includes re-bonded organometallics with focus on its bonding and structure. Preparation, physical and chemical properties of metallocenes have been detailed. Ferrocene, its physciai and chemical characteristics have[1] been elaborated-particularly substitution reaction like Friedel-Craft's and other addition reactions. Structure and bonding based on Valence Bond, Fisher's bonding molecular orbital have been discussed at length approach with a focus on pictorial representation of molecular orbitals of metal and ligand.

EXERCISE

1. Define organometallic compounds.
2. What are the three main classes of organometallic compounds? Give one example of each.
3. Describe three general methods of preparation of organometallic compounds.

4. Give the structural aspect of Lithium alkyls and aryls.
5. In tetramethyl ethylene diamine complexes LiR are more reactive than alkyls themselves.
6. How are grignard reagent prepared? Give metal interchange reaction of organomercury compound.
7. Prepare:
   (*i*) $NaB(C_6H_5)_4$ from BCl_3
   (*ii*) Cyclopropyl mercury bromide from Mercury.
   (*iii*) Diethyl zinc from Zn
   (*iv*) Triethyl aluminium from Al
8. What are ‘Silicones’? How are they prepared?
9. Why is $[Ti(bipy)(CH_3)_4]$ is thermally more stable than $[Ti(CH_3)_4]$?
10. Besides b-hydrogen transfer, what is the other important aspect of decomposition of metal alkyls?
11. Draw the structure of Zeise’s salt. Emphasize the important features on which metal-alkene bonding is based.
12. Give 1UPAC names:

(*i*) 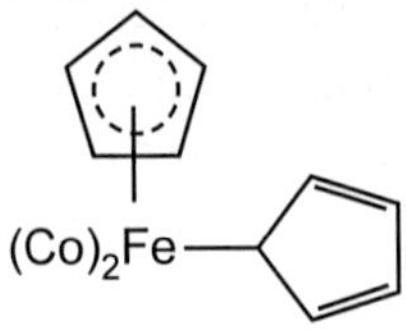

(*ii*) 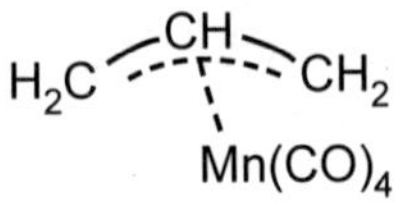

13. Give two step preparation of $(\eta^5\text{-}C_5H_5)_2Ni$ from C_5H_6, Wa and $NiCl_2$.
14. Show how h^3-alkyl complexes can be obtained?
15. Starting from metal carbonyls how will you get carbene complexes.
16. Describe the bonding in (*i*) dimethyl beryllium and (*ii*) Trimethylaluminium in terms of multicenter bonds.
17. Why the bond lengths of A1—C different in $A1_2(CH_3)_6$?
18. Why Zeise’s salt adds to other ligands readily? Explain.
19. Synthesize:
   (*i*) $Na[B(C_2H_5)_4]$ (*ii*) $Li[Al(C_2H_5)_4]$
   (*iii*) C_2H_5 Mg Br (*iv*) $P(C_2H_5)_3$
20. Describe the preparation of ferrocene from $C_5H_6$1. Na and $FeCl_2$ and give Friedel Craft’s reactions.
21. Discuss structure and bonding in ferrocene with the help of Valence Bond Theory and MOT.
22. The two cyclopentadienyl rings are staggered and eclipsed in ferrocene and vuthenocene respectively. Explain.

■ ■ ■

PART–II

COORDINATION CHEMISTRY

Transition Metal Chemistry

1

General Characteristics of *d*-block Elements

1.1 INTRODUCTION

Going strictly by definition 'Transition elements are those in which electrons in degenerate '*d*' and '*f*'-subshells are in the process of filling (incomplete either in the elemental form or in one of the important oxidation states.) According to this definition, Zn, Cd and Hg should not be a part of transition series. These three elements have completely filled '*d*' orbitals both in the element form and in the stable oxidation state. They have striking similarity in properties with other transition elements in the series. Most importantly, their complex formation tendency is similar to "*d*' block elements hence they form a part of the transition elements. Rationally, the three transition series with the elements Zn, Cd and Hg, mark the completion of *d*-orbitals with electrons.

There exist three transition series each comprising of 10 elements *e.g.,* 3*d,* 4*d* and 5*d* involving filling of *d*-orbitals and ending with Zn, Cd and Hg, respectively. The three series occur after the alkaline earth metals Ca, Sr and Ba. The transition series from Sc_{21} to Zn_{30} is called the first or 3*d* transition series. From Y_{39} to Cd_{48} is the second or 4*d* series and from La_{57} to Hg_{80} is the *5d* or third transition series. Thus, a total of 30 elements (3*d,* 4*d* and 5*d*) plus the elements in the 6*d* series which is incomplete, comprise the transition elements. There is marked similarity in the properties of elements within each series which is not seen in the *p*-block elements.

1.2 PLACEMENT IN THE PERIODIC TABLE

The 3*d,* 4*d* and 5*d* series of elements are positioned in between alkaline earth metals, Ca, Sr and Ba on the left and group 13 elements like Ga, In and T1 on

Chemical Periodic Table

Key:

element name **atomic number** **symbol** atomic weight

hydrogen 1 **H** 1.00794																	helium 2 **He** 4.002602
lithium 3 **Li** 6.941	beryllium 4 **Be** 9.012182											boron 5 **B** 10.811	carbon 6 **C** 12.0107	nitrogen 7 **N** 14.00674	oxygen 8 **O** 15.9994	fluorine 9 **F** 18.9984	neon 10 **Ne** 20.1797
sodium 11 **Na** 22.98977	magnesium 12 **Mg** 24.3050											aluminium 13 **Al** 26.981538	silicon 14 **Si** 28.0855	phosphorus 15 **P** 30.97376	sulphur 16 **S** 32.065	chlorine 17 **Cl** 35.453	argon 18 **Ar** 39.984
potassium 19 **K** 39.0983	calcium 20 **Ca** 40.078	scandium 21 **Sc** 44.95591	titanium 22 **Ti** 47.867	vanadium 23 **V** 50.9415	chromium 24 **Cr** 51.9961	manganese 25 **Mn** 54.93805	iron 26 **Fe** 55.845	cobalt 27 **Co** 58.9332	nickel 28 **Ni** 58.6934	copper 29 **Cu** 63.546	zinc 30 **Zn** 65.409	gallium 31 **Ga** 69.723	germanium 32 **Ge** 72.64	arsenic 33 **As** 74.9216	selenium 34 **Se** 78.96	bromine 35 **Br** 79.904	krypton 36 **Kr** 83.798
rubidium 37 **Rb** 85.4678	strontium 38 **Sr** 87.62	yttrium 39 **Y** 88.90585	zirconium 40 **Zr** 91.225	niobium 41 **Nb** 92.90638	molybdenum 42 **Mo** 95.94	technetium 43 **Tc** [98]	ruthenium 44 **Ru** 101.07	rhodium 45 **Rh** 102.9055	palladium 46 **Pd** 106.42	silver 47 **Ag** 107.8682	cadmium 48 **Cd** 112.411	indium 49 **In** 114.818	tin 50 **Sn** 118.710	antimony 51 **Sb** 121.760	tellurium 52 **Te** 127.60	iodine 53 **I** 126.9045	xenon 54 **Xe** 131.293
caesium 55 **Cs** 132.90545	barium 56 **Ba** 137.327	lutetium 71 **Lu** 174.967	hafnium 72 **Hf** 178.49	tantalum 73 **Ta** 180.9479	tungsten 74 **W** 183.84	rhenium 75 **Re** 186.207	osmium 76 **Os** 190.23	iridium 77 **Ir** 192.217	platinum 78 **Pt** 195.078	gold 79 **Au** 196.96655	mercury 80 **Hg** 200.59	thallium 81 **Tl** 204.3833	lead 82 **Pb** 207.2	bismuth 83 **Bi** 208.980	polonium 84 **Po** [209]	astatine 85 **At** [210]	radon 86 **Rn** [222]
francium 87 **Fr** [223]	radium 88 **Ra** [226]	lawrencium 103 **Lr** [262]	rutherfordium 104 **Rf** [261]	dubnium 105 **Db** [262]	seaborgium 106 **Sg** [266]	bohrium 107 **Bh** [264]	hassium 108 **Hs** [269]	meitnerium 109 **Mt** [268]	darmstadtium 110 **Ds** [271]	roentgenium 111 **Rg** [272]	ununbium 112 **Uub** [285]		ununquadium 114 **Uuq** [289]				

lanthanum 57 **La** 138.9055	cerium 58 **Ce** 140.116	praseodymium 59 **Pr** 140.90765	neodymium 60 **Nd** 144.24	promethium 61 **Pm** [145]	samarium 62 **Sm** 150.36	europium 63 **Eu** 151.964	gadolinium 64 **Gd** 157.25	terbium 65 **Tb** 158.9253	dysprosium 66 **Dy** 162.50	holmium 67 **Ho** 164.930	erbium 68 **Er** 167.259	thulium 69 **Tm** 168.934	ytterbium 70 **Yb** 173.04
actinium 89 **Ac** [227]	thorium 90 **Th** 232.038	protactinium 91 **Pa** 231.0359	uranium 92 **U** 238.0289	neptunium 93 **Np** [237]	plutonium 94 **Pu** [244]	americium 95 **Am** [243]	curium 96 **Cm** [247]	berkelium 97 **Bk** [247]	californium 98 **Cf** [251]	einsteinium 99 **Es** [252]	fermium 100 **Fm** [257]	mendelevium 101 **Md** [258]	nobelium 102 **No** [259]

Notes: Elements with atomic weights in square brackets have no stable isotopes. Different sources list different atomic weights for elements. The difference arises from the differing atomic weights of various isotopes. We have tried to list the most stable isotope. For example, some sources list the atomic weight of seaborgium as 263 and others 266. The most stable isotope appears have an atomic weight of 266 so we list that weight here. Roentgenium is still the unofficial name of element 111 but it is the one recommended by the IUPAC so we list it here instead of the generic 'unununium'. Aluminum, cesium, and sulfur are the American spellings for aluminium, caesium, and sulphur.

the right and finally ending with noble gases *i.e.* Kr, Xe and Rn. The 6*d* series remains incomplete and the place has been left vacant in the series for the newer elements as and when they are discovered. Besides, the bottom most element in the vertical column of each p-block from p^1 to p^6 are mostly radioactive in nature. In the periodic table, the filling of electrons is on the basis of energy order in *s, p, d, f* sub-shells. The order $s > p > d > f$ is not maintained because the outermost electron penetrates the charge cloud of low lying electrons. The penetration power in a principal quantum shell decreases in the order $s > p > d > f$. Furthermore, it is attributed to the inefficient shielding of electrons in a particular orbital. The order of shielding effect of an electron in different orbitals is in the order $s > p > d > f$.

Due to the opposite effect of shielding and penetration power, the outermost 4s orbital drops in energy in comparison to 3*d* orbitals in the first and subsequent transition series. Therefore, after filling the outermost $ns^{1,2}$, the *d* orbital elements which have next higher energy are placed. The filling of electrons continues till '*d*' orbital gets completely filled. Scandium which has the valence shell configuration $3d4s^24/p^{\circ}$, has $3d^1$ screening the 4*p* orbitals effectively and not the remaining *d* orbitals. Hence, 3*d* filling gets preference over 4*p*. When 3*d* gets completely filled, the 4*p* orbital becomes next lowest in energy till the atomic number of 36 Krypton. The second and third transition series follow the same pattern.

Therefore, the placement of elements having electrons $3d^{1-10}$ precedes the elements having electrons in $4p^{1-6}$. Krypton has the configuration of $(n-1)d^{1-10}np^6$. Same follows in 4*d* and 5*d* series.

When filling of electrons in 5*d* begins, after La, from Ce to Eu, the electrons start filling 4*f* orbitals which acquire lower energy and then elements starting from H_f get the electrons filled. From Ce to Lu, we have 14 elements called lanthanides or 4/-block elements.

La being in 5*d* series resembles more in properties to 4-block elements.

In the 6*d* transition series after the element, Actinides *i.e.*, $6d^1$, 5*f* orbitals become lower in energy and from Th to Lr, 14 elements will observe the filling of electrons in $5f^{1\text{ to }14}$. The elements Ac has $6d^1$ configuration, but resembles in properties to 5*f*-block elements hence, the series of elements are called Actinides. Some authors include La_{57} in 4-block elements and Ac_{89} in 5-block elements.

It is obvious that transition elements are always one shell behind and inner transition elements, 4*f* and 5*f* series are two shell behind.

From the figure it becomes obvious that energy between different sub-shells drops in irregular ways.

The sequence of filling the sub-shells is represented in the following figure.

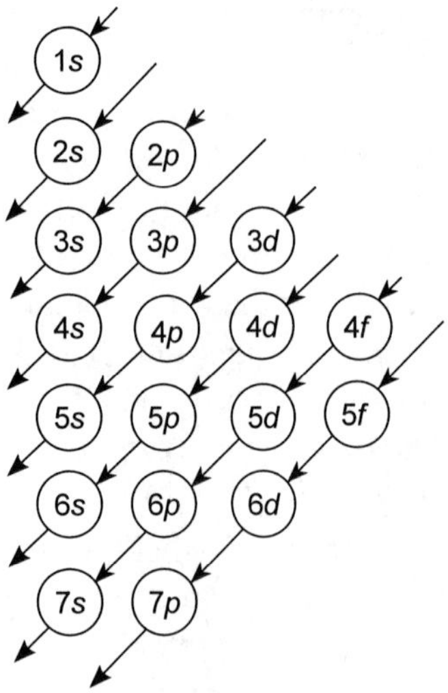

1.2.1 Electronic Configuration

The valence shell configuration of transition elements is $(n\text{–}1)d^{1-10}ns^{1,2}$, where *n* indicates the outermost shell and (*n*–1) indicates the penultimate shell. The *d*-block elements are classified in 3*d*, 4*d*, 5*d* and 6*d* series.

In **3d series** (first transition series), it contains 10 elements from Sc_{21} to Zn_{30} where the last electron fills in 3d orbital and these elements are kept in 4th period. After filling two electrons in 4*s* orbital of Ca *i.e.* $1s^2, 2s^22p^6, 3s^23p^64s^2$ (4*s* drops in energy in comparison to 3*d*), the electrons start filling all the 10 electrons in 3*d* orbital. During the filling of electrons in 3*d*, it is observed that half filled and completely filled 3*d* orbitals have extra stability owing to the lowering of energy due to exchange energy in half filled configuration and neutralisation of spins, lowering the energy in completely filled configuration. This fact comes into light in the configuration of Cr_{24} and Cu_{29} in particular. The general configuration of elements is formalised by (Ar)$3d^{1-10}4s^{1,2}$. In case of Cr, so as to gain extra stability, an electron shifts from 4*s* to 3*d* sub-shell to acquire half filled configuration *i.e.* (Ar) $3d^54s^1$. Similarly, in the configuration of Cu, the completely filled arrangement of electrons is observed. This result in Cu acquiring (Ar) $3d^{10}4s^1$ configuration.

Element	Symbol	Atomic number	Electronic configuration
Scandium	Sc	21	[Ar] $3d^14s^2$
Titanium	Ti	22	[Ar] $3d^24s^2$
Vanadium	V	23	[Ar] $3d^34s^2$
Chromium	Cr	24	[Ar] $3d^54s^1$
Manganese	Mn	25	[Ar] $3d^54s^2$
Iron	Fe	26	[Ar] $3d^64s^2$
Cobalt	Co	27	[Ar] $3d^74s^2$
Nickel	Ni	28	[Ar] $3d^84s^2$
Copper	Cu	29	[Ar] $3d^{10}4s^1$
Zinc	Zn	30	[Ar] $3d^{10}4s^2$

Owing to the interplay of the following forces, it is difficult to predict correct configuration of elements in the second, third, and fourth transition series.

(*i*) Electron-electron repulsion
(*ii*) Nuclear-electron attraction
(*iii*) Shielding outermost electron by inner electrons
(*iv*) Exchange forces.

1.2.2 Second Transition Series (4*d*-series)

The general configuration of these elements is governed by (Kr) $4d^{1-10}5s^{0-2}$. The series starts from Y_{39} and ends at the element Cd_{48}. There occur many irregularities in the configuration of these elements *e.g.,* Nb_{41} should have the configuration (Kr)$4d^35s^2$, but actually it has (Kr) $4d^45s^1$; Ru_{44} should have (Kr) $4d^65s^2$, but it has (Kr) $4d^75s^1$; Rh_{45} should have (Kr) $4d^75s^2$, but it has (Kr) 4d®5s’. Similarly, Palladium (Pd_{46}) should have (Kr) $3d^84s^2$, but in actual, it has (Kr) $3d^{10}4s^0$. The cause for this may be attributed to four factors mentioned earlier.

Element	Symbol	Atomic number	Electronic configuration
Yttrium	Y	39	[Kr] $4d^15s^2$
Zirconium	Zr	40	[Kr] $4d^25s^2$
Niobium	Nb	41	[Kr] $4d^45s^2$
Molybdenum	Mo	42	[Kr] $4d^55s^1$
Techentium	Tc	43	[Kr] $4d^55s^2$
Ruthenium	Ru	44	[Kr] $4d^75s^2$
Rhodium	Rh	45	[Kr] $4d^85s^2$
Palladium	Pd	46	[Kr] $4d^{10}5s^2$
Silver	Ag	47	[Kr] $4d^{10}5s^2$
Cadmium	Cd	48	[Kr] $4d^{10}5s^2$

1.2.3 Third Transition Series or 5*d*-series

The general configuration for these elements is given by (Xe $4f^{0 \text{ or } 14}$ $5d^{1-10}$ $6s^{1,2}$. The series starts from La_{57} to Hg_{80} where the electrons fill 5*d* orbitals Some irregularities in the electronic configuration are observed *e.g.*, Tungsten (W_{74}) should have the elctronic configuraton as (Xe) $4f^{14}5d^56s^1$, but actually it has (Xe) $4f^{14}5d^6s^2$; also, Pt_{78} should have the electronic configuration (Xe) $4f^{14}5d^86s^2$, but in reality it has (Xe)$4f^{14}5d^96s^1$.

Element	Symbol	Atomic number	Electronic configuration
Lanthanum	La	57	[Xe] $4f^05d^16s^2$
Hafnium	Hf	72	[Xe] $4f^{14}5d^26s^2$
Tantalum	Ta	73	[Xe] $4f^{14}5d^36s^2$
Tungsten	W	74	[Xe] $4f^{14}5d^46s^2$
Rehenium	Re	75	[Xe] $4f^{14}5d^56s^1$
Osmium	Os	76	[Xe] $4f^{14}5d^66s^2$
Iridium	Ir	77	[Xe] $4f^{14}5d^76s^2$

Platinum	Pt	78	[Xe] $4f^{14}5d^{9}6s^2$
Gold	Au	79	[Xe] $4f^{14}5d^{10}6s^2$
Mercury	Hg	80	[Xe] $4f^{14}5d^{10}6s^2$

1.2.4 Fourth Transition Series or 6d-series

This series is incomplete. It does not contain all the 10 elements because all the elements have not been discovered so far, but the place has been left vacant in the periodic table. The names and symbols of some of the elements known till date are not in agreement among the chemists. These elements have radioactivity property, but have short half-life and their chemistry has not been investigated so far.

Element	Symbol	Atomic number	Electronic configuration
Actinium	Ac	89	[Rn] $6d^{1}7s^2$
Rutherfordium	Rf	104	[Rn] $5f^{14}6d^{2}7s^2$
Dubnium	Db	105	[Rn] $5f^{14}6d^{3}7s^2$
Seaborgium	sg	106	[Rn] $5f^{14}6d^{4}7s^2$
Bohrium	Bh	107	[Rn] $5f^{14}6d^{5}7s^2$
Hassium	Hs	108	[Rn] $5f^{14}6d^{5}7s^2$
Meitnerium	Mt	109	[Rn] $5f^{14}6d^{7}7s^2$

1.2.5 Electronic Configuration of Transition Metal Ions

Metals can only lose electrons to form cations. In transition elements, electrons can be lost from the outer most and penultimate shells. At the time of filling, *4s* orbital was filled first as there was drop in the energy of 4*s* and then comes the turn of 3*d* orbital. This pattern holds for filling of electrons. When it comes to the loss the electrons, the orbtials 4*s* and 3*d*, readjust their energies and 4*s* becomes outermost and 3*d* becomes penultimate. As a result, the electrons are first lost from 4*s* and then from 3*d* orbtials.

For example:

Ti^0_{22} (Ar) $3d^2 4s^2$

Ti^{2+} (Ar) $3d^2 4s^0$

Ti^{3+} (Ar) $3d^1 4s^0$

It is noticed that electrons are not lost from 3d in case d-orbital is fully filled.

For example:

Zn^0 (Ar) $3d^{10}$ $4s^2$

Zn^{2+} (Ar) $3d^{10} 4s^0$

Cd^0 (Kr) $4d^{10} 5s^2$

Cd^{2+} (Kr) $4d^{10} 5s^0$

Hg^0 (Xe) $5d^{10}$ $6s^2$

Hg^{2+}(Xe) $5d^{10}$ $6s^0$

Electronic configuration of atoms and ions belonging to first transition series are given in Table:

Atom	Electronic configuration	Ion	Electronic configuration
Sc	[Ar] $3d^1 4s^2$	Sc^{3+}	[Ar] $3d^0$
Ti	[Ar] $3d^2 4s^2$	Ti^{2+}	[Ar] $3d^2$
		Ti^{3+}	[Ar] $3d^1$
V	[Ar] $3d^2 4s^2$	V^{2+}	[Ar] $3d^3$
		v^{3+}	[Ar] $3d^2$
		v^{4+}	[Ar] $3d^1$
Cr	[Ar] $3d^5 4s^2$	Cr^{2+}	[Ar] $3d^4$
		Cr^{3+}	[Ar] $3d^3$
Mn	[Ar] $3d^5 4s^2$	Mn^{2+}	[Ar] $3d^5$
Fe	[Ar] $3d^6 4s^2$	Fe^{2+}	[Ar] $3d^6$
Cr;		Fe^{3+}	[Ar] $3d^5$
Kli	[Ar] $3d^7 4s^2$	Co^{2+}	[Ar] $3d^7$
	[Ar] $3d^8 4s^2$	Ni^{2+}	[Ar] $3d^8$
Cu	[Ar] $3d^{10} 4s^1$	Cu^+	[Ar] $3d^{10}$
		Cu^{2+}	[Ar] $3d^9$
Zn	[Ar] $3d^{10} 4s^2$	Zn^{2t}	[Ar] $3d^{10}$

Some important physical constants of transition elements have been listed in Table for 3*d*, Table for 4*d* and Table for 5*d*, for the first, second and third transition series for reference. The physical properties compared are metallic behaviour, melting point, varying oxidation states, paramagnetism and colour of transition metal ions. By comparison, *s*-block elements are also metals, but they are soft with low melting points and show low oxidation states and are diamagnetic. Within transition series, the elements have similar, but graded properties and they differ in properties with *s* and *p*-block elements.

1.2.6 Atomic and Ionic Radii

The atomic and ionic radii particularly of M^{2+} ions show a general decrease from left to right in the period. This is attributed to the increase in the nuclear charge which pulls the electron cloud inwards. In 3*d*-series, from Sc to Mn, the decrease is gradual, but from Cr to Cu, it is nearly constant. This is due to the opposing effect due to screening of outer 4*s*-electrons from the inward pull of the nucleus. Due to opposite forces acting simultaneously, the atomic radii do not show any change from Cr to Cu. In case of Zn, all the electrons are paired. Hence, there is balancing of forces and thus, a slight increase in the atomic radii.

Table 1: Atomic Radii of d-block elements

Sc	Ti	V	Cr	Mn	Fe	Co	Ni	Cu	Zn
1.60	1.46	1.31	1.25	1.29	1.26	1.25	1.24	1.28	1.33
Y	**Zr**	**Nb**	**Mo**	**Tc**	**Ru**	**Rh**	**Pb**	**Ag**	**Cd**
1.80	1.57	1.41	1.36	1.30	1.33	1.34	1.38	1.44	1 49
La	**Hf**	**Ta**	**W**	**Re**	**Os**	**Ir**	**Pt**	**Au**	**Hg**
1.88	1.57	1.43	1.37	1.36	1.34	1.35	1.38	1.44	1.55

The common ions of M^{2+} and M^{3+} follow the trend of The radii. The radii of M^{2+} is somewhat smaller than that of Ca^{2+} ion. The metal oxides MO of transition elements are less basic and less soluble than CaO. Hydration energy of M^{2+} is greater than the hydration energy of Ca^{2+} ion ($Ca^{2+} \rightarrow$ 395 Kcal) and (Ti^{2+} to Cu^{2+}) varies between 446 to 597 Kcal.

Ionic Radii 3d-series (Ti^{2+} to Cu^{2+})

	Ti	V^{2+}	Cr^{2+}	Mn^{2+}	Fe^{2+}	Co^{2+}	Ni^{2+}	Cu^{2+}
Ionic Radii (Å)	0.90	0.88	0.84	0.80	0.76	0.74	0.72	0.69

For M^{2+} ions, the ionic radii is greater than Ga^{3+} (0.62 Å). Therefore, M_2O_3 oxides are similar, but less acidic and more basic than Ga_2O_3. Their hydration energies are less than hydration energy of Ga which is 1124 Kcal.

The hydration energy values for Sc^{3+} and Fe^{3+} are 947 and 1072 Kcal, respectively.

1.2.7 Metallic Behaviour

Like *s*-block elements, all transition elements are essentially metals. This is because both *s*-block and transition elements have one or two electrons in the outermost *s*-orbital. Transition metals are hard, malleable, ductile and have high melting and boiling point. The unpaired electrons in *d* pair up to form covalent bonds. The multiple bonding gives strength to the bond (both metallic and covalent bonding).

The transition elements exhibit three types of structures: face centred cubic (fee), hexagonal close packed (hep) and body centred cubic (bcc). The crystal lattice further strengthens the bond. Zn, Cd and Hg have fully filled *d*-orbitals therefore, there is no covalent bonding. Hence, they are soft and have low melting points. The transition elements are good conductors of heat and electricity and, exhibit lustre due to delocalisation of *s*-electrons.

1.2.8 Atomic Volume and Density

The density and atomic volume are inversely proportional to one another $\left(D = \frac{M}{V}\right)$. Therefore, units of atomic density and atomic volume are g/cm^3 and mL, respectively. Atomic volumes of *d*-block elements are low and consequently densities are high as compared to *s*-block elements. Low atomic volumes are attributed to the filling of electrons in the penultimate $(n-1)$ *d* orbitals and not *ns* orbitals. Due to poor screening, both types, of electrons *viz.* $(n-1)$d and *ns*, the nucleus shows an increase in nuclear chcarge resulting in powerful attraction by the nucleus, and ultimately low volume results in the high densities of transition elements.

Along the transition series, the atomic volume of the element first shows a decrease and then shows an increase in value for group VIII elements. Further, the value increases for elements in group IB and IIB as well.

On moving down in the subgroup, the atomic density increases in the same subgroup. The densities of second transition series are slightly higher than the elements of first transition series as shown in Table.

	III B	IV B	VB	VI B	VII B	←	VIII	→	I B	IIB
	Sc_{21}	Ti_{22}	V	Cr	Mn	Fe	Co	Ni	Cu	Zn_{30}
mL	15.02	10.60	8.35	7.23	7.39	7.10	6.70	6.60	7.10	9.20
g/cm^3	3.01	4.51	6.10	7.19	7.43	7.86	8.90	8.90	8.96	7.14
	Y_{39}	Zr_{40}	Nb	Mo	Tc	Ru	Rh	Pd	Ag	“cC
mL	19.8	14.1	10.8	9.4	–	8.3	8.3	8.9	10.4	13.1
g/cm^3	4.47	6.49	8.40	10.21	11.50	12.20	12.40	12.00	10.51	8.65
	Y_{71}	Hf_{72}	Ta	W	Re	Os	Ir	Pt	Au	Hg_{80}
mL	20.50	13.60	10.91	9.53	8.85	8.43	8.54	9.10	10.22	14.80
g/cm^3	6.17	13.10	16.60	19.30	21.00	22.60	22.50	22.40	19.30	13.60

The value of atomic densities are double for third transition series, from Hf_{72} to Hg_{80} compared to the elements Zr_{40} to Cd_{48} of 2nd transition series. Whereas, the value for La in not double of that of Y_{39} and the reason for this is explained as:

(*i*) Atomic weights of elements of third transition series are double to that of elements belonging to second transition series.

(*ii*) Lanthanide contraction occurs in the elements of third transition series after La_{57} from Hf_{72} to Hg_{80}, which decreases the atomic size futher. This makes the packing of atoms in the metallic crystals high and thus consequent increase in the densities.

1.2.9 Melting and Boiling Points

The melting and boiling points of transition elements are high with respect to those of s and p-block elements. This is because of strong bonding strength between their atoms. The values of melting and boiling points are given in the Table.

3d-series elements:	Sc	Ti	V	Cr	Mn	Fe	Co	Ni	Cu	Zn
M.P. (°C)	1539	1668	1900	1875	1245	1536	1495	1453	1083	419-5
B.P. (°C)	2730	3260	3450	2665	2150	3000	2900	2730	2595	906.00
4d-series elements:	Y	Zr	Nb	Mo	Tc	Ru	Rh	Pd	Ag	Cd
M.P. (°C)	1509	1852	2415	2610	2200	2500	1966	1552	960.8	320.9
B.P (°C)	2927	1500	3300	5560	–	4900	4500	3980	2210	765
5d-series elements:	La	Hf	Ta	W	Re	Os	Ir	Pt	Au	Hg
M.P <°C)	920	2222	2996	3410	3180	2700	2454	1769	1063	–38.4
B.P. (°C)	3470	5400	5425	5930	5900	5500	5300	4530	2920	357

In the first transition series, when we move from Sc to Cr, the value of melting point increases whereas, V and Cr have almost equal melting point.

Similarly, melting point increases when we move from Y to Mo and La to W in second and third series, respectively and then decrease.

The variation depends upon the strength of metallic bonding. Those elements in which unpaired d-electrons are present give additional covalent bonding which strengthens the metal-metal bond and this indicates high melting points. From left to right in the transition series, the number of unpaired d-electrons rise to five, therefore, the value of melting point tends to become very high. But when the electron pairing begins in case of Zn, Cd and Hg (all paired electrons), the melting points decline so much that Hg exists in liquid state at room temperature. In transition elements, where no order is observed, the cause lies as no order exist in their crystalline structure.

1.2.10 Variable Oxidation States

This is an important property which distinguishes transition elements from *s* and p-block elements. The *s* and *p*-block elements have oxidation state either equal to the group number or 8-group number, but the transition elements exhibit varying oxidation states (more than one oxidation state or several oxidation states).

Most of the transition elements show a variety of oxidation states except a few. This is attributed to the following explanations:

(*i*) There is very little difference in energies of $(n-1)d$ and *ns* orbitals, therefore, electrons from $(n-1)d$ orbitals are easily removed.

(*ii*) The *d*-orbitals are degenerate (same energies) and their electrons can be removed stepwise.

(*iii*) After removing *ns* electrons, the remaining core is unstable and hence, one or more electrons can be removed giving varying oxidation states and thus making the core compact and stable.

Various oxidation states shown by the *3d,* *4d* and *5d*-transition series exhibit a particular pattern. These elements show pyramid type of valencies. The maximum number of oxidation states are exhibited by middle element with the decreasing oxidation states before and after the middle element in the d-series of elements.

Table: Common oxidation states of d-block elements.

First transition series

Sc	Ti	V	Cr	Mn	Fe	Co	Ni	Cu	2n
								+ 1	
(+ 2)	+ 2	+ 2	+ 2	+ 2	+ 2	+ 2	+ 2	+ 2	+ 2
+ 3	+ 3	+ 3	+ 3	+ 3	+ 3	+ 3	+ 3		
	+ 4	+ 4	(+ 4)	+ 4	(+ 4)	(+ 4)			
		+ 5	+ 6	(+ 6)	(+ 6)				
				+ 7					

Second transition series

Y	Zr	Nb	Mo	Tc	Ru	Rh	Pd	Ag	Cd
								+1	
					+ 2		+ 2	+ 2	+ 2
+ 3		+ 3	+ 3		+ 3	+ 3	(+ 3)	(+ 3)	
	+ 4		+ 4	+ 4	+ 4	+ 4	+ 4		
		+ 5	+ 5		+ 5				
			+ 6	(+ 6)	+ 6	+ 6	+ 6		
				(+ 7)	(+ 7)				
					(+ 8)				

Third transition series

La	Hf	Ta	W	Re	Os	Ir	Pt	Au	Hg
								+ 1	+ 1
			(+ 2)		(+ 2)	(+ 2)	+ 2		+ 2
+ 3			(+ 3)	(+ 3)	(+ 3)	+ 3	(+ 3)	+ 3	
	+ 4	(+ 4)	+ 4	+ 4	+ 4	+ 4	+ 4		
		+ 5	+ 5	+ 5	+ 5				
			+ 6	+ 6	+ 6	+ 6			
				+ 7	+ 8				

Observations

(*i*) Minimum Oxidation state of +1 and +2 arises out of loss of S– electrons.

(*ii*) Maximum oxidation state is equal to the group number (from group IIIB to group VIIB). In 3*d*-transition series, Cr in group VIB shows maximum oxidation state of +6. Mn shows oxidation state up +7. Similarly for Tc and Ru in 4*d*-series and Re and Os in 5*d*-series. The maximum oxidation state is due to ready Availability of *s* and *d* electrons.

(*iii*) **Stability of Oxidation States:** The 3*d*-elements show stability in these oxidations states where the electron configuration is $3d^0$, $3d^5$ and $3d^{10}$. Mn^{2+} is $3d^5$, therefore, it is more stable than $Mn^{4+}(3d^4)$; $Ti^{4+}(3d^0)$ is more stable than $Ti^{3+}(3d^1)$. Such a rationalisation is not possible for 4*d*-and 5*d*-series of elements.

In 4*d* and 5*d* series of elements, higher oxidation states are more stable while in 3*d*-series, lower oxidation states are stable. The highest oxidation state in 3*d*-series generally exists as oxidation number (hypothetical charge) in the form of oxoanions or oxocations *e.g.*, $Cr^{VI}O_4^{2-}$, $Mn^{VII}O_4^-$. They are all strong oxidising agents. Their existence depends upon the stability of their structure (tetrahedral). In 4*d* and 5*d*-series of elements, compounds that exist are of the type $Ru^{VIII}O_4$, $Os^{VIII}O_4$, $W^{VI}Cl_6$, $Pt^{VI}F_6$ *etc.* The compounds with oxidation states of +2 and +3 are important in 3*d*-series for **aquation** and complex ions. Down the group, the stability of higher oxidation state increases. Lower oxidation state compounds are ionic, while the higher oxidation compounds are more covalent.

1.2.11 Ionisation Energies

Ionisation energy is the energy required to pull off an electron or electrons from the outermost shell of a neutral gaseous atom in its ground state of configuration. In case of removal of one electron from the outer most shell it is called first ionisation energy. If two electrons are removed, it is second and when three electrons are removed, is third ionisation energy and so on. Across the transition series from left to right, with the increase in the nuclear charge, the ionisation energies increase gradually.

The gradual increase in the ionisation energy along each transition series depends upon the increase in nuclear charge and opposing screening effect of the inner *d*-electrons.

On proceeding from top to bottom, ionisation energy values decrease as projected in table, e.g., $Sc_{21} > Y_{39} > La_{57}$. However, ionisation energies after La_{57} in third transition series show an increase in value compared to first and second transition series.

Groups →	**IIIB (3)**		**IVB (4)**	**VB (5)**	**VIB (6)**	**VIIB (7)**	**(8)**	**VIII (9)**	**→ (10)**	**IB (11)**	**MB (12)**
1st or 3*d* series (4th period) →	SC_{21} 631		Ti_{22} 658	V_{23} 650	Cr_{24} 652.8	Mn_{25} 717.4	Re_{26} 759.4	CO_{27} 758	Ni_{26} 736.6	Cu_{29} 745.5	Zn_{30} 906.4
2nd or 4*d* series (5th period) →	Y_{39} 616		Zr_{40} 660	Nb_{41} 664	Mo_{42} 685	Tc_{43} 702	Ru_{44} 711	Rb_{45} 720	Rb_{46} 805	Ag_{47} 731	Cd_{48} 867.7
3rd or 5*d* series (6th period)	La_{57} 538.1	(Ce_{68}–Lu_{71})	Hf_{72} 654	Ta_{73} 761	W_{74} 770	Re_{75} 760	Os_{76} 840	Ir_{77} 880	Pt_{78} 870	Au_{79} 890.1	Hg_{80} 1007

After La_{57} (Ce_{58} to Lu_{71}) 4*f* orbitals get filled with electrons. The nucleus is ineffectively shielded by 4*f*-electrons. Therefore, elements after La_{57} have greater effective nuclear charge and hence, the ionisation energies are higher than those of elements in the first and second transition series. For comparison of ionisation energy values, invariably the trend in first ionisation energies is considered. The values of first ionisation energy of transition elements lie between *s*-and *p*-block elements. These transition elements are less electropositive than *s*-block elements.

Transition elements do form ionic compounds, but not as readily as alkali and alkaline earth metals. The higher valent compounds of transition elements are covalent.

1.2.12 Reducing Action

The E° (electrode potential) value of *d*-block elements are lower than *s*-block elements, therefore, transition elements are weaker reducing agents.

The reducing potential of the metal depends upon the ease with which it gets oxidised:

$$M \longrightarrow M^{n+} + ne^-$$

The standard electrode potential at equilibrium M^{2-}/M is determined:

$$M_{(S)} \longrightarrow M^{2+}_{(aq)} + 2e$$

$$2H^+ + 2e \longrightarrow H_{2(g)}$$

$$M_{(S)} + 2H^+_{(aq)} \longrightarrow M^{2+}_{(aq)} + H_2$$

The enthalpy change ΔH in the above reaction is determined from Born-Haber cycle for a metal.

The E° value of a metal depends upon type of chemical process, such as sublimation energy, ionisation energy and hydration energy. The enthalpy change ΔH for first ionisation energy can be obtained from the Born-HAber cycle.

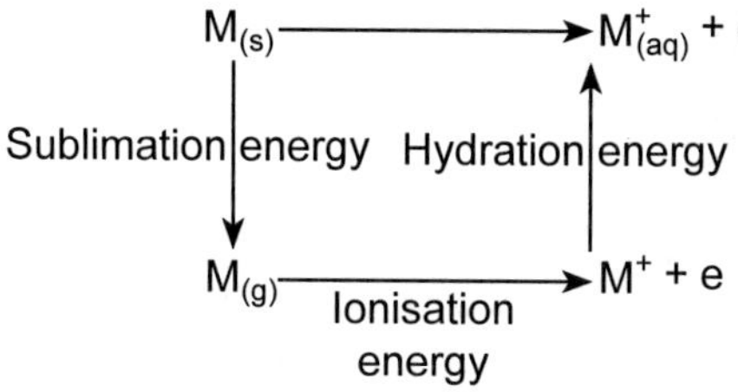

In the above diagram, total ΔH for the process is given by summation of three steps:

$$\Delta H = \Delta H_{sub} + IE + \Delta H_{hyd}$$

(*i*) Sublimation energy is obtained when solid metal changes to gaseous state. In solid metal there are metal-metal bonds. If bonds are stronger then more energy will be required to separate them, therefore. ΔH is positive (endothermic).

(*ii*) Regarding ionisation energy, energy is required to remove an electron from the outermost shell of a gaseous atom. Enthalpy ΔH is positive (endothermic). It depends upon the hold of nucleus on the outermost electron. Smaller the size greater the ionisation energy.

(*iii*) When gaseous ion is surrounded by the water molecules, energy is released. Hence ΔH is –ve. Smaller the ion greater will be hydration energy.

Thus, ΔH is total enthalpy change to bring solid metal to the monovalent ion in aqueous solution.

Similarly, Bom-Haber cycle can be involved to find E° value for the other oxidation states of the metal.

Table 1:

		Ti	V	Cr	Mn	Fe	Co	Ni	Cu
E°, M^{2+}/M	–1.63	–1.20	–0.91	–1.18	–0.44	–0.28	–0.258	+0.34	

For alkali and alkaline earth metals, the E° values at equilibrium are given below:

E°, M^{2+}/M Li	Na	K	Rb	Cs	
	–3.04	–2.71	–2.92	–2.99	–2.02
	Be	Mg	Ca	Sr	Ba
	–1.85	–2.34	–2.84	–2.89	–2.90

The E° value suggest that transition metals are weaker reducing agents than *s*-block elements. In general, transition elements have (*i*) higher sublimation energy, (*ii*) higher ionisation energy and (*iii*) high hydration energy, but the overall E° value remains negative in most elements. In Cu, Ag, Au, Pt, the E° values are positive and these elements are noble. They are unable to displace H_2 from the acids. The negative E° values decrease along each transition series which means their reducing power goes on decreasing.

1.2.13 Colour of Transition Metal Ions

The aqueous solution of most of the transition metal ions are coloured. Even the oxoanions and oxocations are coloured *e.g.*, CrO_4^{2-} is yellow, $Cr_2O_7^{2-}$ is orange, MnO_4^{2-} is purple, CrO_2^{2+} is red. It is well-known that the transition metal ions form aquacomplexes with water molecules. These metal ions form mostly 4-coordinate or 6-coordinate complexes with geometrical structures of tetrahedral, square planar and octahedral geometries. The colour of the solution of the metal ions is listed below:

$[Ti(H_2O)_6]^{3+} \xrightarrow{d^1}$ Purple

$[V(H_2O)_6]^{4+} \xrightarrow{d^2}$ Blue

$[Cr(H_2O)_6]^{3+} \xrightarrow{d^3}$ Dark green

$[Mn(H_2O)_6]^{2+} \xrightarrow{d^5}$ Faintly pink

$[Fe(H_2O)_6]^{2+} \xrightarrow{d^6}$ Faintly green

$[Co(H_2O)_6]^{3+} \xrightarrow{d^7}$ Pink

$[Co(H_2O)_6]^{2+} \xrightarrow{d^6}$ Blue

$[Ni(H_2O)_6]^{2} \xrightarrow{d^8}$ Green

$[Cu(H_2O)_4]^{2+} \xrightarrow{d^9}$ Blue

Some of the transition metal ions give colourless solution in water, such as Sc^{3+}, Ti^{4+}, Cu^+, Ag^+, Zn^{2+}, Cd^{2+} and Hg^{2+}. There seems to be some correlation between the electronic configuration of transition metal ions in the valency shell.

The electronic configuration of those metal ions which give coloured solution in water, generally have unpaired electrons in the degenerate *d*-orbitals while those metal ions which give colourless solutions have either d^0 and d^{10} configuration *i.e.,* either no electron in d-orbital or *d*-orbital is fully filled. In both cases, *d*-orbital is not participating. Therefore, it is confirmed that for getting coloured solutions one must look for the unpaired electrons in degenerate d-orbitals.

From the above feedback it is concluded that crystal field theory (CFT) has a big role to play for the coloured solutions of the transition metal ions.

According to CFT, the magnitude of splitting (Δ_0) decides the wave length to be absorbed from the visible light. Thus, a decrease in the wave length of colour of the complexes change colour from red to violet (V1BGYOR). It has been observed that increase in the number of d-electrons decreases the magnitude of Δ_0 and therefore, the wave length λ of absorbed light increases.

$$\text{Increase in the number of } d\text{-electrons} \propto \frac{1}{\Delta_0} \propto \lambda$$

The increase in the wave length λ changes the colour of the complex from violet to red.

Secondly, the strength of the ligands affects the colour of the complex ions. In the octahedral complexes of $[Co(H_2O)_6]^{3+}$, $[Co(NH_3)_6]^{3+}$ and $[Co(CN)_6]^{3-}$, the strength of the ligands is in order of $H_2O < NH_3 < CN^-$. Due to strong interaction between metal ion and ligands, the amount of splitting Δ_0 between t_{2g} and e_g will be in the same order $\Delta_0(H_2O) < \Delta_0\ (NH_3) < \Delta_0\ (CN^-)$, the wave lengths λ absorbed are in the opposite order since $\Delta_0 \propto \frac{1}{\lambda}$. Thus, $\Delta(H_2O) > \lambda(NH_3) > \lambda(CN^-)$. Accordingly, the light absorbed will be from red to violet and the colour emitted will be blue-green to yellow-green (VIBGYOR). The colour of transmitted light is complementary colour to the absorbed light. The primary colours are red, blue and green. Out of these, light of one colour is absorbed and the complementary colours emitted.

Δ_0:	$[Co(H_2O)_6]^{3+}$	<	$[Co(NH_3)_6]^{3+}$	<	$[Co(CNT)_6]^{3-}$
Absorbed colour	Orange		Green-blue		Violet
Emitted colour	Green-Blue		Orange		Yellow-green

Thirdly, geometry of the complex plays a vital role in the value of Δ_0. The four coordinate complex acquiring tetrahedral and square planar geometry of metal ion Ni^{2+} exhibit different colours. The tetrahedral complexes are blue or green while, square ions that exhibit both 4-coordinate and six coordinate complexes the colour variations are given below:

Complex Ion	**Geometry**	**Colour**
$[Co(H_2O)_6]^{2+}$	Octahedral	Pink
$[CoCl_4]^{2-}$	Tetrahedral	Ink blue
$[Mn(H_2O)_6]^{2+}$	Octahedral	Pink
$[MnCl_4]^{2-}$	Tetrahedral	Green

For Zn, Cd and Hg, the degenerate *d*-orbitals are completely filled with electrons therefore, the splitting of *d*-orbitals or Δ_0 is out of question and there is *no d-d* transition possible. Thus, their complexes are white or colourless.

Exceptions

Generally, Cu_2^{2+} ions and Hg^{2+} acquire d^{10} configuration and these metal ions should give no colour. But Cu_2O is a red compound and HgI_2 is scarlet red. the colour displayed by these compounds is not because of *d-d* transitions. The colour of Cu_2O and HgI_2 may be due to the anions. The colour theory discussed in the complex ions is based on Δ_0 value or *d-d* transition, but in reality colour exhibited by the compound may be due to role of anions, crystal shape, cutting of the crystals, charge transfer spectra *etc.*

1.2.14 Magnetic Properties

The transition metal ions or complexes show different behaviour under the influence of magnetic field. Majority of the compounds of transition metal ions are paramagnetic. Paramagnetism depends upon the unpaired electrons | in *d*-orbitals. But some compounds do show diamagnetism, ferromagnetism, anti ferro-magnetism and ferrimagnetism. The change in the configuration of central metal atom or ion under the influence of applied magnetic field yields the difference in their magnetic properties.

The paramagnetism of a substance is expressed in terms of the magnetic moment μ. Larger the number of unpaired electrons in the complex, greater is its paramagnetism. The magnetic moment μ is measured in the units of Bohr-magneton (BM). The magnetic moment μ is equal to orbital motion (μ_l) and spin motion (μ_s) of an unpaired electron.

$$\mu = \mu_l + \mu_s$$

$$\mu_l = \sqrt{l(l+1)}\,\frac{eh}{4\pi me}$$

$$\mu_s = \sqrt[2]{s(s+1)}\,\frac{eh}{4\pi me}$$

where '*e*' is electronic charge, '*h*' is Planck's constant, '*m*' is mass of an electron and '*c*' is the velocity of light.

$$\mu = \left[\sqrt{l(l+1)} + 2\sqrt{s(s+1)}\right]\frac{eh}{4\pi me}$$

$$1\text{ BM} = \frac{eh}{4\pi me}$$

$$\mu = \left[\sqrt{l(l+1)} + 2\sqrt{s(s+1)}\right]$$

Due to crystal field, number of ligands surrounding the metal ions check the movement of electron cloud round the nucleus therefore, orbital motion is quenched or:

$$\sqrt{l(l+1)} = 0$$

Therefore, the formula of μ is reduced to :

$$\mu = 2\sqrt{s(s+1)}$$

for one electron spin is $= \frac{1}{2}$

$$\mu = 2\sqrt{\frac{1}{2}\left(\frac{1}{2}+1\right)}$$

for 2, 3, 4 ... *n* unpaired electrons

$$\mu_{\text{eff}} = 2\sqrt{n/2(n/2+1)}$$
$$= \sqrt{n(n+2)}$$

for $$\mu_{\text{eff}} = \mu_s = \sqrt{n(n+2)} \text{ BM}$$

In any paramagnetic substance, despite the unpaired δ-electrons the electrons in the inner orbitals arc paired *i.e.,* there is some amount of diamagnetism in all the paramagnetic substances.

		Bohr Magneton (BM)
If	$n = 1$	1.73
	$n = 2$	2.83
	$n = 3$	3.87
	$n = 4$	4.90
	$n = 5$	5.92

These calculated values are in close agreement with the experimental values as given in the Table.

Ion	Valence-shell configuration	*n*	Complex compounds	Shape and hybridisation	Calculated value of $\mu(\mu_{cal})$	Experimental value of
Ti^{3+}	$3d^1$	1	$K_3[Ti^{3+}F_6]$	Octahedral (d^2sp^3)	1.73	1.70
V^{4+}	$3d^1$	1	$[V^{+4}O(acac)_2]$	Square pyra-midal (d^2sp^2)	1.73	1.70
V^{3+}	$3d^2$	2	$NH_4[V^{3+}(SO_4)_2]$ $12H_2O$	Octahedral (d^2sp^3)	2.84	2.80
Cr^{3+}	$3d^3$	3	$[Cr^{3+}(NH_3)_6]Br_3$	Octahedral (d^2sp^3)	3.87	3.77
Cr^{2+}	$3d^4$	4	$[Cr^{2+}(H_2O)_6]SO_4$	Octahedral (sp^3d^2)	4.90	4.80
Fe^{3+}	$3d^5$	5	$[Na_3Fe^{3+}F_6]$	Octahedral (sp^3d^2)	5.92	5.85
Fe^{2+}	$3d^6$	4	$(Et_4N)_2\ [Fe^{2+}Cl_4]$	Tetrahedral (sp^3)	4.90	5.40
Co^{2+}	$3d^7$	3	$Cs_2[Co^{2+}F_4]$	Tetrahedral (sp^3)	3.87	4.60
Ni^{2+}	$3d^8$	2	$[Ni^{2+}(NH_3)_6]Cl_2$	Octahedral (sp^3d^2)	2.84	3.32
Cu^{2+}	$3d^9$	1	$[Cu^{2+}(NH_3)_2^0]Cl_2$	Square planar (dsp^2)	1.73	1.79

Zn^{2+}	$3d^{10}$	0	$[Zn^{2+}(NH_3)_4]^{2+}$	Tetrahedral (d^2sp^3)	0.00	0.00
Fe^{3+}	$3d^5$	5 G)	$K_3[Fe^{3+}(CN)_6]$	Octahedral (d^2sp^3)	5.92	1.73
Co^{3+}	$3d^6$	4 (0)	$[Co^{3+}(NH_3)_6]Cl_3$	Octahedral (d^2sp^3)	4.90	0.00
Co^{2+}	$3d^7$	3	$K_2Ba[Co^{2+}(NO_2)_6]$	Octahedral (d^2sp^3)	3.87	1.88
		presumably in 5*s* orbital				
Ni^{2+}	$3d^8$	2 (0)	$[Ni^{2+}(NH_3)_2^0]Cl_2$	Square planar (dsp^2)	2.84	0.00

Calculated and experimental values of effective magnetic moments of some complexes of 3*d*-series elements. Greater the number of unpaired electrons, greater the μ_{eff} value. The complex ions having finite value of μ_{eff}, are called paramagnetic. There are exceptions in the form of $K_3[Fe(CN)_6]$, $[Co(NH_3)_6]Cl$, $K_2Ba[Co(NO_2)_6]$ where the calculated and observed value shows disagreement. This discrepancy is due to the fact that μ values depend upon the experimental values and not on the valence shell configuration of the central metal cation.

1.2.15 Diamagnetism

The complex ion in which central metal atom or ion contains all paired electrons in 3*d*-orbitals have μ equal to zero. In this, the two electrons present in an orbital have opposite spins, the magnetic moment produced by one electron is cancelled out by the magnetic moment produced by the other electron having opposite spin, so that the net magnetic moment μ_{eff} is zero. Those complexes in which the central atom or ion has zero electron in *d*-orbitals *e.g.,* Ti^{4+}, Sc^{3+} are also called diamagnetic. In Zn^{2+} , Cd^{2+} and Hg^{2+}, all the electrons in *d*-orbitals are fully filled and have no unpaired electron ($3d^{10}$, $4d^{10}$ and $5d^{10}$) and are categorised as diamagnetic.

1.2.16 Ferromagnetism

Those complex ions in which central metal atom or ion has maximum number of unpaired electrons in *d*-orbitals show strong paramagnetism which is termed as ferromagnetism. By ferromagnetism we understand the iron-like attraction towards the magnetic field.

1.2.17 Antiferromagnetism

It is the phenomenon in which the complex compound that has unpaired electrons has its unpaired electrons get paired under the influence of the magnetic field. It shows some paramagnetism or zero value for the magnetic moment.

Ferrimagnetism indicates ferric like configuration in the *d*-orbitals of a complex ion *i.e.,* Fe^{3+} has five unpaired *d*-electron. Under the influence of the magnetic field these unpaired electrons in *d* get paired, but even after pairing one unpaired electron is left.

Fe^3 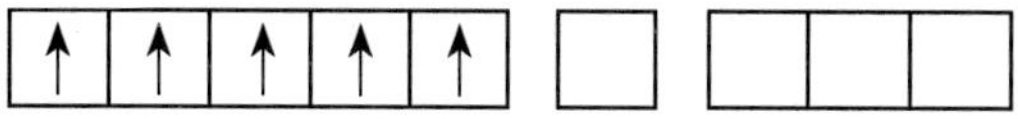

After pairing

Fe^3 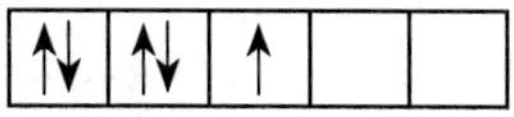

Therefore, ferrimagnetic substances show reduced paramagnetic behaviour.

The experimental magnetic moment value can be used to determine the number of unpaired electrons in *d*-orbitals of metal ion in a complex. The number of unpaired electrons can predict:

(*i*) Geometry of the 4-coordinate complex. It can be tetrahedral sp^3 or square planar dsp^2.

(*ii*) In a 6-coordinate complex, it may be d^2sp^3 (inner orbital octahedral complex) or sp^3d^2 (outer orbital octahedral complex).

For Example: $[Ni(CN)_4]$ has $\mu = 0$ *i.e.,* it is diamagnetic and dsp^2 square planar, $[NiCl_4]^{2-}$ has two unpaired electrons, is paramagnetic sp^3 tetrahedral. Similarly, for 6-coordinate octahedral *e.g.,* $[Fe(CN)_6]^{4-}$ has μ equal to 1.75 BM (one unpaired electron) paramagnetic, d^2sp^3 hybridisation and an inner orbital complex. For $[FeF_6]^{3-}$, μ value is 5.8 BM and it is paramagnetic, outer orbital complex having sp^3d^2 hybridisation.

1.2.18 Complex Formation Tendency

Complex formation tendency is quite dominant in *d*-block elements than in s- and *p*-block elements, although complexes are well known for these elements as well. There is not a single transition element which does not form a complex. The *d*-block elements form complexes with molecules such as, H_2O, CO, NH_3 *etc.* and ions like F^-, Cl^-, CN^-, SCN^-, NO_2^- *etc.* These molecules and ions are called ligands and they can donate a pair of electrons to the transition metal atom or ion.

This tendency of transition elements is based on the properties of transition elements namely:

(*i*) Small cation size

(*ii*) High nuclear charge (effective)

(*iii*) Large number of vacant *d*-orbitals and other *s* and *p* orbitals available for bonding to the ligands.

The vacant orbitals of the metal atoms or ions form a coordinate bond with the ligands ($M \leftarrow L$). Besides, the small size and high positive charge on the metal atoms or ions attract the ligand electrons and enhances the complex formation tendency.

1.2.19 Stability of Complex Compounds

(*i*) It decreases with increase in atomic number of the central metal atom.

(*ii*) Transition metals must be in higher oxidation state to from stable complexes with small highly charged ligands *e.g.,* F^-, Cl^-, NH_3, IO_6^{5-} *etc.* Transition elements with lower oxidation state give stable complexes with π-bonding ligands such as, CO, NO, CN^-, $PC1_3$ *etc.* In lower oxidation states, metal and ligands mentioned above form two **way** bonding $M \rightleftharpoons L$.

(*iii*) In case of transition metal in different oxidation states forming complexes with the same ligand, the complex in which metal ion is of higher oxidation state, is more stable *e.g.,* $[Co(NH_3)_6]^{3+}$ is more stable than $[Co(NH_3)_6]^{2+}$. The metal ion with higher charge interacts strongly with the ligands to give a stable complex.

1.2.20 Catalytic Property

Whenever a catalyst is required for a chemical reaction, it is invariably from the transition elements. This is attributed to the two properties of transition elements namely (*i*) variable oxidation states, (*ii*) large surface area with vacant spaces.

(*i*) Variable Oxidation States: The transition metals have vacant orbitals and variable oxidation states. This property helps in the formation of unstable intermediates. The intermediate compounds are of lower activation energy which increase the rate of reaction. The unstable intermediate compound decomposes to give the final product and the starting substance.

(*ii*) Transition metals provide large area with vacant spaces which help in the absorption of reactants. This brings the reactants closer to one another for the reaction to take place (collision theory).

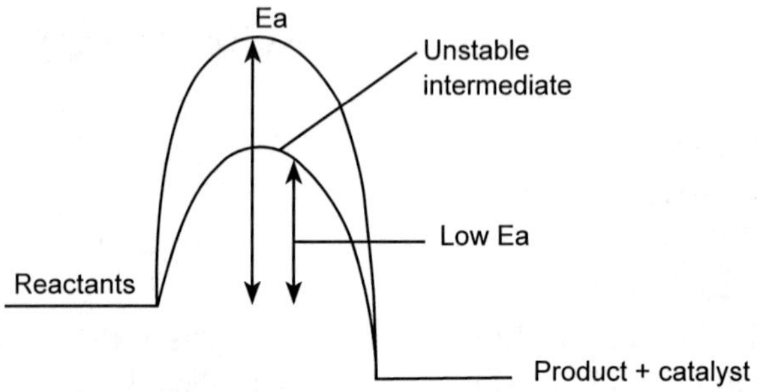

Examples of Catalytic Reactions

(*a*) $2SO_2 + O_2 \xrightleftharpoons{V_2O_5} 2SO_3$

(*b*) $H_2C{=}CH_2{+}H_2 \xrightleftharpoons[410\ K]{Ni} H_3C{-}CH_3$ (Hydrogenation of oil)

(*c*) $2NO + 5H_2 \xrightleftharpoons[Ni]{heated} 2NH_3 + 2H_2O$

(*d*) $2CH_3OH + O_2 \xrightleftharpoons[Black]{Pt} 2HCHO + 2H_2O$

(*e*) Ostwald process: $4NH, + 5O_2 \xrightleftharpoons[800°C]{Pt\ gauge} 4NO + 6H_2O$

(*g*) $N_2 + 3H_2 \xrightleftharpoons[Mo\ 770\ K]{Fe} 2NH_3$

(*h*) $Fe_2O_3 + 3H_2S \xrightleftharpoons[Mo\ 770\ K]{NiO} 2FeS + 3H_2O + S$

(*i*) **Zeigler Natta Catalyst**

$n\,C_2H_4 \xrightleftharpoons{TiCl_4/Al(C_2H_5)_3} (CH_2{-}CH_2)_n$ etc.

1.2.21 Interstitial Compound

Transition metals have a property of Non-metal forming compounds with non-metal atoms atom N like C, H, B and N. The hard transition metals have a closest packed structure showing strong metal-metal bonds. Due to this, a number of small vacant sites are created. In these, atoms of appropriate size get locked to form compounds called interstitial compounds.

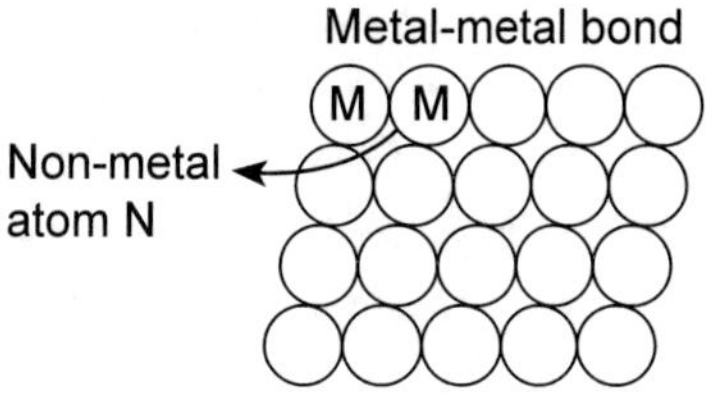

These compounds are non-stoichiometric (bar indicates non-stoichiometry) *e.g.*, $Fe\bar{}O$

1 : 0.84

i.e. the ratio between Fe : O is not 1 : 1, it is 1 : 0.84. The interstitial compounds of metal nitrides, metal oxides, metal carbides, metal hydrides are well known. In actual, steel is an interstitial compound of Fe and carbon. Carbon atom enters the vacant sites.

The non-stoichiometry can be explained by:

(*i*) All the vacant sites are not always occupied by the non-metal atoms, some remain vacant *e.g.*, out of 8 vacant sites only 7 are occupied and hence the fraction.

M : C

1 : 0.41 to 0.59

Carbon occupies octahedral holes.

(*ii*) In case of 4 metal-metal bonds and 4 metal atoms form one bond with the non-metal atom.

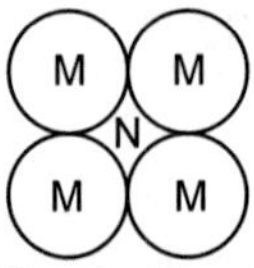

Covalent bond

Therefore, the bond per metal and non-metal atom is ¼ therefore, the non-stoichiometry. The examples of MC, MC, MO and MH are well-known.

1.2.22 Alloy Formation

Atomic size of different transition metals are similar therefore, atoms of one transition element can take up the position meant for the crystal lattice of another. Transition elements are known to form large number of alloys. On heating, transition metals get melted and become miscible. On cooling, the metal solutions form an alloy *e.g.,* Cr dissolved in nickel to form Cr–Ni alloy. Manganese dissolves in iron to form manganese steel. These alloys are harder, have high melting points and are resistant to corrosion unlike the parent metals.

1.2.23 Reactivity

The reactivity of transition elements decreases from 3*d*-series to 5*d*-series. This suggests that elements of 3*d*-series are much more reactive than elements of 4*d*-and 5*d*-series. Moving from left to right in the same transition series shows decrease in level of chemical reactivity *i.e.,* in Cu, Ag, Au and Pt, the behaviour becomes more noble (elements towards the right of transition series).

1.3 GROUP STUDY OF PHYSICAL AND CHEMICAL CHARACTERISTICS OF TRANSITION ELEMENTS

1.3.1 Scandium Group

Transition elements have been put in ten vertical columns or groups. The groups are III_B, IV_B, V_B, VI_B, VII_B, VIII (comprising of three vertical columns) and I_B and II_B. These elements are placed between alkaline earth metals and *p*-block elements in the long form table.

	← *d*-Block Elements (Transition Elements) →										
	III B (3)	IV B (4)	VB (5)	VI B (6)	VI B (7)	← VIII → (8)	(9)	(10)	I B (11)	II B (12)	
	Sc_{21}	Ti_{22}	V_{23}	Cr_{24}	Mn_{25}	Fe_{26}	Co_{27}	Ni_{28}	Cu_{29}	Zn_{30}	***p*-Block Elements**
s-Block Elements	Y_{39}	Zr_{40}	Nb_{41}	Mo_{42}	Tc_{43}	Ru_{44}	Rh_{45}	Pd_{46}	Ag_{47}	Cd_{48}	
	La_{57}	Hf_{72}	Ta_{73}	W_{74}	Re_{75}	Os_{76}	$^{Ir}77$	P^78	Au_{79}	Hg_{80}	
	Ac_{89}	Ku_{104}	Ha_{26}								

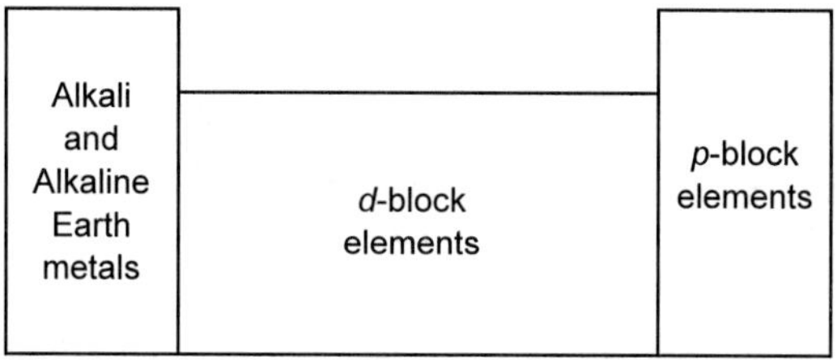

In the long form table, they start from group 3 to 12. Group III includes elements like Scandium, Yttrium, Lanthanum and Actinium and are called as scandium group elements (four elements). The trends in their properties arc quite regular and similar to trends in group I and II elements. They cater to important industrial uses. These elements were earlier called rare earths, but it is a misnomer as many of them are quite abundant on the earth's surface. Due to similarity in their properties it is difficult to separate the individual elements. They are found together with lanthanides in the form of $M^{III}CO_3F$ (bastnaesite) and monazite $M^{III}PO_4$. The extraction of these metals from their compounds is difficult. They are electropositive and react with water.

1.4 PHYSICAL CHARACTERISTICS

1.4.1 Occurrence

These elements occur together with lanthanides. They are found mixed with monazite $M^{III}PO_4$ and other minerals. Their oxides are very stable therefore, they cannot be separated by thermite reaction. So, the metals are obtained by reduction of the chlorides and fluorides with calcium at 1000°C under inert atmosphere of argon.

$$2MCl_3 + 3Ca \underset{1000°C}{\overset{Ar}{\rightleftharpoons}} 3CaCl_2 + 3M_{(metal)}$$

Yttrium is used in red phosphor for TV tubes and to make synthetic garnets used in radar and as gemstones. Lanthanum is produced as Mischmetal (unseparated metal). It is used to improve steel and form alloys with magnesuim. La_2O_3 is used in optical glass to protect from UV light. Actinium is obtained from thorium and uranium in the form of decay product. Ac is radioactive and has half life from 6 hours to 21.8 years. It has $^{228}_{89}Ac$ and $^{227}_{89}Ac$ as radio isotopes.

1.4.2 Atomic Size

The covalent and ionic radii of these elements increase gradually down the group. The second and third elements have almost same size due to lanthanide contraction, but it is observed only after lanthanum.

1.5 CHEMICAL CHARACTERISTICS

These metals are reactive and reactivity increases with increased atomic size down the group. These metals react with water to form basic oxides (MO • OH) and liberate H_2 gas. The reaction is fast in hot water.

(*ii*) Oxides

$$4La + 3O_2 \longrightarrow 2La_2O_3$$
$$2La + 6H_2O \longrightarrow 2La(OH)_3 + 3H_2$$
$$La(OH)_3 \longrightarrow LaO.OH + H_2O$$

Similarly, the basic oxide of ScO.OH is obtained. Sc dissolves in NaOH liberating H_2 gas.

$$Sc + 3NaOH + 3H_2O \longrightarrow Na_3[Sc(OH)_6] + 3/2\ H_2$$

The basic character of hydroxides is in the following order.

$$Sc(OH)_3 < Y(OH)_3 < La(OH)_3$$

The oxides and hydroxides being basic form salt with acids.

$$2Y(OH_3) + CO_2 \longrightarrow Y_2(CO_3)_3 + 3H_2O$$

1.5.1 Electronic Configuration

The valence shell configuration of these elements are $(n{-}1)d^1ns^2$. Their most common oxidations state is +3.

Element		Electronic structure	Oxidation states
Scandium	Sc	[Ar] $3d^14s^2$	III
Yttrium	Y	[Kr] $4d^15s^2$	III
Lanthanum	La	[Xe] $5d^16s^2$	III
Actinium	Ac	[Rn] $6d^17s^2$	III

1.5.2 Oxidation States

All the four elements show +3 oxidation state in solution in which both ns and one electron from $(n-1)$d are removed. The M^{3+} salts have all paired electrons and hence arc colourless and diamagnetic.

Property	Scandium (Sc)	Yttrium (Y)	Lanthanum (La)	Actinium (Ac)
Atomic number	21	39	57	89
Atomic weight	44.96	88.92	138.92	227
Electronic configuration	$[Ar]_{18}\ 3d^14s^2$	[Kr]36 $4d^14s^2$	$[Xe]_{54}\ 4f^05d^16s^2$	$[Rn]_{69}\ 5f^06d^27s^2$
	= 2,8,9,2	= 2,8,18,9,2	= 2,8,18,9,2	= 2,8,18,32, 18,9,2
Atomic radius (pm)	162	180	1$7	
Ionic radius (pm)	81 (+ 2)	–		–
	88.5 (+ 3)	104 (+ 3)	117.2 (+ 3)	118 (+3)
Atomic volume (ml)	15.02	19.8	20.50	–
Density (g/cm3)	3.01	4.47	6.17	–
Melting point (°C)	1539	1509	920	–
Boiling point (°C)	2730	2927	3470	–

Ionisation energy (kJ/mol)	631	616	538.1	–
Oxidation states	+ 2, + 3	+ 3	+ 3	+ 3
Colour of M^{3+} ions	colourless	colourless	colourless	colourless
Heat of fusion (Kcal/mol)	3.85	4.1	2,7	–
Heat of vaporisation (Kcal/mol)	72.85	94.0	95.5	–
Reduction potential (volts $M^3 + 3e^- \rightleftharpoons M$)	–2.1	–2.37	–2.52	–
Electronegativity	1.3	1.2	1.1	1.1

$La(OH)_3$ is a strong base and can liberate NH_3 from ammonium salts. Their hydroxides are decomposed to oxides on heating (similar to Group 2 elements)

$$2Y(OH)_3 \xrightarrow{\Delta} Y_2O_3 + 3H_2O$$

$$Y_2CO_3 \longrightarrow Y_2O_3 + 3CO_2$$

$$Y_2(SO_4)_3 \longrightarrow Y_2O_3 + 3SO_2 + 3/2O_2$$

$$2Y(NO_3)_3 \longrightarrow Y_2O_3 + 6NO_2 + 3/2O_2$$

1.5.3 Halides

They react with halogens to form MX_3 trihalides which resemble Ca halides. The fluorides are insoluble like CaF_2. The chlorides are deliquescent and crystallise as hydrates.

$$ScCl_3.(H_2O) \xrightarrow{\text{Heat}} Sc_2O_3 + 6HCl + 4H_2O$$

$$YCl_3.7H_2O \xrightarrow{\Delta} YOCl + 2HCl + 6H_2O$$

$$LaCl_3.7H_2O \longrightarrow LaOCl + 2HCl + 6H_2O$$

Anhydrous $ScCl_3$ cannot be used in FCR reaction in place of anhydrous $AlCl_3$.

The elements react with hydrogen on heating at 300°C forming MH_3. The hydrides react with H_2O liberating hydrogen. The hydrides are salt like containing H^-. When heated with carbon, Sc forms ScC_2 which in turn reacts with H_2O to form ethyne.

$$Sc_2O_3 + 7C \xrightarrow{1000°C} 2ScC_2 + 3CO$$

$$ScC_2 + H_2O \longrightarrow C_2H_2 + SeO.OH$$

1.5.4 Complexes

The elements in the group do not have a strong tendency to form complexes in +3 state. Sc^{3+} ion is the smallest in the group, hence it readily forms complexes *e.g.*, $[Sc(OH)_6]^{3-}$ and $[ScF_6]^{3-}$ both are octahedral in shape.

$$ScF_3 + 3NH_4F \longrightarrow 3NH_4^+ + [ScF_6]$$

Y and La have larger atomic size, they form complex with coordination number of 7 or 8. The common complexes with complexing ligands oxalic acid, citric acid, acetylacctone and edta *e.g.* $[Y(acac)_3(H_2O)]$ 7-coordinate with monocapped trigonal prism and $\{La(ac\ ac)_3(H_2O)_2\}$ 8 coordinate with square antiprism structure, $\{La\ edta\ (H_2O)_4\}$ is 10 coordinate and $\{La\ (SO_4)_3\ 9(H_2O)\}$ is 12 coordinate.

Similarity in La^{3+} and Ca^{2+} compounds is observed because of similar size [La = 1.032Å and Ca = 1.00Å]. In biological role, La^{3+} appears to replace Ca^{2+} in nerve impulses.

■ ■ ■

2

Titanium Group

2.1 INTRODUCTION

This group consists of three elements *viz.*, titanium, zirconium and hafnium. Titanium is comercially important as:

(*i*) It gives strength, hardness and corrosion resistance when alloyed with other metals.

(*ii*) $TiCl_3$ is used as Zeigler-Natta catalyst for the polymerisation of ethylene.

Zirconium and hafnium are used in nuclear reactors.

2.2 PHYSICAL CHARACTERISTICS

Occurrence

Titanium ore is called as ilmenite and TiO_2 is most abundant on earth. Zirconium is found mainly in the form of zircon $ZrSiO_4$ and small amounts as baddeleyite ZrO_2. Hf and Zr show similar properties and **size** due to the interception of lanthanide contraction. Separation of Zr and Hf is very difficult.

2.2.1 Electronic Configuration

The valence shell electronic configuration is given as:

Titanium	(Ar) $3d^2 4s^1$
Zirconium	(Kr) $4d^2 5s^2$
Hafnium	(Xe) $4f^{14} 5d^2 6s^2$

2.2.2 Oxidation States

The common oxidation states of these three elements are listed ahead:

	Oxidation states
Ti	–1,0, **+2, +3, +4**
Zr	**+2, +3, +4**
Hf	**+2, +3, +4**

(The more important, abundant and stable oxidation states are shown in bold ink.)

2.2.3 Extraction of Metals

Ti is accepted as the modern "wonder metal" of twentieth century. It is harder, stronger, higher melting and lighter than steel. Traces of non-meta!lic impurities like H, C, N and O make Ti, Zr and Hf brittle. Ti and A1 alloy is used in supersonic aeroplanes.

Element	Covalent radius (A)	Ionic radius M^{4+} (A)	Melting point (°C)	Boiling point (°C)	Density ($g\ cm^{-3}$)	Pauling's electro negativity
Ti	1.32	0.605	1667	3285	4.50	1.5
Zr	1.45	0.72	1857	4200	6.51	1.4
Hf	1.44	0.71	2222	4450	13.28	1.3

2.2.4 Physical Constants

When alloyed with steel it makes it harder and tougher. Due to high melting point Ti metal is quite difficult to be extracted. The main reason is its high reactivity with air, 0_2, N, and hydrogen. In the reaction, the formed oxide is not reduced by carbon or CO because carbides are formed.

In the first step, TiO_2 is heated with C and Cl_2 at 900°C to form $TiCl_4$.

$$\text{(Rutile) } TiO_2 + 2C + 2Cl_2 \longrightarrow TiCl_4 + 2CO$$

$$2FeTiO_2 + 6C + 7Cl_2 \longrightarrow TiCl_4 + 6CO + 2FeCl_2$$

$TiCl_4$ is a colourless liquid which is removed by distillation. Ti is obtained from the following methods:

(*i*) **Kroll Process:** $TiCl_4$ is reduced by Mg or Ca.

$$TiCl_4 + 2Mg \longrightarrow 2MgCl_2 + Ti$$

$MgCl_2$ is removed by leaching with H_2O or dilute HCl. With Na metal > reduced in the atmosphere of Argon.

$$TiCl_4 + 4Na \xrightarrow{Ar} 4NaCl + Ti$$

NaCl is easily removed by leaching with water. Zirconium is also obtained by Kroll process.

(*ii*) Van Arkel-de Boer Process

This method is quite useful in order to obtain very pure metal and is used for both metals *viz.* Ti and Zr. With iodine, they are heated in an evacuated vessel and both the metals form their respective iodides. The iodides of these metals at 1400°C using tungsten filament get volatilise and separate metal from the impurities.

$$\underset{\text{(impure)}}{\text{Ti}} + I_2 \xrightarrow{50+250°C} TiI_4 \xrightarrow[\text{W filament}]{1400°C} Ti + 2I_2$$

Zirconuim is also extracted by the similar method on a smaller scale. Zr is more immune to corrosion than Ti. Zr and Hf have very similar properties and can be separated by solvent extraction of their compounds. An alloy made by Zr and Nb is frequently used in superconductors. Hf has an application in nuclear reactors for regulating neutrons.

2.2.5 Atomic Size

When we move down in Group IV from Ti to Hf, the covalent and ionic radii increase, also noticed an identical size in Zr and Hf due to filling of 4*f* level with 14 lanthanide elements. The electronic configuration of Hf and La are quite similar therefore, they show almost identical properties.

2.3 CHEMICAL CHARACTERISTICS

2.3.1 Reactivity and Passivity

The compressed form of metal structures renders them unreactive at low or moderate temperatures. Sometimes formation of their oxide film enables us to check further attack At room temperature, the metals do not react with acid or alkali.

It has been noticed that Ti can be dissolved in hot concentrated HCl resulting in the formation of Ti^{3+} and H_2. Also, hot HNO_3 reacts to give $TiO_2.(H_2O)_n$. Regarding Zr, it can be dissolved in concentrated H_2SO_4 and aquaregia. The HF has been found to be an outstanding solvent in order to dissolve all these three metals.

$$Ti + 6HF \longrightarrow H_2[TiF_6] + 2H_2$$

At 400°C they begin to react and above 600°C they are highly reactive. They form oxides MO_2, halides MX_4, interstitial nitrides MN and interstitial carbides MC by direct combination. Their hydrides MH_2 are not affected by air and water.

Their reduction potential with respect to acid solution is tabulated below:

Oxidation state

$$\overset{+4}{TiO^{2+}} \xrightarrow[+0.10]{} \overset{+3}{Ti} \xrightarrow[-0.37]{} \overset{+2}{Ti^{2+}} \xrightarrow[-1.67]{+1} \overset{+}{Ti}$$

2.3.2 Behaviour of Ti Group Compounds in Different Oxidation States

The most stable and common oxidation state for all the elements is +4. Anhydrous $TiCl_4$ is covalent and has tetrahedral structure in the gaseous state. TiF_4 is ionic and its crystal structure is polymeric F-bridged structure where Ti is surrounded by 6F atoms.

The +4 state compounds are white or colourless and diamagnetic, they have d° configuration with all paired electrons. The TiO^{2+} titanyl ion is formed in solution, but it polymerises $(TiO)_n^{2+}$.

Ti^{+3} ions are stronger reducing agents than Sn^{2+} ions. They exist as solids and are stable.

Ti^{3+} compounds have d^1 configuration and are paramagnetic. Their magnetic moment value is 1.73 BM. There is one *d-d* transition and one band in visible spectrum. In Ti^{3+} solution, the colour is violet.

Ti^{2+} compounds are unstable and strong reducing agents, and exist in solid state. Ti(O), (–1) and (–2) states exist in the dipyradyl complexes.

Oxides

TiO_2 shows +4 oxidation state and is used as a white pigment in paints. It is non-toxic and does not turn black when exposed to H_2S. It is used for whitening paper and nylon. It is used as a filler in plastics and rubber. Due to impurities TiO_2 becomes coloured and in order to obtain pure TiO_2, rutile is heated with chlorine and coke at 900°C. Then formed $TiCl_4$ is heated at 1200°C with oxygen.

$$TiO_2 + 2C + 2Cl_2 \longrightarrow TiCl_4 + 2CO$$

$$TiCl_4 + 2O_2 \longrightarrow \underset{\text{pure}}{TiO_2} + 2Cl_2$$

Ilmenite or titanium iron oxide ($FeTiO_3$) is reacted with concentrated H_2SO_4 to form sulphate cake as (TiO) SO_4 (Titanyl sulphate). $(TiO)SO_4$ is hydrolysed by boiling with H_2O.

The crystal structures of oxides indicate that they are ionic. In TiO_2 structure, Ti is surrounded by 6 atoms octahedrally. Oxides are insoluble in H_2O. Ti^{4+} exists as TiO^{2+} titanyl ions in solution which is polymeric.

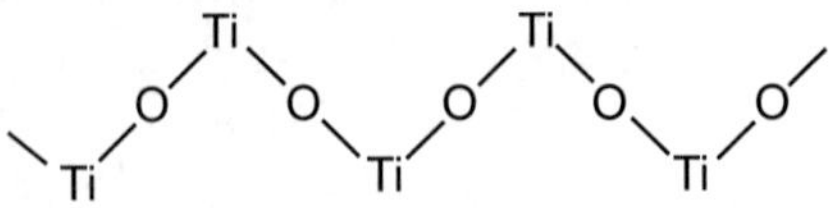

Polymeric $(TiO)_n^{2+}$

Similarly, ZrO^{2+} exists in solution and form polymeric crystals. $ZrO(NO_3)_2$ is soluble in water. Phosphate of Ti, Zr and Hf are insoluble. TiO_2, ZrO_2 and HfO_2 are white solids and non-volatile. The basic property increases with increase in atomic number. TiO_2 is amphoteric and ZrO_2 and HfO_2 are basic in nature.

$$\underset{\text{Titanyl sulphate}}{TiO.SO_4} \xleftarrow[H_2SO_4]{\text{cone.}} TiO_2.(H_2O)_n \xrightarrow[NaOH]{\text{cone.}} \underset{\text{Sodium titanate}}{NaTiO_3.(H_2O)_n}$$

Mixed oxides are formed when $Ti0_2$, $Zr0_2$ or HfO_2 are fused at 2500°C with other metal oxides to form titanates, zirconates and hafnates.

$$TiO_2 + Na_2O \longrightarrow Na_2TiO_3$$

Similar to ilmenite, perovskite ($CaTiO_3$) also occurs naturally. Perovskite structure has a cubic close-packed with O and Ca, with Ti occupying one third of octahedral sites. Ca has a coordination number as 12, also Ca and Ti are as far apart as possible.

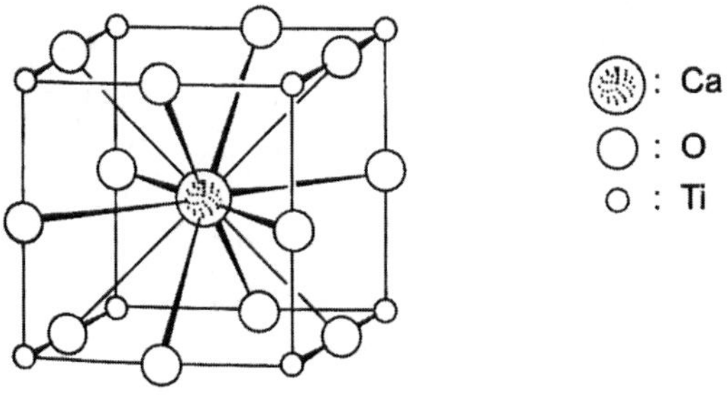

Perovskite Structure

Also, $BaTiO_3$ has a perovskite structure, Ba^{2+} ions are very big in size to fit into close-packed oxide without expanding it. This results in enhancement of the size of octahedral holes in which Ti can rattle. In an electric field, Ti atoms go on one side creating polarisation and making the crystal ferroelectric. It is used in gramophone, pick-ups microphones and ceramic capacitors.

Ti(IV) solution give yellow-orange colour on addition of H_2O_2. The colour is due to the formation of peroxo complex $[TiO_2.\ OH.(H_2O)_n]^+$ below pH 1.0.

2.3.3 Halides

Ti and Zr form all the halides like fluorides, chlorides, bromides and iodides $TiCl_4$ is prepared by passing Cl_2 over TiO_2 and C.

$$TiO_2 + 2C + 2Cl_2 \longrightarrow TiCl_4 + 2CO$$

Fluorides can be prepared by the reaction of

$$TiCl_4 + 4HF \longrightarrow TiF_4 + 4HCl$$

The iodides can be prepared by heating I_2 and Ti. TiI_4 is used in van Arkel-de Boer process for purifying the metals. The compound titanium tetrachloride ($TiCl_4$) is found to be colourless, diamagnetic, fuming liquid whereas, $ZrCl_4$ is a white solid. All halides are tetrahedral and can be hydrolysed vigorously by

H_2O with the formation of fumes in moist air giving polymeric $TiO_2.(H_2O)_n$. When hydrolysed with aqueous HCl they form oxychlorides.

$$TiCl_4 + 2H_2O \longrightarrow TiO_2(H_2O)_n + 4HCl$$

$$TiCl_4 + H_2O \longrightarrow TiOCl_2 + 2HCl$$

TiF_4 is most stable compared to other halides.

Complexes: Octahedral complexes are formed with a variety of ligands.

$$TiF_4 + 2HF \longrightarrow [TiF_6]^{2-}$$

$$TiCl_4 + 2HCl \longrightarrow [TiCl_6]^{2-} \text{ unstable}$$

Complexes of Zr and Hf have a higher coordination number *e.g.*, $Na_3[ZrF_7]$. $Na_3[HfF_7]$, $Na_4[ZrF_8]$ and $Na_4[HfF_8]$.

Ti^{3+} compounds have d^1 configuration and are more basic than Ti^{4+} compounds. Ti^{3+} used as its halide $TiCl_3$ which is violet powder and has its role in Zeigler-Natta catalyst. Ti^4+ forms purple aqua ion $[Ti(H_2O)_6]^{3+}$ when $TiCl_4$ aqueous solution is reduced by Zn which is a powerful reducing agent. $Ti^{(III)}$ Cl_3 shows different colours in aqueous solution due to different number of water molecules and chloride ions surrounding them *i.e.* $[Ti(H_2O)_6]^{3+}3Cl^-$ and $[Ti(H_2O)_5Cl]^{2+}2Cl^-$. These two situations develop different level of crystal field splitting of the *d* level. To determine Fe^{3+} through titration, Ti^{3+} is frequently used.

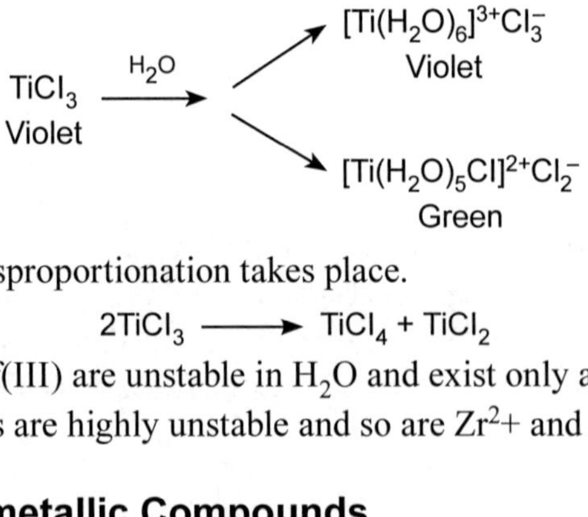

On heating disproportionation takes place.

$$2TiCl_3 \longrightarrow TiCl_4 + TiCl_2$$

Zr(III) and Hf(III) are unstable in H_2O and exist only as solid.

Tf compounds are highly unstable and so are Zr^2+ and Hf^{2+} compounds.

2.3.4 Organometallic Compounds

Cyclopentadienyl compounds of Ti, Zr and Hf are quite stable and their hepta coordinate compounds are also well-known. Several stable compounds of Ti are known such as, $[Ti(\eta\text{-}C_5H_5)_2(CO)_2]$, $[Ti(\eta^5\text{–}C_5H_5)_2(NR_2)_2]$, $[Ti(\eta^5\text{–}C_5H_5)_2(SCN)_2]$ where cyclopentadienyl molecules are pentahaptic *i.e.* 5–C atoms in each ring are attached to Ti. Reduction of these compounds yield $[Ti(C_5H_5)_2.X)]$ and $[Ti(C_5H_5)_2]$ resembling ferrocene, but the Ti compound is dimeric. Only a few Ti alkyl and aryls are known *e.g.*, CH_3TiCl_3, $C_6H_5Ti(C_3H_7O)_3$. Compounds with alkyl group attached to Ti help in the polymerisation of alkenes.

■ ■ ■

3

Vanadium Group

3.1 INTRODUCTION

Vanadium is the parent metal along with niobium and tantalum metals. Vanadium enjoys commercial importance specially by forming alloy as ferrovanadium and its significant role as a catalyst in chemical reactions in the form of V metal and V_2O_5. The role of oxoanion of vanadium called vanadates occupy importance in solution chemistry. The other two metals of vanadium group *viz.* niobium and tantalum are useful though in little requirement. Vanadium forms a number of complexes of interest in its lower oxidation states.

3.2 PHYSICAL CHARACTERISTICS

3.2.1 Occurrence and Extraction

An odd atomic numbers are observed in V, Nb and Ta, and in comparison to the elements of neighbouring group therefore, they are not much abundant. Regarding abundancy of metals, vanadium occupies 19th position by weight in the earth's crust. Vanadium can be found in some ores like vanadite $PbCl_2.3Pb_3(PO_4)_2$ with lead ore, carnotite $K_2(UO_2)_2(VO_4)_2.3H_2O$ with uranium ore and also in crude oil. When vanadates are heated with Na_2CO_3 or NaCl at 800°C it gives $NaVO_4$ which forms a red precipitate of sodium polyvanadate with H_2SO_4. Sodium polyvanadate on heating at 700°C gives black V_2O_s (black colour is due to impurities).

In order to get pure vanadium metal VCl_5 is reduced with H or Mg, or by reducing V_2O_5 by Ca. Vanadium can also be obtained by the process of electrolysis of fused halide complex. V_2O_s is used as a catalyst in the manufacture of H_2SO_4 by Contact process. In the Contact process for making sulphuric acid the catalyst Pt has been replaced by V_2O_5 as it is cheaper and also reduces the poisoning level created by impurities due to arsenic. V_2O_5 is used in the oxidation reactions for the conversion of naphthalene to phthalic acid and toluene to benzaldehyde. V_2O_5 plays its role as an essential catalyst.

The other two elements *viz.* niobium and tantalum occur together. The common ores for Nb are pyrochlorite $CaNaNb_2O_6F$ and columbite (Fe, Mu) Nb_2O_6. For tantalum, the ore is tantalite (Fe, Mn)Ta_2O_6. Nb is used in stainless steel (Nb/steel is used in nuclear reactors). Ta has its role in electronic industry for making capacitors. Ta rod is also used to repair fractured bones by using it in form of wires or screws. TaC has highest melting point of 3800°C.

3.2.2 Electronic Configuration and Oxidation States

The important oxidation states are underlined and the less important in normal type. Unstable or doubtful oxidation states are given in brackets.

	Electronic structure	Oxidation states
Vanadium	V (Ar) $3d^34s^2$	(–1), (0), (I), (II), IV V
Niobium	Nb (Kr) $4d^35s^2$	(–1) (0) (I) (II) m (IV) V
Tantalum	Ta (Xe) $4f^{14}5d^36s^2$	(–1) (0) (I) (II) III (IV) V

Behaviour of Compounds in Different Oxidation States: For all the three elements in the group, the oxidation states vary between –1 to +5. In case of vanadium, +II and +III states are reducing agents, +IV is stable and +V is slightly oxidising. Regarding Nb and Ta, +V is the most stable state and most common although, these two elements also exist in lower oxidation states. Zinc and acid can reduce vanadium +V to vanadium +II. Nb +V can be reduced to Nb^{3+} while, Ta +V cannot be reduced. This shows stability of +V state and instability of lower oxidation states.

3.2.3 Atomic Size

Atoms are smaller than the titanium group due to poor shielding by *d*-electrons. The atomic size of Nb and Ta are identical due to lanthanide contraction. Therefore, Nb and Ta occur together, show similar properties and are difficult to separate.

3.2.4 Graded General Properties

The three metals of vanadium group *viz.* V, Nb and Ta are silvery metals having high melting points. Vanadium has the highest melting point in 3*d*-series of metals. This is due to maximum *d* electrons participation in metallic bonding, Nb and Ta also show high melting points. Impurities make these metals hard and brittle, otherwise they are soft and ductile. They are non-corrosive owing to the oxide film on the surface and with HF they form complexes. These metals are not affected by air, water and acids at room temperature. V dissolves in concentrated H_2SO_4, HNO_3 and aquaregia (3 volume of concentrated HCl and one volume of concentrated HNO_3), but unaffected by alkali, whereas Nb and Ta dissolve in NaOH.

	Covalent radius (A)	Ionic radius (A)		Melting point (°C)	Boiling point (°C)	Density (g cm^{-3})	Pauling's electro-negativity
		M^{2+}	M^{3+}				
V	1.22	0.79	0.640	1915	3350	6.11	1.6
Nb	1.34	–	0.72	2468	4758	8.57	1.6
Ta	1.34	–	0.72	2980	5534	16.65	1.5

At high temperature, these three metals form non-stoichiometric interstitial compounds with non-metals C, H, N, O *etc.* Comparatively, vanadium is stable in different oxidation states than Nb and Ta which are stable in +5 oxidation state. V^{2-} and V^{3+} are reducing agents. In +4 oxidation state V forms VO^{2+} ion which is very stable in solids and solutions. Vanadium +5 state is covalent such as, VF_5, VO_2^+ and VO_4^{3-} (vanadates). The basic property of +5 oxides increases when we move down the group. V_2O_5 is amphoteric. V dissolves in water giving pale yellow solution and with NaOH, it forms vanadates which are colourless solutions and wide ranging vanadates that are formed depend upon the pH. At high pH, orthovanadates (VO_4^{3+}) are obtained. Niobates and tantalates are also formed like Nb_2O_5 and Ta_2O_s which can be dissolved in NaOH. V^{2+} and V^{3+} are well known ionic compounds whereas, Nb(II) (III) Ta(II) and Ta(lll) are not ionic, but are complexes.

Colour: Colour in transition metal compounds may be due to:

(*i*) *d-d* transitions

(*ii*) Crystal defects

(*iii*) Charge transfer spectra

Below +5 oxidation state the colour is attributed to *d-d* transition due to incomplete *d*-orbital electrons. In +5 oxidation state, the configuration is d^o therefore, the compounds are expected to be white or colourless *e.g.* VF_5, NbF_5 etc. However, V_2O_5 is red or orange, $NbCl_5$ is yellow, $NbBr_5$ is orange and NbI_5 is brass coloured owing to factors other than *d-d* transition.

3.3 COMPOUNDS OF V, Nb AND Ta

Compounds of these elements with non-metal C, H and N are interstitial and non-stoichiometric. At high temperature, they form metal nitrides with formula MN. Carbides form hard interstitial compounds like MC and MC_2, NbC and TaC. VC_2 reacts with water to liberate ethyne. With H_2 all the three elements form:

VH	, NbH	and	TaH
1 : 0.71	1 : 0.86		1 : 0.76

3.3.1 Halides

All the three metals form pentahalides when heated with halogens like chlorine, bromine, iodine. The remaining halides of vanadium are VCl_4, VBr_3 and VI_3. All halides are volatile, covalent and are hydrolysed by water.

The list of halides formed in different oxidation states is given in the table.

Oxidation state			
(+ II)	**(+ III)**	**(+ IV)**	**(+V)**
VF_2 blue	VF_3 yellow-green	VF_4 green	VF_6 colourless
VCl_2 pale green	**VCl_3** red-violet	VCl_4 reddish brown	– (liquid)
VBr_2 orange-brown	**VBr_3** brown	VBr_4 magenta	–
VI_2 red-violet	VI_3 black-brown	–	–
–	(NbF_3) *blue	NbF_4 black	**NbF_5** white
–	$NbCl_3$ black	$NbCl_4$ violet	**$NbCl_6$** yellow
–	$NbBr_3$ brown	$NbBr_4$ brown	**$NbBr_5$** orange
–	NbI_3	NbI_4 grey	**NbI_5** brass
–	(TaF_3) *blue	–	**TaF_5** white
–	$TaCl_3$ black	$TaCl_4$ black	**$TaCl_5$** white
–	$TaBr_3$	$TaBr_4$ blue	**$TaBr_6$** yellow
–		TaI_4	**TaI_5** black

The most stable oxidation states are shown in bold.
*may be fluorides.

Pentahalides

Vanadium can form only pentafluoride, but Nb and Ta are able to form full range of halides like fluorides, chlorides, bromides and iodides.

$$M_2O_5 + F_2 \longrightarrow MF_5$$

$$NbCl_5 \text{ or } TaCl_5 + F_2 \longrightarrow MF_5$$

$$NbCl_5 \text{ or } TaCl_5 + HF \longrightarrow MF_5$$

$$VF_5 \text{ disproportionates at 600°C: } 2VF_5 \longrightarrow VF_5 + VF_3$$

VF_5 is a colourless liquid, while other pentahalides are solids. Fluorides are made of octahedral structures MF_6. While studying the geometry, VF_5 forms a long chain of octahedral, but $NbCl_5$ and $TaCl_s$ form cyclic tetramers with four octahedral joined together.

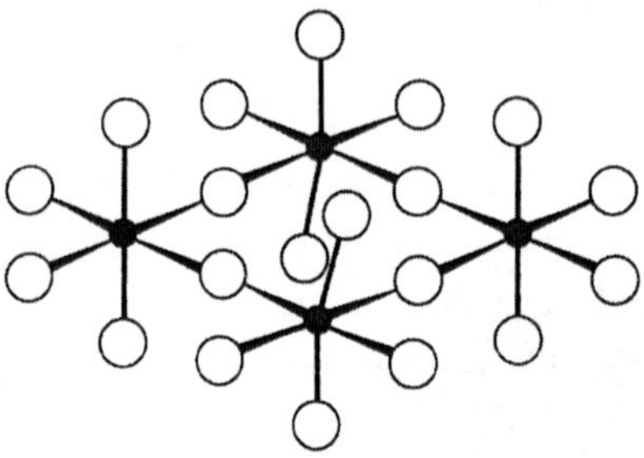

In vapour phase, the pentahalides exist as monomers. The pentafluorides react with F^- to form octahedral complexes $[MF_6]^-$. The heavier Nb and Ta with higher concentration of F^- form 7-coordinate complexes with monocapped trigonal prism structures, *e.g.,* $[NbF_7]^{2-}$ and $[TaF_7]^{2-}$. $[TaF_8]^{3-}$ is a square antiprism structure. The oxohalides can be formed by heating V_2O_5 and Cl_2 or when pentahalides are heated in air *e.g.,* $VOCl_3$. $VOCl_3$ is tetrahedral in shape.

The tetrachloride of vanadium disproportionates at room temperature.

$$2VCl_4 \longrightarrow 2VCl_3 + Cl_2 \text{ (VCl}_5\text{ does not exist)}$$

VCl_4 hydrolyses with water

$$VCl_4 + H_2O \longrightarrow VOCl_2 + 2HCl$$

Trihalides of V, Nb and Ta are known except TaI_3, VX_3 are polymeric. VCl_3 and VBr_3 are made by direct combination. VF_3 is made from VCl_3 and HF. They have d^2 configuration and are reducing agents. The colour is black to brown. VI_3 decomposes as given.

$$2VI_3 \longrightarrow VI_2 + VI_4$$

Dihalides

Dihalides of vanadium are quite familiar and can be prepared from trihalides hy the action with zinc and acid. VF_2 has a rutile (TiO_2) structure. The +2 oxidation state compounds are reducing and are hygroscopic in nature. The Nb and Ta show complex formation tendency at lower oxidation state.

Standard reduction potential of various oxidation states in acid solution has been given in volts.

Standard Reduction Potentials (Volts)

Acid solution Oxidation state +V		IV		III		+II		+I 0
$V(OH)_4^+$	+1.0	V^{2+}	+0.34	V^{3+}	+0.26	V^{2+}	+1.18	V
				−0.26				
Nb_2O_5	0.05			Nb^{3+}	−1.10			Nb
				−0.64				
Ta_2O_5				−0.75				Ta

3.3.2 Oxides

When metals of this group react with dioxygen at high temperatures they form pentaoxides *e.g.* V_2O_5. V_2O_5 is an orange red solid, while Nb_2O_5 and Ta_2O_5 are white in colour. To get pure V_2O_5, acidified ammonium metavanadate is reacted with sulphuric acid or by strong heating

$$2NH_4VO_3 + H_2SO_4 \longrightarrow \underset{\text{(hydrated)}}{V_2O_5} + (NH_4)_2SO_4 + H_2O$$

$$2NH_4VO_3 \xrightarrow{\Delta} V_2O_5 + 2NH_3 + H_2O$$

A list of oxides formed in different oxidation states are given

+2	+3	+4	+5
VO	V_2O_3	VO_2	V_2O_5
Mb		NbO_2	Nb_2O_5
TaO		TaO_2	Ta_2O_5

Pentaoxides by far are the most stable. V_2O_5 is considered as amphoteric and mainly acidic in nature and it forms orthovanadates VO_4^{3-} when reacted with NaOH. When pH is raised, it polymerises to give a range of isopolyacids. V_2O_5 dissolves in strong acid forming pale yellow VO_2^+ ion.

Some reactions of V_2O_5 are :

$$V_2O_5 + NaOH \longrightarrow \text{range of vanadates with varying pH}$$

$$V_2O_5 + H_2O_2 \longrightarrow \text{peroxovanadates (red)}$$

$$V_2O_5 + 3Cl_2 \longrightarrow 2VOCl_3$$

$$V_2O_5 + SO_2 \longrightarrow 2VO_2 + SO_3$$

$$V_2O_5 + H_2 \longrightarrow VO_2 + V_2O_3$$

V_2O_s is used as a catalyst in Contact Process during manufacture of sulphuric acid. Catalytic activity is because of its tendency to lose or gain oxygen when heated.

+4 state oxides VO_2 has d^1 configuration and has dark blue colour. With acids it forms a blue solution containing VO^{2+} vanadyl ion. Vanadyl ion forms a large number of compounds $VO.SO_4$, VOS_2 and complexes of the type $[VOX_4]^{2-}$, $[VO(Oxalate)]^{2-}$ $[VO(NCS)_4]^{2-}$ and $[VO\ (acetylacetone)_2]$

+3 Oxides V_2O_3 is a non-stoichiometric oxide. It can be produced by reduction of V_2O_5 by carbon and hydrogen. In acids, it forms blue or green hydrated ion $[V(H_2O)_6]^3$. The oxide has d^2 configuration. It is a black solid and blue in hydrated form.

The octahedral complexes formed by it are $[V(H_2O)_6]^{3+}$, $[VF_6]^{3-}$, $[V(CN)_6]^{3-}$, $[V(CN)_6]^{3-}$ and $[V^{III}\ (oxalate)_3]^{3-}$. It also forms dimeric triacetate complex $[V_2^{in}\ (CH_3COO)_6]$

+2 Oxides

VO is non-stoichiometric VO $V_{0.94} \cdot O_{1.12}$

It is ionic in solid state and has NaCl type structure. Its configuration is d^3 with grey-black colour. It is prepared by reducing V_2O_5 with H., at 1700°C. It is basic and soluble in H_2O. Its solutions are strongly reducing as it is oxidised by air and water. The complexes known are $K_4[V(CN)_6].7H_2O$ and $[V(en)_3]Cl_2$.

3.3.3 Vanadates

V_2O_5 dissolves in strong NaOH to form colourless orthovanadates VO_4^{3-}. They are tetrahedral in shape. They polymerise at lower pH to form large number of isopolyacids in solution. The oxoions form dimers, trimers and pentamers. In acids, $V_2O_5(H_2O)_n$ is precipitated which can dissolve in very strong acid giving complex ions and finally VO_2^+ ion is formed. At different pH values, various solids crystalline out, but they do not have the same structure.

Nb_2O_5 and Ta_2O_5 are white solids and are not attacked by acids except HF. M_2O_5 disolves in NaOH to form niobates and tantalates.

Flow diagram of VO_4^{3-} is acid solution.

$$\underset{\text{colourless}}{[VO_4]^{3-}} \xrightarrow{\text{pH12}} \underset{\text{colourless}}{[VO_3.OH]^{2-}} \xrightarrow{\text{pH10}} \underset{\text{colourless}}{[V_2O_6.OH]^{3-}} \xrightarrow{\text{pH9}} \underset{\text{orange}}{[V_3O_9]} \xrightarrow{\text{pH7}}$$

$$\underset{\text{red}}{[V_5\dot{O}_{14}]^{3-}} \xrightarrow{\text{pH6.5}} \underset{\text{brown precipitate}}{[V_2\dot{O}_5.H_2O]_n} \xrightarrow{\text{pH2.2}} [V_{10}O_{28}]^{6-} \xrightarrow{\text{pH<1}} \underset{\text{pale yellow}}{[VO_2]^+}$$

*deluded because solids of this formula have been precipitated

3.3.4 Organometallic Compounds

The main examples are $V(CO)_6$. These form M—C a bonds.

V forms bis (cyclopentadienyl) compounds like $[V(\eta^3—(C_5H_5)_2Cl_2]$, $[V\eta^3—(C_5H_5)_2Cl]$ and $[V(\eta^5—(C_5H_5)_2]$. The last one is vanadocene. It is paramagnetic, dark violet and air sensitive. Nb and Ta form similar organometallic compounds.

■ ■ ■

4

Chromium Group

4.1 INTRODUCTION

The metals chromium, molybdenum amd tungsten exist in group 6. Chromium metal is largely used in the process of electroplating and in forming various alloys with ferrous and non-ferrous metals. In addition, they also form important compounds *i.e.* dichromates of sodium and potassium. $Cr0_3$ and Cr_20_3 are commonly used in industry. Tungstate and molybdate ions contribute in the formation of iso and heteropoly acid. Chromium acetate has an unusual dimeric structure and are ruby red crystals. Mo and W form octahedral metal clusters $[M_6X_g]^{4-}$. Mo plays an important role in dinitrogen fixation in plant world.

4.2 PHYSICAL CHARACTERISTICS

4.2.1 Abundance, Occurrence and Extraction

Chromium occurs in plenty in earth's crust whereas, molybdenum and tungsten are scanty. Cr is mainly obtained from chromite ore called iron chromium oxide ($FeCr_2O_4$) which is slightly magnetic. Chromium can be produced as pure Cr and ferrochrome alloy which consists of Fe, Cr and C. It is used in the production of several ferro-alloys *i.e.* stainless steel and hard chromium steel.

$$FeCr_20_4 + 4C \xrightarrow[\text{Furnace}]{\text{Electric}} \underset{\text{ferrochrome}}{4CO + Fe + 2Cr}$$

In order to get pure chromium, the chromite ore is fused with NaOH to get sodium chromate and ferric oxide. The sodium chromate is soluble whereas, ferric oxide is insoluble.

$$3FeCr_2^{III}O_4 + 8NaOH + 3\frac{1}{2}O_2 \xrightarrow[1100°C]{air} 4Na_2Cr^{VI}O_4 + Fe_2O_3 + 4H_2O$$

The yellow solution of Na_2Cr0_4 is acidified to give sodium dichromate ($Na_2Cr_2O_7$) which is then reduced to chromium oxide (Cr_2O_3) by heating with C.

$$Na_2Cr_2O_7 + 2C \longrightarrow Cr_2O_3 + Na_2CO_3 + CO$$

Finally, to obtain pure Cr metal, the chromium oxide is reduced by aluminium through thermite process.

$$Cr_2O_3 + 2Al \longrightarrow Al_2O_3 + \underset{\text{pure}}{2Cr}$$

The metal is brittle, hence it is used when alloyed with other metals. The metal is non-corrosive and protects other metals from corrosion by electro-deposition.

Mo and W are rare metal elements. Mo is used as a catalyst in petrochemical industry. Molybdenum is extracted from molybdenite (MoS_2) ore, which is also obtained as by-product from CuS ore. By the process of roasting, the MoS_2 is converted to MoO_3 in air. Then MoO_3 is dissolved in dilute NH_4OH to get ammonium molybdate which is then reduced by hydrogen to obtain pure Mo.

Tungsten occurs in the form of tungstate as wolframite $FeWO_4$. $MnWO_4$ and scheelite $CaWO_4$. By fusing wolframite with Na_2CO_3, sodium tungstate is obtained which gives tungstic acid on acidification. Whereas, with HCl, scheelite is acidified at the time of precipitation of tungstic acid. Then its anhydrous oxide is obtained by heating which is then finally reduced with H_2 at 850°C to get pure tungsten metal.

4.2.2 Electronic Configuration and Oxidation States

The electronic configuration of the elements of chromium group have been tabulated below:

Element	Electronic structure	Oxidation states
Chromium	$(Ar)3d^5 4s^1$	(–11),(–1), (0) **+2**, **+3**, (+4), (+5) **+6**
Molybdenum	$(Kr)4d^5 5s^1$	(–11)(–1)(0)(+2)(+3)(+4)(+5) **+6**
Tungsten	$(Xe)\ 4f^{14} 5d^4 6s^2$	(–11)(–1)(0)(+1)(+2)(+3) +4 +5 **+6**

Most stable and abundant oxidation states have been shown in bold, less important in normal type and unstable in bracket.

4.2.3 Oxidation States

The metals Cr and Mo, have $(n{-}1)d^5\ ns^1$ configuration with stable half filled configuration and W has $d^4 s^2$ configuration. The expected oxidation states vary from +1 to +VI for Mo and W. Their lower oxidation states occur as complexes *e.g.* dipyradyl complexes, carbonyl complexes and carbonyl ions. For chromium, three states are known *i.e.* Cr(II); Cr(III) and Cr(VI). The Cr(II) is well know for reducing, Cr(III) is the most stable form and Cr(VI) has very strong oxidising property. Regarding molybdenum and tungsten, Mo(+VI) and W(+VI), they are most stable oxidising state. Whereas, W(+III) and Mo(+III) are strongly reducing. On going down the group, the higher oxidation states become more stable and lower oxidation states less stable. The trend of higher oxidation state becoming more stable and lower oxidation states less stable down the group is maintained as shown in the table.

Table: Oxides and halides of Cr, Mo and W

Oxidation states				
(+ II)	(+ "I)	(+ IV)	(+ V)	(+ VI)
–	Cr_2O_3	CrO_2	–	CrO_3
–	–	MoO_2	Mo_2O_5	MoO_3
–	–	WO_2	(W_2O_5)	WO_3
CrF_2	CrF_3	CrF_4	CrF_5	(CrF_6)
$CrCl_2$	$CrCl_3$	$CrCl_4$	–	–
$CrBr_2$	$CrBr_3$	$CrBr_4$	–	–
CrI_2	CrI_3	CrI_4	–	–
–	MoF_3	MoF_4	MoF_5	$M°F_6$
$MoCl_2$	$MoCl_3$	$MoCl_4$	$MoCl_5$	$(MoCl_6)$
$MoBr_2$	$MoBr_3$	$MoBr_4$	–	–
MoI_2	MoI_3	MoI_4?	–	–
–	–	WF_4	WF_5	WF_6
WCl_2	WCl_3	WCl_4	WCl_5	WCl_5
WBr_2	WBr_3	WBr_4	WBr_5	WBr_6
WI_2	WI_3	WI_4?		

The most stable oxidation states are bold, unstable are bracketed.

4.3 GENERAL CHARACTERISTICS

These metals are found to be hard with high melting point and low volatility, Tungsten has highest melting point of all the non-alloyed metals as well as Second highest of all the elements after carbon. Regarding chromium, being protected by surface coating of oxide at low temperature, it is unreactive. Cr is non-corrosive and therefore, it is used for electroplating iron. HCl and H_2SO_4 are the acids where Cr can be dissolved whereas, in concentrated HNO_3 or aqua regia, chromium is rendered passive owing to the formation of Cr_2O_3 transparent and non-reactive film on the surface. However, Mo and W are found to be relatively inert. Also, Mo and W can be dissolved in the mixture of HNO_3 and HF and fused with KNO_3/NaOH.

With HCl gas, Cr forms anhydrous $CrCl_2$ and H_2. When reacted with halogens in heating conditions, Cr forms trihalides ($CrCl_3$) whereas, Mo and W form hexahalides ($MoCl_6$) even at room temperature.

$$2Cr + 3Cl_2 \longrightarrow 2CrCl_3$$

$$Mo + 3Cl_2 \longrightarrow MoCl_6$$

There is close similarity in the properties of Mo and W due to lanthanide contraction.

Cr and Mo on heating with O_2, form Cr_2O_3 and MoO_3, respectively

$$2Cr + 3O_2 \longrightarrow 2Cr_2O_3;\ 2Mo + 3O_2 \longrightarrow 2MoO_3$$

Molybdates are reduced by H_2S to precipitate MoS, in group II analysis

Standard Reduction Potentials (Volts)

Acid Solution

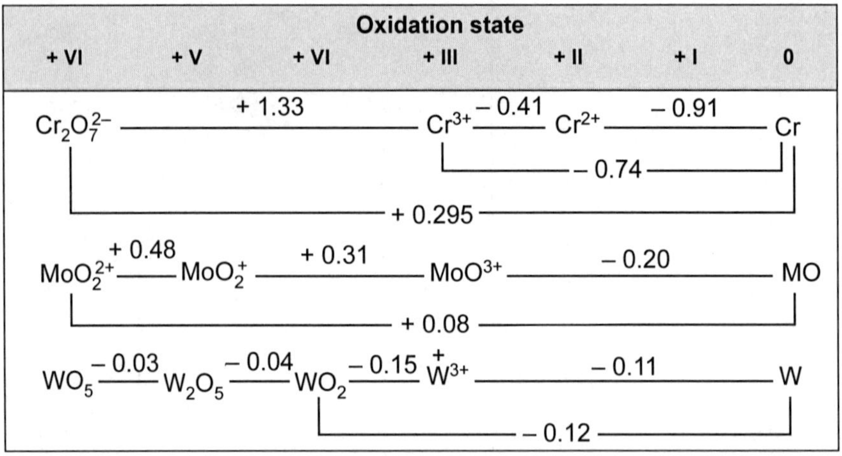

Basic Solution

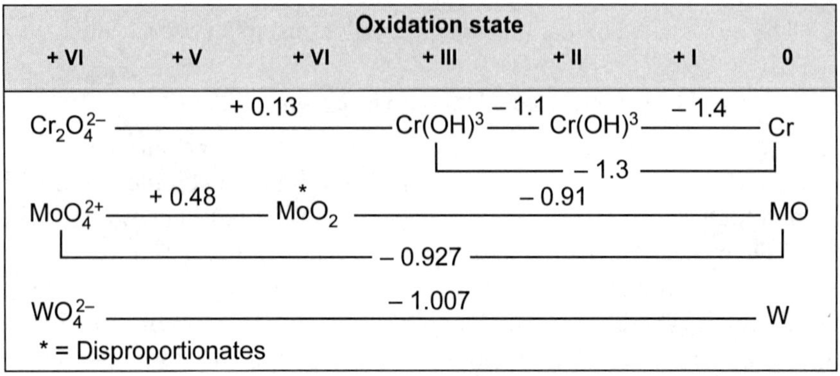

Many of these potentials have been calculated from thermodynamic data, and the existence of species such as Mo^{3+}, W^{3+} and W_2O_5 is questionable.

4.3.1 Compounds in (+6) State

The Cr(+6) compounds are known to be strong oxidising agents *e.g.* CrO_4^{2-} chromates, $Cr_2O_7^{2-}$ dichromates. CrO_3 chromic anhydride and CrO_2Cl_2 chromyl chloride, $CrOX_4$ and CrF_6.

Sodium Chromate and Potassium Dichromate

Sodium chromate is prepared from chromite ore by fusing with NaOH or Na_2CO_3 in the presence of air.

$$4FeCr_2O_3 + 8\,Na_2CO_3 + 7O_2 \longrightarrow 8Na_2CrO_4 + 2Fe_2O_3 + 8CO_2$$

Sodium chromate is a yellow solid and a strong oxidising agent.

Among dichromates, potassium dichromate ($K_2Cr_2O_7$) is preferred over sodium dichromate ($Na_2Cr_2O_7$) to use in several chemical reactions because sodium dichromate is hydroscopic in nature. The potassium dichromate is an orange coloured compound and used as moderately strong oxidising agent in volumetric titrations and in some organic reactions. It is prepared by acidified chromate solution.

$$\underset{\text{yellow}}{2CrO_4^{2-}} + \underset{\text{(acid)}}{2H^+} \longrightarrow \underset{\text{orange}}{Cr_2O_7^{2-}} + H_2O$$

$$Na_2Cr_2O_7 + 2KCl \longrightarrow K_2Cr_2O_7 + 2NaCl$$

$K_2Cr_2O_7$ is used as a primary standard. It forms blue-violet peroxo compounds when H_2O_2 is added to acidified potassium dichromate.

Its acidified reaction can be explained by the ionic equation.

$$Cr_2O_7^{2-} + 14H^+ + 6e^- \longrightarrow 2Cr^{3+} + 7H_2O \quad E° = 1.33V$$

$$Cr_2O_7^{2-} + 2H^+ + 4H_2\,O_2 \longrightarrow \underset{\text{blue violet peroxo compound}}{2CrO(O_2)_2} + 5H_2O$$

The peroxo compounds can be extracted with ether. It reacts with pyridine to form $[CrO(O_2)(OH)]^-$ which is explosive in less acidic conditions. Its structure is pentagonal pyramid where the ether group reacts with pyridine.

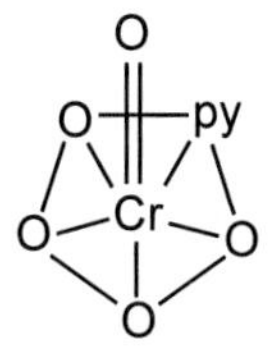

In alkaline solution, the reaction gives red-brown K_3CrO_8 which is a tetraperoxo species $[Cr(\,O_2)_4]^{3-}$. Dark red-brown $(NH_3)_3CrO_4$ compound with pentagonal bipyramid structure is formed with ammonia solution.

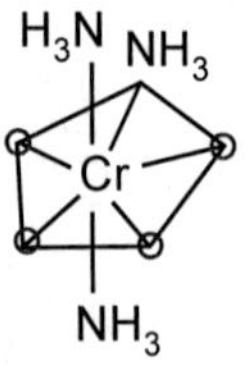

Pentagonal bipyramid

4.3.2 Chromium Trioxide or Chromic Anhydride

Chromium trioxide is prepared from saturated solution of sodium dichromate (Na_2Cr,O_7) by adding concentrated H_2SO_4.

$$Na_2Cr_2O_7 + H_2SO_4 \longrightarrow \underset{\text{orange solid}}{2CrO_3} + Na_2SO_4 + H_2O$$

CrO_3 is toxic and corrosive. It dissolves in water to form red chromic acid which acts as a strong oxidising agent. Also, it dissolves in NaOH to give CrO_4^{2-} ion and loses oxygen on heating at 250°C to form Cr_2O_3 green.

$$2CrO_3 \longrightarrow 2CrO_2 + O_2$$

$$2CrO_2 \longrightarrow Cr_2O_3 + \frac{1}{2}O_2$$

CrO_3 reacts with F_2 and Cl_2, to form CrO_2F_2, and $CrOF_4$. At 170°C and 25 atmospheres, it forms CrF_6. Chromium also form chromic acid (H_2CrO_4), useful to clean glass apparatus. When dissolved in acetic acid it acts as an oxidising agent for organic reactions.

MoO_3 and WO_3 are formed by heating the metal in air. They are not attacked by acids. They dissolve in NaOH forming MoO_4^{2-} and WO_4^{2-} but are insoluble in water and have oxidising power. They form mixed oxides with alkali and alkaline earth metal oxides.

Chromyl chloride can be prepared from the following reactions:

(*i*) $CrO_3 + 2HCl \xrightarrow[\text{H2SO}_4]{\text{Cone.}} CrO_2Cl_2 + H_2O$

(*ii*) $K_2Cr_2O_7 + 6HCl \xrightarrow[H_2SO_4]{\text{Cone.}} 2CrO_2Cl_2 + 2KCl + H_2O$

$\downarrow$ Red vapours

With NaOH, CrO_2Cl_2 forms Na_2CrO_4. Chromyl and molybdenyl chlorides are covalent and are decomposed by water.

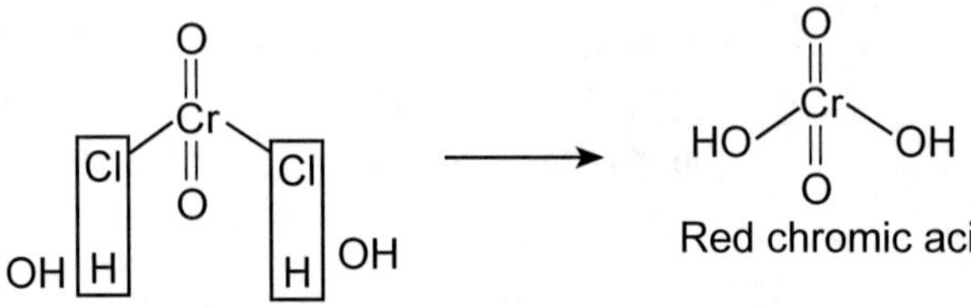

Red chromic acid

4.3.3 Halides

CrF_6 is prepared by heating the elements under pressure and cooling. It is an unstable yellow solid and decomposes into $CrF_5 + F_2$. On the other hand, MoF_6 and WF_6 are quite stable. They are easily hydrolysed. They are diamagnetic and colourless having d° configuration. $MoCl_6$ and WCl_6 are black and WBr_6 is dark blue. WCl_6 reacts with water to form tungstic acid.

Cr(+5) compounds are unstable and decompose to Cr(+3) and Cr(+6) *e.g.* K_3CrO_8 contains tetraperoxo species $[Cr(O_2)_4]^{3-}$ and is dark brown. Heating Mo and Cl_2 gives Mo_2Cl_{10}. It is dimeric and hydrolyses in water as well as paramagnetic (μ = 1.6 BM) with one unpaired electron and therefore, no metal-metal bonding.

+4 State: Cr(+4) compounds are rare and formed by heating the elements together. CrF_4 is prepared by heating the elements at 350°C. $MoCl_4$ has two polymeric forms, one form shows chain of octahedra with metal-metal bond, whereas other form do not show metal-metal bond. CrO_2 and TiO_2 have a rutile structure. CrO_2 is made from CrO_3 by reduction. MnO_2 and WO_2 are also made by reducing with hydrogen. They are brown-violet, insoluble in reducing solvents, but soluble in concentrated HNO_3 making MoO_3 and WO_3.

+3 State: Cr(+3) compounds are ionic and most stable. They are very stable in acidic solution and can be oxidised to (+6) state in aquaregia.

Cr_2O_3 in a green solid obtained by heating $(NH_4)_2Cr_2O_7$.

$$(NH_4)_2Cr_2O_7 \xrightarrow{\Delta} N_2 + Cr_2O_3 + 4H_2O$$

$$4Cr + 3O_2 \longrightarrow 2Cr_2O_3 \text{ (green solid)}$$

$$4CrO_3 \longrightarrow 2Cr_2O_3 + 3O_2$$

NaOH added to Cr^{3+} gives the following reaction

$$Cr^{3+} + 3OH^- \longrightarrow Cr(OH)_3 \longrightarrow Cr_2O_3(H_2O)_n$$

On heating strongly, the oxide becomes inert to acids and bases. When heated strongly, it gives $[Cr(H_2O)_6]^3$+ with acids and chromites with strong alkali. Chromite solution is available in the form of $[Cr(OH)_6]^{3-}$, $[Cr(OH)_5H_2O]^{2-}$. The Cr_2O_3 has commercial use as a pigment in paint and catalyst in the manufacture of polythene and butadiene.

$CrCl_3$ is a solid compound forming red-violet flakes as it has a layer structure. In aqueous solution, it forms violet coloured $[Cr(H_2O)_6]^{3+}$ ions and other related complexes, *i.e.*, $[Cr(H_2O_4)Cl_2]^+$, $[Cr(H_2O)_4Cl^+$ and $[Cr(H_2O)_3Cl_3]$, It can also form chrome alum $K_2SO_4Cr_2(SO_4)_324H_2O$. The hexa-aqua ion is acidic and it may give bridged complexes with hydroxo bridges.

$$[Cr(H_2O)_6]^{3+} \rightleftharpoons H^+ + [Cr(H_2O)_5(OH)]^{2+} \rightleftharpoons$$

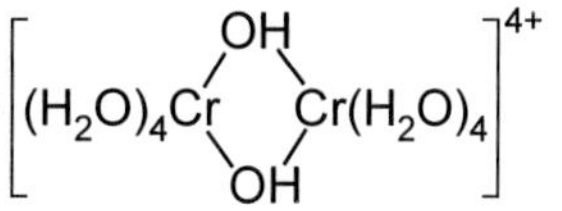

The 6-coordinate complexes have octahedral structures. The configuration of these complexes is d^3 with three unpaired electrons and high CFSE. The ammine and oxalate complexes may show different isomerism.

Complexes	Isomerism
$[Cr(NH_3)_6]^{3+}$	one form
$[Cr(NH_3)_5Cl]^{2+}$	one form
$[Cr(NH_3)_4Cl_2]^{+}$	*cis* and *trans* isomers
$[Cr(NH_3)_3Cl_3]$	mer and fac isomers
$[Cr(C_2O_4)_3]^{3-}$	*d* and / isomers

Reinecke's salt $NH_4[Cr(NH_3)_2(NCS)_4].H_2O$ is used to precipitate large positive ions. The other complexes with reduced magnetic moment are

$$[(H_3N)_5Cr\text{—}OH\text{—}Cr(NH_3)_5]^{+5} \text{ and } [(H_3N_2)\text{—}N\text{—}O\text{—}Cr(NH_3)_5]^{4-}$$

Mo(+3) and W(+3) ions have no oxides, but their halides are known. Mo(III) compounds are stable, oxidise in air and hydrolyse in water. They form octahedral complexes $[MoCl_6]^{3-}$. It has been found that Cr(III) may form rarely basic acetate having formula $[Cr_3O(CH_3COO)_6L_3]^+$, where L stands for H_2O or some other ligand. The structure consists of triangle of 3 Cr atoms where six acetate groups act as bridges between the Cr atoms. The six acetate groups are arranged in such a way that two acetate groups are across each edge of the triangle.

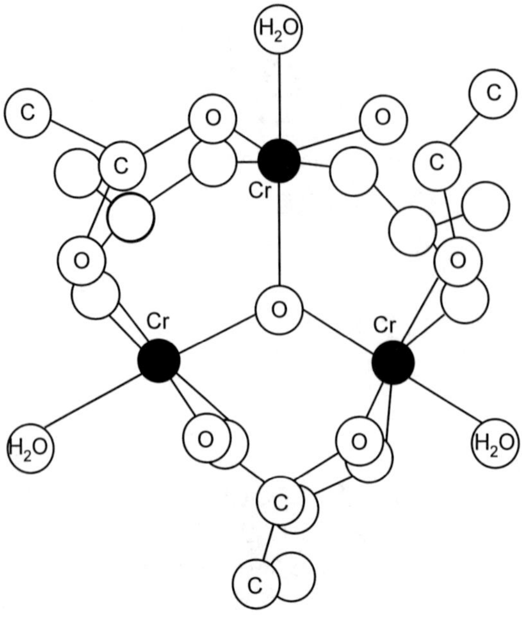

Cr (II) Compounds: Cr^{2+} aqueous solution exists as $[Cr(H_2O)_6]^{2+}$. It is obtained by reducing Cr^{3+} solutions.

$$Cr^{3+} + e^- \longrightarrow Cr^{2+} \quad E° = -0.41V$$

Cr^{2+} reduces water to hydrogen in acidic solution and Cr^{2+} compounds are easily oxidised in air to Cr^{3+}. Cr^{2+} may be stabilised by forming complexes, such as $[Cr(NH_3)_6]^{2+}$ or $[Cr(dipyridyl)_3]^{2+}$ The Cr^{2+} forms several complexes with N ligands and other chelating ligands.

One of the most stable Cr^{2+} compounds is Cr(II) acetate dihydrate $Cr_2(CH_3COO)_4.2H_20$. It is prepared by reacting Cr^{2+} compounds in solution with sodium acetate. It is red solid and is dimeric. The molecule shows an octahedral geometry. Each Cr^{2+} is surrounded by octahedron made up of four O atoms from bidentate acetate groups, one O from H_2O and the other from Cr^{2+} ion as shown in figure.

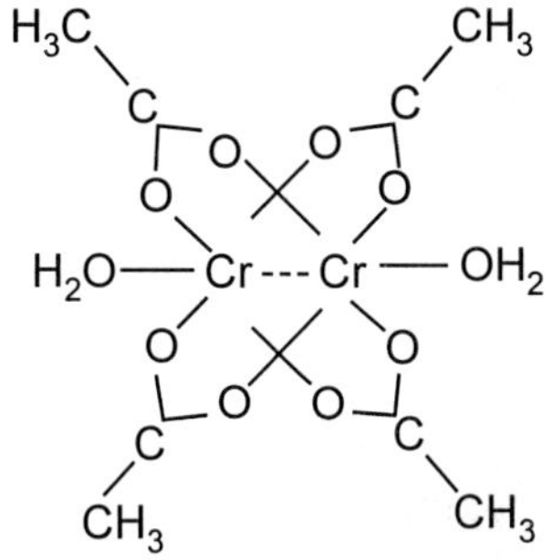

There is strong Cr-Cr bond because of short distance between two Cr atoms (2.36Å). Cr^{2+} has d^4 configuration with four unpaired electrons and Cr(II) acetate is diamagnetic. This suggests that the four unpaired electrons are in metal-metal bond. Mo(II) and W(II) form higher halides in the form of cluster compounds *e.g.,* $[Mo_6Br_8]Br_4.2H_2O$. Similarly, W(II) forms $[W_6Cl_8]^{4-}$ complexes.

(+1) State: Compounds with d^5 configuration are doubtful. They may exist in the form of complexes in which they are further stabilised *e.g.* $[Cr(dipy)_3]^+ClO_4^-$. Mo and W do form sandwich type structured compounds like $(C_6H_6)_2Mo^+$ and $C_5H_5M_QC_6H_6$.

4.3.4 Zero, (–I) and (–II) State Compounds

Zero oxidation state exists in metal carbonyls $M(CO)_6$. In these compounds, *a* bonding exists due to electron pair donated by CO group to the metal and strong *dπ-pπ* back donation occurs from filled metal orbitals to CO. All the three metals form octahedral compounds soluble in organic solvents. E.O. Fischer (1955) prepared dibenzene chromium which occurs as dark brown crystals and have sandwich type structure.

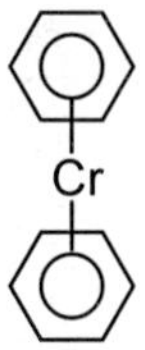

Dibenzene Chromium

It is similar to ferrocene. Here Cr has a coordination nuumber of 12.

(*i*) $2CrCl_3 + 2Al + AlCl_3 + 6C_6H_6 \longrightarrow 3[AlCl_4]^- + 3[Cr(\eta^6\text{-}C_6H_6)_2]^+$

(*ii*) $CrCl_3 + 2C_6H_5MgBr \longrightarrow [Cr(\eta^6\text{-}C_6H_6)_2]^+Cl^- + MgBr_2 + MgCl_2$:

Complexes can be formed with one benzene ring as well *e.g.*,

$$[Cr(\eta^6\text{-}C_6H_6)(CO)_3].$$

In a similar way, Mo and W also form such complexes. It has metal with a coordination number 9 and are yellow solids. Among these, the benzene compounds are more reactive than $\eta^5\text{-}C_5H_5$ compounds.

The lower oxidation states also occur in carbonyl ions *e.g.*,

$$[M_2(CO)_{io}]^{2-} \text{ indicates } (-I)$$

$$[M(CO)_5]^{2-} \text{ indicates } (-II)$$

4.3.5 Chromates, Molybdates and Tungstates

CrO_3, MoO_3 and WO_3 are acidic in nature. They dissolve in NaOH forming tetrahedral CrO_4^{2-}, MoO_4^{2-} and WO_4^{2-} ions. Chromates, molybdates and tungstates exist both in solution and solid form. The CrO_4^{2-} ions are strong oxidising agents, whereas MoO_4^{2} and WO_4^{2-} are relatively weak oxidising agents.

On acidification, $CrO_4{}^{2-}$ forms condensates of $Cr_2O_7^{2-}$ Dichromate ions have two tetrahedral units with a common oxygen atom at one corner.

$$2CrO_4^{2-} + 2H^+ \longrightarrow Cr_2O_7^{2-} + H_2O$$

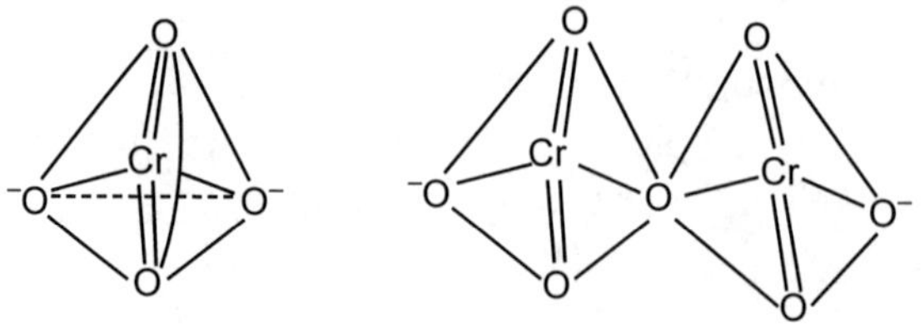

Fig. Chromate and dichromate Ions

The most important $Na_2Cr_2O_7$ or $K_2Cr_2O_7$ compound is a strong oxidising agent in acid medium. It is moderate oxidising agent for many organic reactions.

Molybdates and tungstates on acidification give a range of polymolybdates and polytungstates.

$$[MoO_4]^{4-} \xrightarrow[H^+]{pH_6} [Mo_7O_{24}]^{6-} \xrightarrow[1.5 \text{ to } 3.0]{pH} [Mo_8O_{26}]^{4-} \xrightarrow[<1.0]{pH} MoO_3.2H_2O$$

Para molybdate — Octa molybdate — Hydrated oxide

The condensation products of tungstates are shown as:

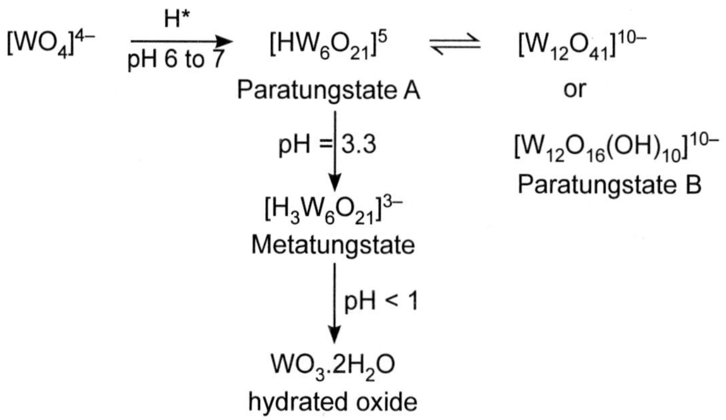

On heating, phosphate solution with ammonium molybdate and nitric acid, yellow precipitates of ammonium phosphomolybdate are formed.

e.g., $(NH_4)_3\,[PO_4.\,Mo_{12}O_{36}]$
yellow precipitate

4.3.6 Tungsten Bronzes

When Na_2WO_4 is strongly heated with WO_3 and H_2, the tungsten bronzes are formed. Presently, they are made from Na_2WO_4 and W metal resulting in the production of blue, purple, red or yellow tungsten bronzes which are semi-metallic solids. In addition, they have metallic lustre and are able to conduct electricity. These tungsten bronzes are non-reactive to acids and alkalis and are useful in the production of metallic paints.

4.4 THE MANGANESE GROUP

Manganese metal exists along with Fe in iron ores. Its principal ore is pyrolusite MnO_2 and resembles Fe in properties. It is used for manufacturing manganese steel and large quantities of MnO_2 are used for making 'Dry' batteries. $KMnO_4$ is a very common oxidising agent.

Mn(III) acetate has an uncommon structure. In biolgoical world, Mn is important and very essential element in process of photosynthesis. The other elements in the same group are Tc (Technetium) and Re (Rhenium), which are rare elements. Their lower oxidation states stabilise by forming cluster compounds (+2, +3 and +4) and higher oxidation states of Tc and Re are stable.

4.4.1 Electronic Configuration and Oxidation States

A brief summary of these three elements i.e. Mn, Tc and Re show their electronic configurations and oxidation states have been tabulated as below:

Element	**Electronic configuration**	**Oxidation states.**
Mn	(Ar) $3d^5 4s^2$	–1, 0, +1, +2, +3, +4, +5, +6, +7
Tc	(Kr) $4d^5 5s^2$	0, 2, 3, 4, –5, +6, +7
Re	(Xe) $5d^5 6s^2$	0, 1, 2, 3, 4, 5, 6, 7

4.4.2 Occurrence and Uses

Manganese is widely distributed in the terrestrial crust on earth in large quantity. Pyrolusite is the principal manganese ore and pure Mn can be obtained by reducing MnO_2 or Mn_3O_4 with the help thermite process. It can also be obtained by electrolysing aqueous $MnSO_4$ solutions. Manganese is used for making important alloys *i.e.,* ferromanganase and other non-ferrous alloys where Mn amount is used in little amount. Manganin, a non-ferrous alloy contains 84% Cu, 12% Mn and 4% Ni. Manganin is used in making electrical instruments because its electrical resistance is not dependent upon temperature.

Tc is the first man-made chemical element and does not occur in nature. Among the uranium fission products, ^{99}Tc is one of them which is metastable nuclear isomer. At high temperature of 400-450°C, technetium oxidizes to form pale-yellow heptoxide (Tc_2O_7) which is a volatile oxide. Tc can be obtained by neutron bombardment of Mo. Re (Rhenium) comes under the category of very rare elements and occurs in molybdenum sulphide ores in very small amount. Re is obtained in the form of Re_7O_7 by roasting process which is then dissolved in NaOH to get ReO_4^- perrhenate ions. KCl is added in the concentrated solution ofperrhenate ions to get potassium perrhenate ($KReO_4$). Later, it is reduced with hydrogen to get metal. Re is mainly used to obtain an alloy Pt-Re which is used as catalyst for making lead free petrol.

4.4.3 Oxidation States

The valence shell configuration of $(n-1)d^5ns^2$ suggests wide ranging oxidation states. Manganese shows wide range of oxidation states and Mn^{2+} is most stable. In alkaline solutions, Mn^{2+} is oxidised to Mn^{4+}. $KMnO_4$ is the strongest oxidising agent in solution. Mn(+VII) is well known in $KMnO_4$ solution. The lower oxidation states also exist as carbonyl compounds *i.e.* $Mn_2(CO)_{10}$. The +VII state is most common and stable in Tc and Re. Re (III) is also stable in complexes.

Table: Oxides and halide

Oxidation states						
(+II)	**(+ III)**	**(+ IV)**	**(+ V)**	**(+ VI)**	**(+ VII)**	**Others**
MnO	Mn_2O_3	MnO_2	–	–	Mn_2O_7	Mn_3O_4
–	–	TcO_2	–	TcO_3	Tc_2O_7	
–	Re_2O_3h	ReO_2	(Re_2O_5)	ReO_3	Re_2O_7	
MnF_2	MnF_3	MnF_4	–	–	–	
$MnCl_2$	–	–	–	–	–	
$MnBr_2$	–	–	–	–	–	
MnI_2	–	–	–	–	–	
–	–	–	$TcCF_5$	TcF_6	–	
–	–	$TcCl_4$	–	($TcCl_6$?)	–	
–	–	–	–	–	–	

–	–	–	–	–	–	
–	–	ReF_4	ReF_5	ReF_6	ReF_7	
($ReCl_2$)	$ReCl_3$	**$ReCl_4$**	$ReCl_5$	($ReCl_6$?)	–	
($ReBr_2$)	ReBr,	**$ReBr_4$**	$ReBr_5$	–	–	
(ReI_2)	ReI_3	**ReI_4**	–	–	–	

The most stable oxidation states are shown in bold, unstable ones in brackets, h = hydrous oxide.

Table: Reduction Potential in Volts

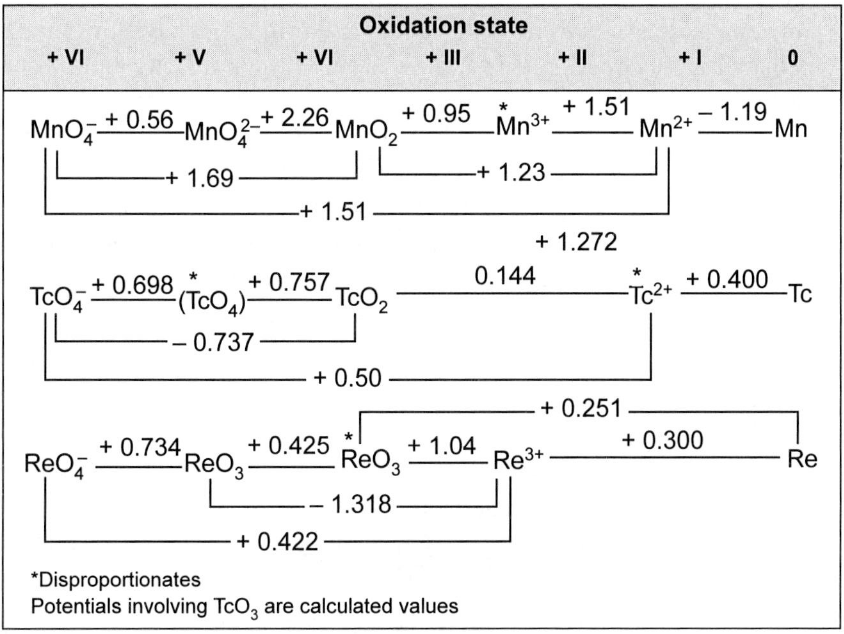

4.5 GENERAL CHARACTERISTICS

Regarding reactivity of the elements of group 7 (*d* block elements) in the periodic table, Mn is more reactive than other neighbouring elements. It slowly reacts with water and hydrogen is liberated. It dissolves in dilute acids and reacts in air when finely divided. It reacts with non-metals *i.e.* O_2, N_2 and Cl_2 and form Mn_3O_4, Mn_3N_2 and $MnCl_2$, respectively. With F_2, it forms a mixture of MnF_2 and MnF_4. Its melting point is lower than the preceding elements in the row *e.g.,* Ti, V and Cr. It also differs by having four different solid structures.

Tc and Re are less reactive metals in comparison to Mn as they do not react with water and do not dissolve in HCl, HF and other oxidising acids. They react with O_2 to form Tc_2O_7 and Re_2O_7 when heated. Whereas, when Tc and Re are heated with F_2, then TcF_5, TcF_6 and ReF_6 and ReF_2, respectively are formed.

Mn^{2+}, Mn^{3+} and MnO_4^{2-} and MnO_4^- are ionic compounds of Mn. Whereas, Tc and Re do not show ionic chemistry, but form oxoions *e.g.* TcO_4^- and ReO_4^-.

Tc and Re prefer to form metal-metal bonds in lower oxidation states. In general, lower oxidation state compounds are basic and higher oxidation state acidic.

Colour: Mn^{2+} compounds are pale pink in colour. MnO_2 is black due to *d-d* transition. Whereas, Mn(+VII) compounds are colourless due to d^0 configuration, but MnO_4^- compounds are purple black. The colour is attributed to charge transfer spectra.

In comparison with iron, Mn is harder and brittle, but have lower melting point than iron. In Group 7 *i.e.* Mn, shows a very little similarity with Group 17 elements *i.e.* halogens, by having 7 outer electrons. On heating, it reacts with most non-metals. Mn_2O_7 and Cl_2O_7 are similar and, MnO_4^- and ClO_4^- both are isomorphous. It is also important to note that chromates CrO_4^{2-}, manganates MnO_4^{2-} and ferrates FeO_4^{2-} are similar and occurring in the same row.

Compounds in Lower Oxidation States zero and (–I) oxidation state: The compounds exist in the form of carbonyls *e.g.*

$$[Mn(CO)_5]^- \longrightarrow \text{(–1) oxidation state}$$

$[Mn_2(CO)_{10}]$ and $[Re_2(CO)_{10}]$ $\longrightarrow$ zero valent state whtereas, Mn with ammonia in the form of cyanide *e.g.* $K_6[Mn(CN)_6]$: $2NH_3$ is unstable and highly reducing.

It is difficult to obtain the elements Mn(+I) and Re(+I) at low oxidation state with strong reducing character as follows:

$$K_3[Mn(CN)_6] \xrightarrow[\text{liq. } NH_3]{\text{K in}} K_5[Mn^1(CN)_6]$$

$$[Re_2(CO)_{10}] \xrightarrow[\text{under pressure}]{Cl_2} [Re_2^1(CO)_{10}Cl]$$

+II Oxidation State: It can be obtained from pyrolusite (MnO_2) by action with acids.

$$MnO_2 + 4HCl_3 \longrightarrow MnCl_2 + 2H_2O + Cl_2$$

$$2MnO_2 + 2H_2SO_4 \longrightarrow 2MnSO_4 + O_2 + 2H_2O$$

In addition to the above method, they can also be produced by reducing the higher oxidation state compounds. If NaOH or NH_4OH is added to $MnCl_2$ or $MnSO_4$, a pale pink precipitate is formed which finally turns brown-black due to oxidation of $Mn(OH)_2$.

While observing the electronic configuration of Mn^{2+}, there is half filled *d* shell as $3d^5$ configuration and shows greater stability in comparison to Cr^{2+} and Fe^{2+}. The Mn^{2+} complexes are difficult to oxidise.

In Mn^{2+}, the formed complexes are octahedral with high-spin arrangement of unpaired electrons and such arrangement gives zero CFSE (crystal field stabilization energy) *i.e.* $[MnCl_6]^{4-}$ and $[Mn(NH_3)_6]^{2+}$, are not stable in solution. Whereas, chelated complexes are more stable and exist as solids *e.g.* $[Mn(en),]^{2+}$, $[Mn(C_2O_4)_3]^{4-}$ and $[Mn\ edta]^{4-}$. These complexes show very pale colours due to *d-d* transitions. The *d-d* transition not only promote electron from t_{2g} level to e_g level, but also it may change spins for pairing unpaired e_g electrons.

Table: Conversion of higher oxidation states to lower

Oxidation state						
+ VII	+ VI	+ V	+ IV	+ III	+ II	0
Acid solution						
$MnO_4^- \xrightarrow{+0.56} MnO_4^{2-} \xrightarrow{+2.26} MnO_2 \xrightarrow{+0.95} \overset{*}{Mn}^{3+} \xrightarrow{+1.51} Mn^{2+} \xrightarrow{-1.19} Mn$						
Basic solution						
$MnO_4^- \xrightarrow{+0.56} MnO_4^{2-} \xrightarrow{+2.27} \overset{*}{Mn}O_4^{3-} \xrightarrow{+0.93} MnO_2 \xrightarrow{+0.1} Mn^{3+} \xrightarrow{-0.2} Mn(OH)_2 \xrightarrow{-0.55} Mn$						
* = Disproportionates						

$[Mn(CN)_6]^{4-}$ is a complex of Mn and has one unpaired electron due to the effect of strong field ligand (spin pairing). $K_4[Fe(CN)_6].3H_20$ has blue colour and has the same structure as $K_4[Fe(CN)_6]$, While studying $[Mn(CN)_5NO]^{3-}$, has shown a low spin arrangement which is slightly more stable than high spin arramgement. In low spin complexes, CFSE is five times Δ_0 which is cancelled out due to spin-pairing. The low spin complexes are more reactive, therefore, can (be oxidised easily by removing the electron which affects the unit of pairiing energy and ultimately increases CFSE. Also, they can be reduced quickly because the addition of electron fills the orbitals proportionately.

$$Mn^{I}(CN)_6]^{5-} \xleftarrow[\text{Zn reduction}]{} Mn^{II}(CN)_6]^{4-} \xrightarrow[\text{Air oxidation}]{} Mn^{III}(CN)_6]^{2-}$$

The low spin d^5 complexes constitute *d-d* transitions and the compounds are very strongly coloured.

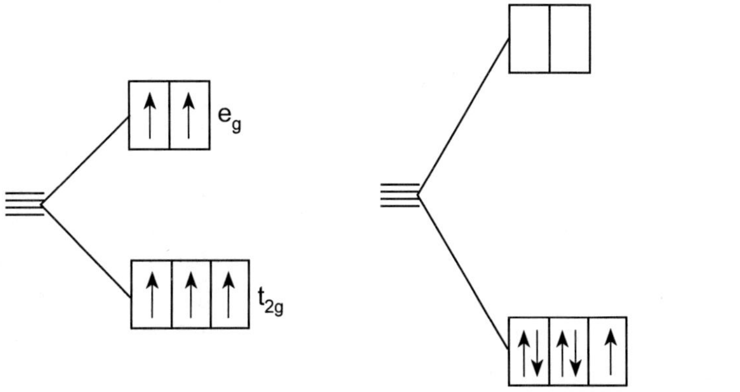

(a) high-spin d^5 arrangement (spin free) (weak ligands)

(b) low-spin *cf* arrangement (spin paired) (strong ligands)

Fig. High and low-spin arrangements in octahedral crystal field.

Mn^{2+} square planar complexes are few *e.g.* $MnSO_4.5H_2O$ contains $[Mn(H_2O)_4]^{2+}$ square planar unit. The halide complexes *i.e.* $[MnCl_4]^{2-}$, $[MnBr_4]^{2-}$ and $[MnI_4]^{2-}$, are tetrahedral with green-yellow colour. These halides show a distinct character in solution, by picking up two water molecules or two halide ions, form pink octahedral complexes. It is noted that Re(+II) is not ionic and forms some complexes *i.e.* $[Re(py)_2Cl_2]$ which exist as *cis* and *trans* isomers and hence, is not tetrahedral, but is square planar.

4.5.1 Mn(III) Oxidation State Compounds

The most stable oxide is Mn_3O_4 even at high temperature and is black in colour. It is prepared by heating any oxide of Mn or $Mn(OH)_2$ to 1000°C and it is a mixed oxide of Mn^{II} and $Mn^{III}{}_2O_4$ having a spinel structure. In these oxides, the 0-atoms are closely packed in octahedral holes and in tetrahedral holes with Mn(Ill) and Mn(II), respectively.

With process of electrolysis, hydrated manganic ion $[Mn(H_2O)_6]^{3+}$ can be obtained in solution by oxidising Mn^{2+} with potassium peroxodisulphate or by reducing MnO_4^-. In concentrated solution, Mn^{3+} reduces to more stable d^5 Mn^{2+} ion. Very few Mn^{3+} salts are known, namely MnF_3, $Mn_2(SO_4)_3$ and Mn_2O_3. These disproportionate in acid and hydrolyse in water:

$$2Mn^{3+} + H_2O \longrightarrow Mn^{2+} + MnO_2 + 4H^+$$

$$Mn^{3+} + 2H_2O \longrightarrow MnO.OH + 3H^+$$

4.5.2 Mn(III) Complexes

Mn(+III) complexes are stable in aqueous solution and are octahedral with high spin complexes with value of μ = 4.9BM indicating four unpaired electrons, *e.g.* $[Mn(H_2O)_6]^{3+}$, $Cs^IMn^{III}(SO_4)_2.12H_2O$ and $[Mn(acac)_3]$,

In $KMnO_4$ and oxalic acid titration, $[Mn^{III}(C_2O_4)_3]^{3-}$ may be formed which may hinder the reduction of MnO_4^- to Mn^{2+}. Therefore, the titration is performed at 60°C to decompose the $[Mn^{III}(C_2O_4)_3]^{3-}$ complex. If air is bubbled through a solution having Mn^{2+} and KCN, the complex $K_3[Mn^{III}(CN)_6]$ is formed. CN^- being a strong field ligand causes spin pairing to give low spin complex.

While comparing technetium and rhenium, Tc(Ill) is unstable, but $Re_2O_3.(H_2O)_n$ and the heavier halides are known. These halides are trimeric and dark ted $(ReCl_3)_3$ and $(ReBr_3)_3$ solids, whereas $(ReI_3)_3$ is black in colour. With rhenium and halide complexes, a different situation has been observed when perrhenates are reduced by H_2 or NaH_2PO_2 in acid solution and as a result, dimeric $[Re_2Cl_8]^{2-}$ and $[ReBr_8]^{2-}$ halide complexes are obtained. This is isostructural with $[Mo_2CL_8]^{4-}$.

4.5.3 +IV Oxidation State Compounds

MnO_2 is a commercially important oxide specially in dry cell batteries *e.g.* alkaline batteries and zinc-carbon batteries, also used in brick and glass industry to colour them with red or brown. It decomposes to Mn_3O_4 on heating at 530°C which is a mixed-valence compound. It occurs as black mineral pyrolusite having rutile structure. MnO_2 is prepared by:

(*i*) By Heating Mn and O_2

(*ii*) By oxidising $Mn(NO_3)_2.6H_2O$ in air

(*iii*) By Electrolytic oxidation of $Mn^{II}SO_4$

(*iv*) By precipitation of hydrated MnO_2 from the titration of $KMnO_4$ carried out in alkaline solution

$$MnO_4^- + 2H_2O + 3e^- \longrightarrow MnO_2 + 4O^-;\ E = 1.23$$

MnO_2 does not react with acids unless heated and oxidises Cl^- to Cl_2 when reacted with concentrated HCl. Even today, this is the commercial method of preparation of chlorine.

$$MnO_2 + 4HCl \longrightarrow Mn^{2+} + Cl_2 + 2Cl^-$$

In the similar way, MnO_2 is used to oxidise hot concentrated H_2SO_4 to liberate O_2.

$$2MnO_2 + 2H_2SO_4 \longrightarrow 2MnSO_4 + O_2 + 2H_2O$$

Also, Mn is reacted with flourine to produce blue coloured MnF_4 which is unstable. Other halide complexes formed by Mn are $K_2[MnF_6]$, $K_2[MnCl_6]$, $K_2[Mn(CN)_6]$ and $K_2[Mn(IO_3)_6]$, This is the highest oxidation state of manganese where these complexes are formed. MnO_2 is used in dry batteries.

The other important use of MnO_2 is in the preparation of $KMnO_4$, as an oxidising agent in organic reactions, use as catalyst in the preparation of dioxygen as given in the following reactions:

(*a*) Preparation of potassium permanganate

$$MnO_2 + KNO_2 \xrightarrow{NaOH} K_2MnO_4 + NO$$

$$K_2MnO_4 + H_2O \xrightarrow[\text{Oxidation}]{\text{Electrolytic}} KMnO_4 + KOH + H_2$$

(*b*) As an oxidising agent in organic reactions

$$C_6H_5.CH_3 + 2Mn_2SO_4 + 3H_2O \longrightarrow C_6H_5.CHO + 2Mn_2SO_4 + 3H_2O$$

$2C_6H_5.NH_3 + 4MnO_2 + 5H_2SO_4 \longrightarrow$ 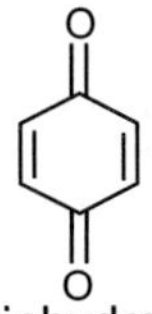 $+ (NH_4)_2SO_4 + 4MnSO_4 + 4H_2O$

Quinhydrone

(*c*) As catalyst for preparation of dioxygen

$$2KClO_3 \xrightarrow{MnO_2} 2KCl + 3O_2$$

About technetium and rhenium, TcO_2 and ReO_2 exhibit stable +4 oxidation state and are black, and brown colour, respectively. They are produced by the following reactions:

(*i*) By burning metal in a limited quantity of oxygen
(*ii*) By heating Tc_2O_7 with T_c
(*iii*) Thermal decomposition of meta-molybdates NH_4MO_4(M=Tc or Re)
(*iv*) Hydrated oxides are prepared by reducing TcO_4^- and ReO_4^- by Zn/HCl

TcO_2 is insoluble in alkali, but ReO_2 reacts to form ReO_3^{2-} rhenites. TcS_2 and ReS_2 are effective catalysts for hydrogenation and are preferred over Pt as they are not poisoned by sulphur compounds.

Regarding reactions with halides, technetium forms tetra and hexa halides. The $TcCl_4$ is a solid crystal structure and $TcCl_6$ consists of distorted edge-sharing octahedral groups linked into endless *zig-zag* ***chains***. It is paramagnetic and has no metal-metal bond. The rhenium complex $[ReCl_6]^{2-}$ is octahedral.

$$ReO_4 \xrightarrow[+KI]{\text{Con. HCl}} [Re^{IV}Cl_2]^{2-}$$

$[ReCl_6]^{2-}$ is hydrolysed

$$[ReCl_6]^{2-} + H_2O \longrightarrow ReO_2.(H_2O)_n$$

Other halide complexes are $[MF_6]^{2-}$, $[MBr_6]^{2-}$ and $[MI_6]^{2-}$ $[Tc^{IV}(CN)_6]^{2-}$ are prepared from $[MI_6]^{2-}$ and KCN. Re is oxidised by CN to form $[Re^{V}(CN)_8]^{3-}$.

4.5.4 +V State Compounds

The known compound is manganate (V) ion MnO_4^{3-} occurs as bright blue salt K_3MnO_4. It is produced by reducing $KMnO_4$ solution with excess sodium sulphite. It tends to disproportionate. Tc +5 state is rare, while Re(+5) is readily hydrolysed by water and disproportoinate.

$$ReCl_5 + 8H_2O \longrightarrow H\,Re^{VII}O_4 + Re^{IV}O_2 + (H_2O)_n$$

The $ReCl_2$ is dimeric and without metal-metal bonding. The oxide of rhenium is also known as Re_2O_5.

4.5.5 +VI State Compounds

The compounds of +6 state of Mn are called manganates. They are dark green and prepared by oxidising MnO_2 in fused KOH in the presence of air as well as by treating $KMnO_4$ with alkali:

$$4MnO_4^- + 4OH^- \longrightarrow 4MnO_4^{2-} + O_2 + H_2O$$

They are stable in strong alkali, whereas in dilute alkali, it disproportionates water or acid solution. They act as strong oxidsing agents.

$$3MnO_4^{2-} + 4H^+ \longrightarrow 2MnO_4^- + MnO_2 + 2H_2O$$

Regarding rhenium and technetium, Re(+6) oxide is known as ReO_3 (red oxide) and TcO_3 is doubtful.

The hexahalides of technetium and rhenium are well-known *i.e.* TcF_6, ReF_6 and $ReCl_6$. These hexahalides are prepared by direct combination of elements. These compounds have low melting point therefore, are sensitive to water and can be hydrolysed.

$$3ReF_6 + 10H_2O \longrightarrow 2HReO_4 + ReO_2 + 18HF$$

4.5.6 +VII State Compounds

Though +7 state compounds are not much in number, but the most commonly known compound is potassium permanganate *i.e.,* $KMnO_4$ comes under this category. It is widely used as oxidising agent in analytical chemistry. Also, volumetric tlritamations of $KMnO_4$ are carried out in acidic medium where MnO_4 is reduced to Mn^{2+}.

$$MnO_4^- + 8H^+ + 5e^- \longrightarrow Mn^{2+} + 4H_2O \quad E° = 1.51\ V$$

In alkaline solution, $Mn^{IV}O_:$ is the product,

$$MnO_4^- + 2H_2O + 3e^- \longrightarrow MnO_2 + 4OH^- \quad E° = 1\ 23\ V$$

The course of reaction depends upon the pH value of medium. $KMnO_4$ acts as a self indicator. Acid solution of $KMnO_4$ decomposes slowly.

Its decomposition is catalysed by sunlight therefore, it is stored in coloured bottles.

$$4MnO_4^- + 4H^+ \longrightarrow 4MnO_2 + 2H_2O + 3O_2$$

Addition of $KMnO_4$ to concentrated H_2SO_4 results in the formation of MnO_3^+ ions which is green solution.

$$KMnO_4 + 3H_2SO_4 \longrightarrow K^+ + MnO_3^+ + 3HSO_4^- + H_3^+O$$

If large quantity of $KMnO_4$ is used then Mn_2O_7 is formed which is an explosive oil. For manufacturing $KMnO_4$ in large quantity, the following reaction is performed:

$$\underset{}{MnO_2} \xrightarrow[\text{with KOH + KNO}_3]{\text{Fused}} \underset{\text{Manganate Ion}}{MnO_4^{2+}} \xrightarrow[\text{in alkalin solution}]{\text{Electrolysed}} \underset{\text{Permanganate Ion}}{MnO_4^- + e}$$

In addition to the above methodology, strong oxidising agents like PbO_2 or $NaBiO_3$ are used with Mn^{2+} ions to form permanganates. Qualitatively presence of Mn^{2+} can be confirmed by treating the Mn^{2+} solution with $NaBiO_3$ when purple colouration of $KMnO_4$ is observed.

Being a good oxidising agent, $K.MnO_4$ is frequently used in several organic reactions for the preparation biological products *i.e.* saccharin, vitamin C, niacin *etc.* It is used as effective disinfectant and used extensively in water treatment industry. It kills bacteria due to oxidation and leaves no bad taste in water. About technetium and rhenium, Tc(+7) and Re(+7) occur in the form of several compounds *e.g.,* hepta oxides M_2O_7, hepta sulphides M_2S_7, MO_4^-. They are relatively stable. Tc_2O_7 and Re_2O_7 are stable yellow solids. $HTcO_4$ and $HReO_4$ are colourless acids having tetrahedral structures. Hydride complexes $K_2[TcH_9]$ and $K_2[ReH_9]$ are well known.

4.5.7 Biological Use

Mn^{2+} is an important constituent and activator of multiple enzymes in animal and plants. It converts nitrogeneous waste into urea. Urea is carried by blood to the kidneys where it is filtered and excreted as urine. It is an essential trace element for plant growth and added along with the fertilizers.

4.6 THE IRON GROUP

Iron finds greater use than other elements. The major use of iron is in steel making which is important all over the world. Regarding biological world, it is important constituent which has several important uses *i.e.,* forms a part of haemoglobin which carries oxygen in the blood, as, myoglobin for oxygen storage, scavenging and storage in nitrogenase enzyme *etc.* It also forms many unusual complexes. The other eight elements in group VIII are ruthenium (Ru), osmium (Os), cobalt (Co), rhodium (Rh), iridium (Ir), nickel (Ni), palladium (Pd), platinum (Pt).

4.6.1 Electronic Structures and Oxidation States

Element	Electronic structure	Oxidation states
Fe	[Ar] $3d^6 4s^2$	0, +$\underline{2}$, +$\underline{3}$, (+4), (+5), (+6)
Ru	[Kr] $4d^7 5s^1$	0, +$\underline{2}$, +$\underline{3}$, +4, (+5), +6, (+7), (+8)
Os	[Xe] $4f^4 5d^6 6s^2$	2, (+1), (+2), (+3), +$\underline{4}$, (+5), +6, (+7), + 8

The most stable oxidation states are underlined, less stable in normal type and unstable in ( ) bracket.

4.6.2 Occurrence and Extraction

After O, Si and Al, Fe is the fourth most abundant element in the earth's surface. Also, most of the earth's core consists of iron and nickel. The major ores of iron are haematite (Fe_2O_3), magnetite (Fe_3O_4), limonite (FeO(OH)), siderite ($FeCO_3$) and pyrites FeS_2. India has about 6 per cent reserves in the world. The other two elements of this group Ru and Os are very rare, and are found together with Pt, Cu, Ag and Au. Main source is NiS/Cus ores, where it is present in trace amounts in these ores.

Oxides of iron are the major source for iron extraction and it takes place in blast furnace. These blast furnace is huge structure which is heated by blasts of hot air and with burning coke inside for further rise in temperature. The temperature at the entry of the furnace is 2000°C and 1500°C at the bottom, 200°C at the top of furnace. The top of furnace is charged with iron ore, reducing agent in the form of coke and slag ($CaCO_3$). The furnace functions with blowing of air at the bottom and the coke burns to produce heat and CO. As a result, the iron oxide is reduced by CO. The formed molten iron dissolves 3 to 4 per cent carbon from coke and form pig iron. Pure iron melts at 1535°C, whereas pig iron melts at 1015°C due to impurities present in it. Finally, the molten iron collects through the bottom of the furnace in the hearth.

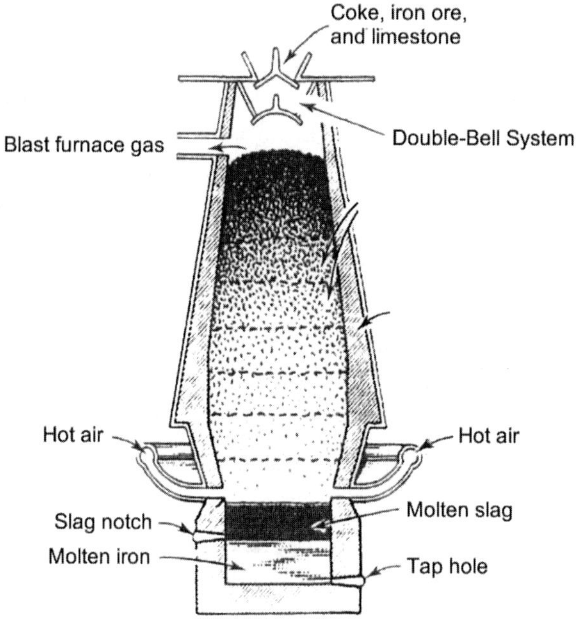

Fig. Diagram of a blast furnace

Due to high temperature, $CaCO_3$ decomposes to CaO which in turn reacts with sand or clay to form $CaSiO_3$ slag. This is molten and collects at the bottom. The slag floats over molten metal protecting metal from oxidation. Molten iron and molten slag have separate outlets. The molten metal is allowed to solidify into pig iron which is hard and brittle. Pig iron cannot be machined due to brittleness. It contains 4.3 per cent carbon and impurities like Si, P, S and Mn.

In order to reduce the brittleness, the non-metal impurities must be removed. Iron in the form of steel can be machined into desired shape *i.e.* roll and is ductile because it has less than 2 per cent carbon. Presence of low carbon in iron is called soft steel and with high carbon called as hard steel. Many times, Mn is present as in iron ore and Mn as impurity is beneficial for steel as it enhances the quality of steel. The slag is used for making building material (cement).

The reactions involved in the overall process for extraction of Fe are

$$Fe_2O_3 + 3C \longrightarrow 3CO + 2Fe$$

$$CaCO_3 + SiO_2 \longrightarrow CaSiO_3 + CO_2$$

With respect to temperature at different stages of the process, the reactions are identified:

At 400°C

$$4Fe_2O_3 + CO \longrightarrow 2Fe_3O_4 + CO_2$$

$$Fe_3O_4 + CO \longrightarrow 2FeO + CO_2$$

At 500-600°C

$$2CO \longrightarrow C + CO_2$$

With deposition of C as soot, reduces FeO at 800°C.

$$FeO + CO \longrightarrow C0_2 + Fe$$

At 900°C

$$CaCO_3 \longrightarrow CaO + CO_2$$

At 1000°C

$$FeO + CO \longrightarrow Fe + CO_2$$

$$CO_2 + C \longrightarrow 2CO$$

By combining the above two reactions we get

$$FeO + C \longrightarrow CO + Fe$$

At 1800°C

$$CaO + SiO_2 \longrightarrow CaSiO_3$$

$$FeS + CaO + C \longrightarrow CaS + Fe + CO$$

$$MnO + C \longrightarrow Mn + CO \text{ (Mn in iron)}$$

$$SiO_2 + 2C \longrightarrow Si + 2CO \text{ (Si in iron)}$$

4.6.3 Steel Manufacture

Steel is obtained by removing impurities like C, Si, P and Mn from pig iron. The impurities are removed as gases or converted into slag.

Wrought iron is the purest form of the metal. It can be obtained by burning off all the carbon and other impurities in pig iron. Wrought iron is extremely soft and malleable and can be converted into steel by adding requisite amount of carbon.

4.6.4 Bessemer Process

This was the first industrial process for the production of steel from molten pig iron. The process is carried out in a large pear shaped container called Bessemer converter made up of steel and has a lining of silica and clay or of dolomite. From its narrow opened upper end, the iron ore, coke and limestone are introduced and the finished product is poured out from the bottom. The wide bottom has a number of perforations from where the blast of compressed air is forced upward into the converter during operation. The converter is tilted at an angle to get the charge and during blow of air, it is turned upright. Later, it in inclined for pouring the molten steel after the completion of operation. During operation, when the air passes upwards through the molten pig iron, the impurities like silicon, manganese and carbon form their respective oxides. There is conversion of Si and Mn to form $MnSiO_3$ slag and these reactions produce excess heat to convert C to CO and CO_2. The low carbon content can be examined by the colour of flame produced, then molten steel is poured into cast iron moulds. This gives bars or blocks of steel which can be rolled or forged. The iron ores contain Si and P as impurities which are oxidised and dolomite is used when the amount of phosphorous is high.

Reactions in the Bessemer process are:

$$P_4O_{10} + 6Fe + 3O_2 \longrightarrow 2Fe_3(PO_4)_2$$

$$Fe_3(PO_4)_2 + 2Fe_3C + 3Fe \longrightarrow 2Fe_3P + 6FeO + 2CO$$

$$FeO + SiO_2 \longrightarrow \underset{\text{Ferrous silicate}}{FeSiO_3}$$

In phosphorous-rich ores, basic Bessemer convertor is used instead.

(*i*) It is lined with basic dolomite or limestone. It reduces reaction between $Fe_3(PO_4)_2$ slag and the lining.

(*ii*) $CaCO_3$ is added to facilitate slag forming. These are basic and react-with P_4O_{10} to form $Ca_3(PO_4)_2$ basic slag. This is used as a phosphatic fertiliser.

4.6.5 Common Oxidation States

The iron has maximum oxidation state as (+VI) which is rare. Common oxidation states of oxides and halides of the Fe group metals listed in the following Table.

Table Oxides and halides of Fe, Ru and Os

Oxidation states							
(+II)	**(+III)**	**(+IV)**	**(+V)**	**(+VI)**	**(+VII)**	**(+VIII)**	**Others**
FeO	Fe_2O_3	–	–	–	–	–	Fe_3O_4
–	Ru_2O_3*	**RuO_2**	–	$(RuO_3)^h$	–	RuO_4	
–	–	**OsO_2**	–	$(OsO_3)^h$	–	**OsO_4**	
FeF_2	FeF_3	–	–	–	–	–	
$FeCl_2$	$FeCl_3$	–	–	–	–	–	
$FeBr_2$	$FeBr_3$	–	–	–	–	–	
FeI_2	–	–	–	–	–	–	
–	RuF_3	RuF_4	**RuF_s**	(RuF_6)			
–	$RuCl_3$	$RuCl_4$	–	–	–	–	
–	$RuBr_3$	–	–	–	–	–	
–	RuI_3	–	–	–	–	–	
–	–	OsF_4	OsF_5	OsF_6	(OsF_7)	–	
–	–	**$OsCl_4$**	$OsCl_5$	–	–	–	$OsCl_{3\,5}$
–	–	$OsBr_4$	–	–	–	–	
–	–	OsI_4	–	–	–	–	

Note: The metal compounds with most stable oxidation state are bold, while the unstable are shown in brackets.

Fe(II) and Fe(III) are slightly oxidising. In Ru and Os, +VIII state is shown in RuO_4 and OsO_4. Ru(III) and Os(IV) are their most stable oxidation states although Ku(V), Os(VI) and Os(VIII) are reasonably stable. It has been observed that the higher oxidation states become more stable, while descending the group elements.

4.7 GENERAL CHARACTERISTICS

The pure iron is a bright silvery-white metal and rapidly oxidizes (rusts) in moist air or in water. It is malleable, ductile, strongly magnetic and not very hard, but quite reactive. While oxidising, the metal converts to $Fe_2O_3.xH_2O$ hydrated ferric oxide or rust. It dissolves in cold dilute non-oxidising acids forming Fe^{2+} salt and liberating hydrogen. It forms Fe^{3+} and Fe^{2+} salts in warm acids and air.

With oxidising agents like concentrated HNO_3 or $K_2Cr_2O_7$ the metal forms a protective coating of metal oxide. Fe is slightly amphoteric. It is also affected by concentrated NaOH.

Regarding ruthenium and osmium, both elements are noble and resistant to acids. But Os can be oxidised with aqua regia to form OsO_4. Rusting is very common in iron and it is a special type of corrosion and in simple way for understanding, it is the conversion of Fe° to Fe^{2+} ions and electrons. In aqueous or moist condition, electrons discharge H^+ ions present in water as H_2 which reacts with atmospheric oxygen to produce H_2O.

$$Fe^{\circ} \longrightarrow Fe^{2+} + 2e^-$$

$$2H^+ + 2e^- \longrightarrow 2H + \underset{\text{(air)}}{\frac{1}{2}O_2} \longrightarrow H_2O$$

Being positive, Fe becomes anode and the impurity of noble metal in Fe becomes cathode to form electrochemical cell. As a result, Fe^{2+} ions get oxidised to form FeO.OH, Fe_2O_3 or Fe_3O_4. The oxides do not form adhering protective film therefore, corrosion continues. In order to protect Fe from corrosion, O_2, H_2O and other impurities should be cut off from Fe. This can be achieved by (*i*) electroplating (*ii*) tinning or (*iii*) galvanising. Also, painting Fe surface with Pb_3O_4 or treating with H_3PO_4 to form $FePO_4$ is effective. Alternatively, sacrificial anode and impressed current method may be used.

The effect of lanthanide contraction is less in Ru and Os as compared to other groups. The horizontal similarity is more pronounced which is attributed to similar size Fe^{2+} (0.61 Å) low spin, Co^{2+} is 0.65 Å low spin and Ni^{2+} is 0.69 Å.

4.7.1 Low Oxidation State Compounds

The compounds of low oxidation (–11) state are not very common in this group, though, Fe and Ru do form carbonyl ions like $[Fe(CO)_4]^{2-}$ ion and $[Ru(CO)_4]^{2-}$.

There are examples of carbonyl compounds in which zero valent state occur and they may be mononuclear *e.g.,* [$Fe(CO)_5$], [$Ru(CO)_5$] and [$Os(CO)_5$]; or dinuclear *e.g.,* [$Fe_2(CO)_9$] and [$Os_2(CO)_9$]; or trinuclear *e.g.,* [$Fe_3(CO)_{12}$]. [$Ru_3(CO)_{12}$] and [$Os_3(CO)_{12}$],

[$Fe(CO)_5$], [$Ru(CO)_5$] and [$Os(CO)_5$] occur in liquid state at room temperature, whereas other carbonyls are volatile solids. [$Fe(CO)_5$] has commercial availability. The di-and trinuclear carbonyls compounds have M–M bonds.

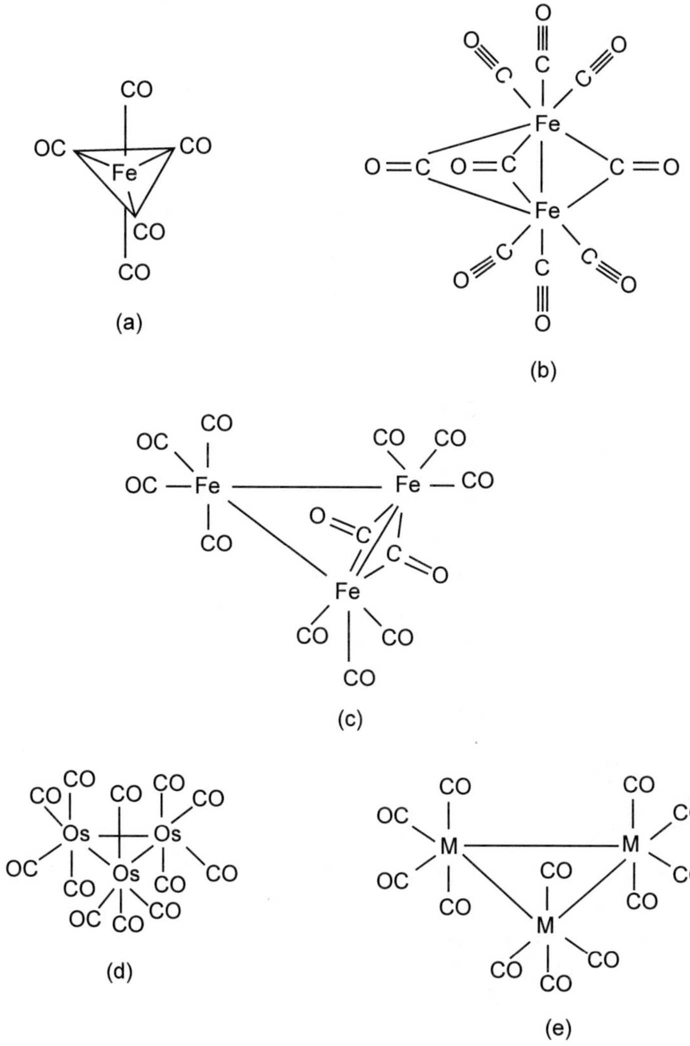

Fig. Structures of [$Fe(CO)_5$], [$Fe_2(CO)_9$]. [$Fe_3(CO)_{13}$], [$Os_3(CO)_{12}$], [$Ru_3(CO)_{12}$]

4.7.2 $[Fe(CO)_5]$

It is also called as Carbonyl Iron Powder and chemically iron pentacarbonyl. Iron pentacarbonyl is one of the homolytic metal carbonyls. $Fe(CO)_5$ is volatile most frequently encountered liquid metal complexes. $Fe(CO)_5$ is used in magnetic coils of high-frequency for radios and televisions, active ingredients of radar absorbent materials and also as a chemical precursor for the synthesis of various iron-based nanoparticles. It can be easily oxidised and its gas forms explosive mixture. By reacting with aqueous alkali or water give carbonylate anions.

$$[Fe(CO)_5] + 3NaOH \longrightarrow Na[HFe(CO)_4] + Na_2CO_3 + H_2O$$

The carbonyl compounds undergo substitution reactions, CO ligand is replaced by other ligands *e.g.* PF_3, PCl_3, PPh_3 $AsPh_3$ and benzene.

$$[Fe(CO)_5] + PF_3 \longrightarrow [(PF_3)Fe(CO)_4] + CO$$

$$[Fe_2(CO)_9] + 6PPh_3 \longrightarrow Fe_2(CO)_3(PPh_3)_6 + 6CO$$

These carbonyl compounds can also be used as catalysts in many reactions.

$$2HC{\equiv}CH + 3CO + H_2O \xrightarrow{[Fe(CO)_5]} \underset{\text{p-Quinol}}{HO{-}C_6H_4{-}OH} + CO_2$$

$$C_6H_5NO_2 + 2CO + H_2 \xrightarrow{[Ru_3(CO)_{12}]} C_6H_5NH_2 + 2CO_2$$

Formation of nitrosyls compounds by substituting CO by NO *e.g.*

$$Fe(CO)_5 + 2NO \longrightarrow [Fe(CO)_2(NO)_2] + 3CO$$

3CO groups are substituted by 2NO groups as 2NO donates 6 electrons.

The formed complex $[Fe(H_2O)_5NO]$ is the brown ring test for nitrates in salt analysis. The colour is due to charge transfer spectra. The complex carries 'Fe' as +1 and NO^+ nitrosyl ion.

4.7.3 +11 Oxidation State Compounds

Among all the oxidation states, this is the most important oxidation state. Its salts are called ferrous salts which are pale green crystalline compounds and contain $[Fe(H_2O)_6]^{2+}$ ion. *e.g.,* $FeSO_4.7H_2O$, $FeCl_2.6H_2O$, $Fe(ClO_4)_2.6H_2O$. They are readily oxidised to Fe(III), and it is difficult to obtain them in pure state. They are reasonably pure in double salt $FeSO_4.(NH_4)_2SO_4.6H_2O$ which is primary standard in volumetric titrations with $KMn0_4$, $K_2Cr_2O_7$ and ceric solutions. $FeSO_7$ and H_2O_2 are used as Fenton's reagent for oxidising alcohols to aldehydes.

4.7.4 Oxides

The iron oxide (FeO) is non-stoichiometric with chemical formula $Fe_{095:1}O$. It is obtained in the form of black powder by strong heating of ferrous oxalate in vaccum.

$$Fe(C_2O_4) \longrightarrow FeO + CO_2 + CO$$

$$4FeO \longrightarrow \underset{(FeO.Fe_2O_3)}{Fe_3O_4} + Fe(II)$$

FeO is basic in nature and can be easily dissolved in acids. It has cubic close-packed structure with Fe^{2+} ions occupying all octahedral holes. The ferrous hydroxide ($Fe(OH)_2$), white solid can be precipitated from the solutions having Fe(+II). But when it comes in contact of O_2 from air and turns dark green and later brown colour because it oxidised to a mixture of $Fe(OH)_2$ and $Fe(OH)_3$ and finally $Fe_2O_3.xH_2O$. $Fe(OH)_2$ dissolves in acids as well as in NaOH. In NaOH, it forms blue-green complex $Na_4[Fe(OH)_6]$ which can be crystallised.

4.7.5 Halides

Iron reacts with halogen acids to form hydrated dihalides like $FeF_2.8H_2O$ (white), $FeCl_2.4H_2O$ (pale green) and $FeBr_2.4H_2O$ (pale green). The dihalide $FeCl_2.4H_2O$ has an octahedral unit in which Cl-atoms are in *trans* positions. When iron is heated with gaseous HF or HCl, it forms anhydrous FeF_2 and $FeCl_2$, respectively. On heating, there is a possibility of formation of FeF_3 and $FeCl_3$ also. $FeBr_2$ and FeI_2, can also be prepared by heating the elements.

4.7.6 Complexes

Iron Fe^{2+} forms a large number of complex compounds. The most important complex of Fe^{2+} is haemoglobin (red pigment of blood). The complexes of Fe^{2+} are either tetrahedral or octahedral. $K_4[Fe(CN)_6]$ is an important complex. It is a yellow solid. It can be produced by the action of Fe^{2+} ions and CN^- ions. Potassium ferrocyanide is used for the test of Fe^{2+} and Fe^{3+} ions in solution.

$$K_4[Fe(CN)_6] + FeCl_2 \longrightarrow K_2Fe^{II}[Fe^{III}(CN)_6] \quad + \quad 2KCl$$

white precipitate of potassium ferroferricyanide

Fe^{3+} ions give prussian blue precipitate or colouration.

$$K_4[Fe(CN)_6] + FeCl_3 \longrightarrow KFe^{III}[Fe^{II}(CN)_6] \quad + \quad 3KCl$$

blue precipitate of potassium ferriferrocyanide

If Fe^{2+} ions are reacted with $K_3[Fe(CN)_6]$, it gives deep blue colour. This is called Turnbull's blue.

$$K_3[Fe(CN)_6) + FeCl_2 \longrightarrow KFe^{II}[Fe^{III}(CN)_6]$$

Turnbull's blue

Both Prussian blue and Turnbull's blue are dissolved in oxalic acid to get fountain pen ink and used in paint. Reduction potential value shows that it is easier to oxidise $[Fe(CN)_6]^{4-}$ in aqueous solution to $[Fe(CN)_6]^{3-}$ than to oxidise $[Fe(H_2O)_6]^{2+}$ to $[Fe(H_2O)_6]^{3+}$.

$$[Fe^{III}(H_2O)_6]^{3+} + e \longrightarrow [Fe^{II}(H_2O)_6]^{2+}\ E° = +0.77V$$

$$[Fe^{III}(CN)_6]^{3-} + e \longrightarrow [Fe^{II}(CN)_6]^{4-}\ E° = +0.36V$$

$K_4[Fe(CN)_6]$ can precipitate Cu(II) ions from solution to form red-brown complex of Cu(II) $[Fe(CN)_6]$, The $K_4[Fe(CN)_6]$ is kinetically inert therefore, its CN^- group is not easily removed or substituted. The salt is non-poisonous but care should be taken as there is evolution of HCN with dilute acids. Single CN^- group may be substituted by H_2O, CO, NO_2^- or NO^+. The well known salt is sodium nitrosopentacyanoferrate (II) having formula as $Na_2[Fe(CN)_5NO].2H_2O$. It reacts with sulphide ion to forrn purple complex.

$$2[Fe(CN)_5(NO)]^{2-} + S^{2-} \longrightarrow 2[Fe(CN)_5(NOS)]^{4-}$$

Fe(II) form stable complexes with ethylene diamine, $2–2^+$ dispyridyl and 1,10 phenanthroline. Complex with phenanthroline is bright red. It is used in colorimetric determination of iron. In comparison to oxidising $[Fe(H_2O)_6]^{2+}$ to $[Fe(H_2O)_6]^{3+}$, it is difficult to oxidise $[Fe(phen)_3]^{2+}$ (red colour) to $[Fe(phen)_3]^{3+}$ (blue colour) due to π bonding between metal and the ligand.

4.7.7 Test for NO_3^- or NO_2^-

The brown ring test for nitrates depends upon the brown complex $[Fe(H_2O)_5NO]^{2+}$. The test reveals the mixing of freshly prepared $FeSO_4$ solution with the solution of NO_2^- or NO_3^- ions in a test tube. By adding concentrated H_2SO_4 along the side of the test tube, it forms a thick layer at the bottom. Chemically H_2SO_4 reacts with NO_3^- ions to form NO which combines with Fe^{2+} ions to give brown coloured $[Fe(H_2O)_5NO]^{2+}$ complex at the interface of the two liquids. The brown colour disappears if the mixture is heated or shaken. NO is evolved and yellow solution of $Fe_2(SO_4)_3$ is obtained. Regarding nitrites (NO_2^-) give brown colour before the addition of H_2SO_4.

Besides the octahedral complexes, heavier halides give tetrahedral complexes of the type $[FeX_4]^{2-}$. They are formed with large cations and large anions which give high lattice energy.

Simple compounds of Ru(II) and Os(II) are very few, but $[Ru^{II}(H_2O)_6]^{2+}$ is well known. The tetrahedral complexes are formed with ligands which are able to perform back bonding, *e.g.,* CO and PR_3 (phosphine ligand). Complexes with CN^-, Cl^-, NH_3 and amines are also common. Regarding ruthenium, it forms complex which is binuclear carboxylate $Ru_2(CH_3COO)_4(H_2O)_2$ and contain M–M bond. The binuclear dinitrogen complexes of Ru(II) where N_2 ligand may act as a bridging group was reported.

$$[(H_3N)_5RuN_2]^{2+} + [Ru(NH_3)_5H_2O]^{2+} \longrightarrow [(H_3N)_5—Ru—N—N—Ru(NH_3)_5]^{4+}$$

4.7.8 Compounds with Oxidation State (III)

Iron in the +III state compounds are called ferric salts and are obtained by oxidising the corresponding Fe(+II) compounds which are usually yellow-brown in colour. The yellow-brown colour is due to colloidal iron oxide or FeO.OH. Except I^-, all the common anions form salt with ferric ions and salts exist in both forms *i.e.,* anhydrous and hydrated. The Fe(+III) salts exist in various colours *i.e.* $FeCl_3.6H_2O$, $FeF_3.4\frac{1}{2}\ H_2O$ and $Fe_2(SO_4)_3.9H_2O$ (all yellow); $FeBr_3.6H_2O$ and FeF_3 are green in colour; $Fe(NO_3)_3.6H_2O$ is colourless, $Fe(NO_3)_3.9H_2O$ is pale purple. In aqueous solution, $[Fe(H_2O)_6]^{3+}$ ions exist. $Fe_2(SO_4)_3$ is a very common salt and exist in six different hydrates. Ferric sulphate is a very useful salt in our life for many purposes like as a coagulant for purifying drinking water and in the treatment of industrial effluent and sewage. There are some double and important salts called alums, e.g. ammonium ferric alum $(NH_4)_2SO_4.Fe_2(SO_4)_3.24H_2O$ and chrome alum $K_2SO_4.Cr_2(SO_4)_3.24H_2O$, and can be crystallized. They are used as mordants in dyeing.

4.7.9 Oxides

$FeCl_3$ on hydrolysis gives reddish brown precipitate of hydrous oxide as $Fe_2O_3.(H_2O)_n$ and a part of this precipitate is FeO.OH. By heating this hydrated oxide at 200°C it forms brown α-Fe_2O_3 having hexagonal close packed lattice. It has been found that by oxidising magnetite (Fe_3O_4), it gives γ-Fe_2O_3 which has cubic close-packed arrangement of ions of O^{2-} and Fe^{3+}. The physical nature of Fe_3O_4 is black solid and obtained by igniting Fe_2O_3 at 1400°C. It is mixed or double oxide of $FeO.Fe_2O_3$. All the oxides like $\overline{FeO}$, $\overline{Fe_2O_3}$ and $\overline{Fe_3O_4}$ are non-stoichiometric. Fe_2O_3 is basic and difficult to dissolve in acid, while FeO.OH dissolves in acids to form pale violet $[Fe(H_2O)_6]^{3+}$ ion and in concentrated NaOH, $[Fe(OH)_6]^{3-}$ is formed. This proves slightly amphoteric character.

$$Fe_2O_3 + 2NaOH \longrightarrow \underset{\text{sodium ferrite}}{2NaFeO_2} + H_2O$$

Also, there is formation of red purple FeO_4^{2-} ferrate ions by passing Cl_2 through an alkaline solution of hydrated Fe_2O_3. Ferrates are stronger oxidising agents than $KMnO_4$.

4.7.10 Halides

By combining directly the metal and halogen, the reaction gives FeF_3, $FeCl_3$ and $FeBr_3$. It has been observed that Fe(III) salts are less ionised than Fe(II) salts. The formed ferric fluoride (FeF_3) can form complexes by combining with alkalic metal fluorides as $Na_3[FeF_6]$ complexes. The Fe^{3+} complexes are of high spin and octahedral because the orbitals are singly occupied *i.e.* $3d^5$ configuration and of F^- is a weak field ligand. Due to the small size of Fe^{3+}, Fe(+III) complexes are more stable than that of Fe(+II). The colour in Fe(+III) complexes arise due to charge transfer spectra. $FeCl_3$ dissolves in ether and water, giving solvated monomeric species.

$FeCl_3.6H_2O$ is used as an oxidising agent and as a mordant in dyeing.

It is used as a catalyst for the manufacture of CCl_4.

$$CS_2 + 3Cl_2 \xrightarrow[\text{at } 60°C]{FeCl_3 \text{ catalyst}} CCl_4 + S_2Cl_2$$

FeCl, catalyst + 6S

$$CS_2 + S_2Cl_2 \xrightarrow[\text{at } 60°C]{FeCl_3 \text{ catalyst}} CCl_4 + 6S$$

$FeCl_3$ forms high spin complexes with weak field ligands and spin-paired complexes with strong field ligands.

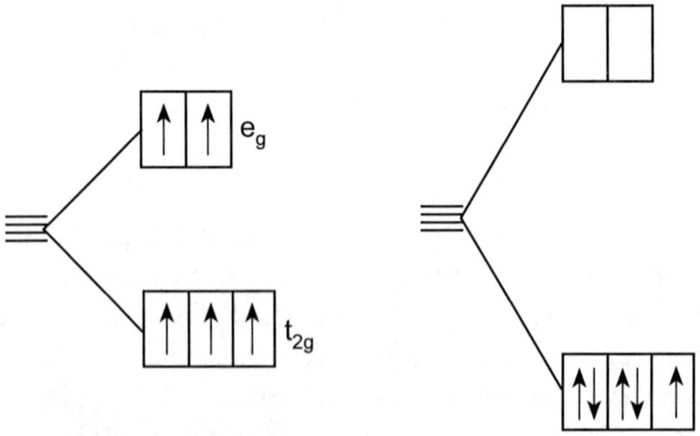

(a) weak ligand high spin (b) strong ligand spin paired

$FeCl_3$ structures:

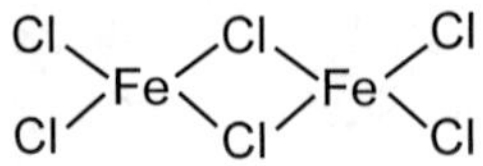

[i] $(FeCl_3)_2$ gas dimer

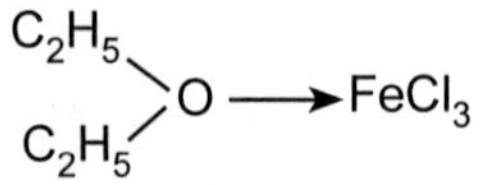

[ii] solvated $FeCl_3$

$[Fe(H_2O)_4Cl_2]Cl.\ (H_2O)_2$

(*ii*) Hydrated $FeCl_3$

The hydrated $FeCl_3$ normally consists of *trans* $[Fe(H_2O)_4Cl_2]Cl.2H_2O$, whereas it forms tetrahedral $[FeCl_4]^-$ ion in strong HCl.

4.7.11 Complexes

The ferric (Fe^{3+}) form complexes preferably with those ligands through O atoms than those where N is donor. The complexes having NH_3 ligand complexes are unstable in water *i.e.* $[Fe(H_2O)_6]^{3+}$ is stable, while $[Fe(NH_3)_6]^{3+}$ is unstable. Also, it has been noted that complexes with chelating N ligands like dipyridyl and 1,10-phenanthroline, are less stable than the complexes formed with Fe(II). These ligands results in spin pairing.

At pH 2 to 3, the hydrated ion $[Fe(H_2O)_6]^{3+}$ pale purple has a tendency to hydrolyse into yellow solution

$$[Fe(H_2O)_6]^{3+} \rightleftharpoons [Fe(H_2O)_5(OH)]^{2+} + H^+$$

On increasing the pH value to 4–5, polymerization of the above formed hydroxy species takes place and develop into a dimer which is a brown solid.

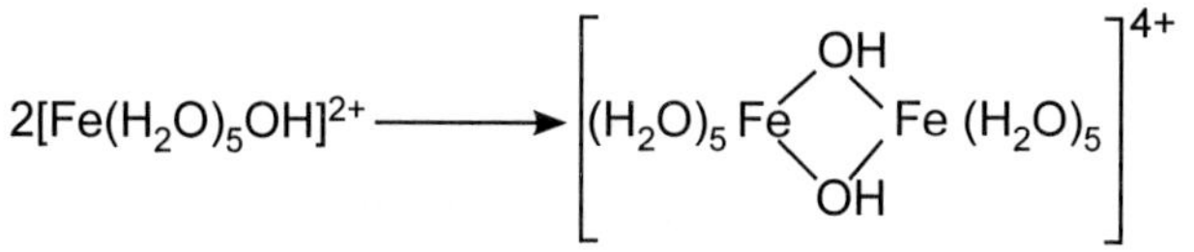

At still higher pH, the precipitates of hydrated oxide are formed which are reddish brown in colour.

The common complexes with O donors are $[Fe(PO_4)_3]^{6-}$, $[Fe(HPO_4)_3]^{3-}$ both colourless and $[Fe(C_2O_4)_3]^{3-}$ dark green are formed. When ferric is treated with oxalic acid for removing the rust stains it produces soluble $[Fe(C_2O_4)^{3-}$ complex. Another complex $[Fe(acac)_3]$ is red in colour, $[Fe(C_2O_4)_3]^{3-}$ is red and $[Fe(phen)_3]^{3+}$ has blue colour. These complexes can be obtained by oxidising Fe(+II) with $KMnO_4$ or by electrolysis.

When the ions of Fe^{3+} are mixed with ions of SCN^- in an aqueous solution, results in blood red complex as $[Fe(SCN)(H_2O)_5]^{2+}$ along with some $Fe(SCN)_3$ and $[Fe(SCN)_4]^-$. Such a complex is useful in the estimation of Fe(III) and this colour is destroyed by adding F^- ions as it forms $[FeF_6]^{3-}$ which is colourless. Fe^{3+} and Mn^{2+} have d^5 high spin configuration with weak field ligands. Due to more charge of Fe^{3+} intense charge transfer spectra is obtained, but not found in Mn^{2+}. With strong field ligands CN^-, SCN^- and oxalate complexes show intense colour, because of *d-d* spectra. Fe(III) and Ru(III) form basic acetates of the formula $[Fe_3O(CH_3COO)_6L_3]^+$.

Regarding ruthenium and osmium, they form stable $[M^{III}(NH_3)_6]^{3+}$ complexes. Ru can also form mixed halogen ammonia complexes as well.

Compounds with +IV state: When Ru and Os bum in the presence of air they form oxides of blue black colour (RuO_2) and copper coloured (OsO_2), and both are solid. The osmium also forms stable tetra fluorides and tetra chloride.

Compounds with (+V) state: The (+V) state exists but it is unstable tetrameric pentaflourides with bridging F groups are known. Also, complexes are known.

$$RuCl_3 + KCl + F_2 \longrightarrow K[Ru^VF_6]$$

Compounds with (+VI) state

The Ferrate FeO_4^{2-} ions come in this category and are the most import compounds where Fe has (+VI) oxidation state. They can be prepared by passing Cl_2 into an alkaline solution of hydrated ferric oxide $Fe_2O_3.xH_2O$ or by oxidising Fe(III) with NaOCl or by electrolytically. Ferrate is purple coloured and a stronger oxidising agent than $KMnO_4$. The sequence of stability of

(+VI) decreases in the order along period $CrO_4^{2-} > MnO_4^{2-} > Fe_4^{2-} >> Co_4^{2-}$. The ferrates decompose in water or acid solution liberating dioxygen, but are stable in alkaline solution.

$$FeO_4^{2-} + 5H_2O \longrightarrow Fe^{3+} + 1\tfrac{1}{2}\, 2O_2 + 10OH^-$$

Na_2FeO_4 and K_2FeO_4 are highly soluble, while $BaFeO_4$ is a precipitate. FeO_4^{2-} ion has a tetrahedral structure like CrO_4^{2-} ion.

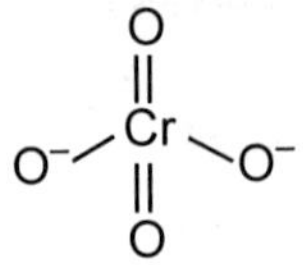

RuF_6 is well known and is prepared by heating Ru and Cl_2 together. OsF_6 is more stable than-RuF_6.

Compounds of (+VIII) state

The maximum oxidation state for the compounds of ruthenium and osmium. Both form tetra oxides as RuO_4 and OsO_4 which are y ellow solids melting at 25°C and 40°C, respectively. OsO_4 is prepared by burning finely divided metal with O_2 or by treating the metal with concentrated HNO_3 whereas, RuO_4 is obtained by oxidation with $KMnQ_4$ and also RuO_4 is less stable. These oxides are toxic and strong oxidising agents. They contain tetrahedral structure and are slightly soluble in H_2O, but highly soluble in CCl_4. The aqueous solution o OsO_4 can be reduced to OsO_2 (black) or Os. The vapours of OsO_4 are injurious to eyes. OsO_4 is important in organic chemistry as it is used for addition to the double bond to give *cis* glycols. Though OsO_4 is not basic, but HCl reduces to *trans*-2-$[Os^{VI}O_2Cl_4]^{2-}$, $[Os^{IV}Cl_6]^{2-}$ and $[Os_2{}^{IV}OCl_{10}]$. Regarding RuO_4, it dissolves in NaOH and liberates O_2 to give perruthennate and ruthennate ion.

$$RuO_8 + 4OH^- \longrightarrow 4Ru^{VII}O_4^- + 2H_2O + O_2$$

perruthennate ion

Ru(+8) is reduced to perruthenate and ruthennate (+6) ion

$$4RuO_4^- + 4OH^- \longrightarrow 4Ru^{VI}O_4^{2-} + 2H_2O + O_2$$

ruthennate ion

RuO_4^{2-} analogous to FeO_4^{2-} ion. OsO_4 is relatively more stable and is not reduced by NaOH and instead, octahedral *trans*-osmate (+8) ion is formed:

$$Os^{VIII}O_4 + 2OH^- \longrightarrow [Os^{VIII}O_4.(OH)_2]^2$$

Perosmate

Osmate (+8) ion can react with NH_3 to produce $[OsO_3N]^-$ complex which t,e reduced by ethyl alcohol to give *trans* $[OsO_2.(OH)_2]$.

4.7.12 Cyclopentadienyl Compounds

Wilkinson first reported the iron-hydrocarbon derivative called di-p-cyclopentadienyliron. It is produced by the following reaction:

$$2C_5H_5MgBr + FeCl_2 \longrightarrow Fe(C_2H_5)_2 + MgBr_2 + MgCl_2$$

Presently it is called Ferrocene. It forms orange crystals stable in air and is diamagnetic, it is soluble in organic solvents and insoluble in H_2O, NaOH solution and concentrated HCl. It is thermally stable up to 500°C. It has a sandwich structure with two parallel planar cyclopentadienyl rings

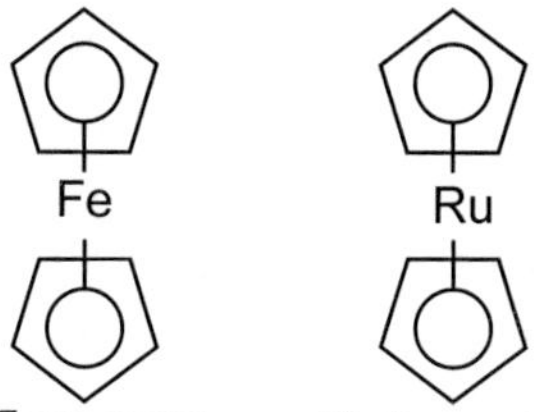

Ferrocene Ruthenocene

Since ail the five C atoms in the cyclopentadienyl ring are equidistant from Fe and it is indicated as η^5-C_5H_5. All the first transition scries elements form similar compounds, but ferrocene is more stable. Ferrocene displays aromatic character, therefore, they undergo FCR as illustrated:

$$(\eta^5\text{-}C_5H_5)_2Fe + CH_3COCl \longrightarrow (\eta^5\text{-}C_5H_4COCH_3)(\eta^5\text{-}C_5H_5Fe) + HCl$$

With excess acetyl chloride, we get disubstituted compound. This confirms that C_5H_5 is a real ligand. The general method of preparation is from cyclopentadiene in THF solution

(i) $C_5H_6 + Na \xrightarrow{THF} C_5H_sNa + ½H_2$

$$2C_5H_5^- + FeCl_2 \longrightarrow (\eta^5\text{-}C_5H_5)_2Fe + 2Cl$$

(ferrocene)

$$2C_5H_5^- + NiCl_2 \longrightarrow (\eta^5\text{-}C_5H_5)_2Ni$$

Nickelocene

(*ii*) Another method uses C_5H_5Tl

$$C_5H_6 + Tl(OH) \xrightarrow{H_2O} C_5H_5Tl + H_2O$$

$$2C_5H_5Tl + FeCl_2 \longrightarrow 2TlCl + (\eta^5\text{-}C_5H_5)_2Fe$$

(*iii*) A more convenient preparation may use a strong base to remove proton from C_5H_6

$$2C_5H_6 + FeCl_2 + 2Et_2NH \longrightarrow Fe(C_5H_5)_2 + 2Et_2NH_2Cl$$

The bonding in these sandwich compounds is π-bonding. The detailed bonding is too complicated to be described, some cyclopentadeinyl compound have two rings at an angle, then they form tetrahedral structures, *e.g.*,

$(\eta^5\text{–}C_5H_5)_2TiCl_2$, $[Ti(\eta^5\text{–}C_5H_5)_2(CO)_2]$ $[Ti(\eta^5\text{-}C_5H_5)_2(NR_2)_2]$, $[Ti(\eta^5\text{-}C_5H_5)_2(SCN)_2]$ and $[V(\eta^5\text{-}C_5H_5)_2Cl_2]$

4.7.13 Bioinorganic Chemistry of Iron

Small amount of iron is necessary in biological world of plant and animal whereas its large amount may become toxic much similar to Cu and Se. Overall it is one of the most important in various biological processes:

1. In animal world, iron is oxygen carrier in the blood.
2. Iron in myoglobin which is an iron- and oxygen-binding protein found in the muscle tissue of vertebrates in almost all mammals and acts as oxygen storage in muscle tissue.
3. Iron is important as an electron carrier in plants, animals and bacteria (cytochromes) as well as involved in electron transfer in plants and bacteria (Ferrodoxines).
4. Storage and scavenging of Fe in animals in the form of ferretin and transferrin complexes that are present in all cells, but most common in bone marrow, liver, and spleen.
5. It is involved in the process of N_2 fixation by bacteria through the enzyme nitrogenase.
6. The role of iron is very important in several enzymes *i.e.* aldehyde oxidase, catalase and peroxidase and succinic dehydrogenase (aerobic oxidation of carbohydrates).

Approximately, 4 gm of iron is reported in human body and out of which 70 per cent of iron is found in haemoglobin pigment of red blood cells, which major amount of the rest is in the form of ferretin. The function of haemoglobin is very important in order to pick up dioxygen from the lungs and through blood it supplies the oxygen to all the different parts of the body where it is transferred to myoglobin molecule and stored. The O_2 is stored in myoglobin till it is used for release of energy from glucose molecules. O_2 from haemoglobin is replaced by H_2O molecules. The protein part of haemoglobin can absorb H^+ and this helps in removing the CO_2 molecules from the tissues, since CO_2 is converted to HCO_3^- and H^+. By this physiological process, the blood removes the more soluble HCO_3^- and the reduced haemoglobin removes the H^+. Then the blood returns to heart through the veins and again pumped into the lungs where HCO_3^- is converted back into CO_2. This CO_2 is exhaled into the air through lungs. So, the blood picks up O_2 again and the whole process is repeated. The important point to notice is the important feature of haemoglobin which acts as an oxygen carrier is the reversibility of the process. Therefore, it is interesting to note that if a very stable complex is formed with O_2 then large energy would have been released in lungs and little energy is released in muscles when O_2, is released in the muscles. When haemoglobin carries oxygen it termed as oxyhaemoglobin and its reduced form is called deoxyhaemogobin.

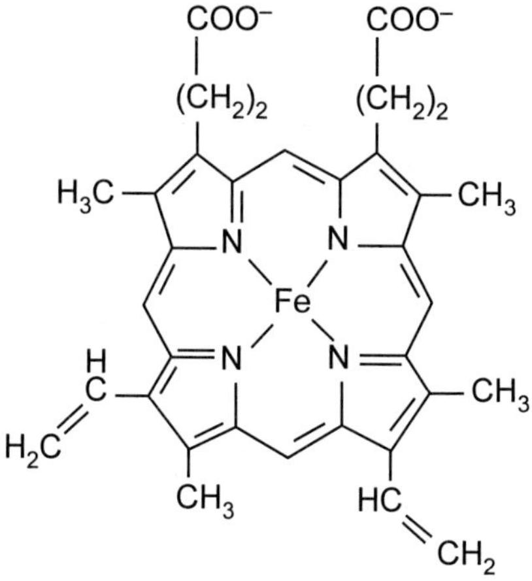

Structure of Haem b

The transfer of O_2 from haemoglobin is also very interesting physiological process in which Fe(II) involves instead of Fe(III). Though, the ligands like CO, CN^- and PF_3 are also present and may get engaged with the O_2 site. Such coordination is also reversible, but very difficult. In case of CN^- ligand, the coordination is irreversible. Therefore, such ligands reduce or prevent the oxygen transfer and cause death of the organism. In addition, the CN^- ligand also interferes with cytochrome enzyme system which causes extreme toxicity. That is why nature has designed our physiological process for respiration in such a way that the complex formed with iron and oxygen in blood is of appropriate stability.

■ ■ ■

5

The Cobalt Group

5.1 INTRODUCTION

Occurrence, Extraction and Utility

The group IX includes elements, such as cobalt, rhodium and iridium which are relatively rare in earth's crust. Among these elements, cobalt is most abundant which accounts for 23 ppm by weight, whereas rhodium and iridium are among the rarest in the earth. These elements have odd atomic numbers.

The main commercial cobalt ores are cobaltite $CoAsS_3$ smaltite $CoAs_2$ and linnacite Co_3S_4. The cobalt occurs in combination with Ni, Cu and Pb ores.

Regarding metallurgical process, the ore is processed through roasting and a mixture of oxides called speisses is obtained. During the process, the byproducts As_4O_{10} and SO_2 gas are produced. The next step is to treat oxides with H_2SO_4 so that Fe, Co and Ni get dissolve and are separated from Cu and Pb. In order to precipitate the iron as hydrous oxide $Fe_2O_3.(H_2O)_n$ lime is added to the solution. Then with addition of NaOCl, $Co(OH)_3$ is precipitated which is ignited to get Co_3O_4. Finally, Co_3O_4 is further reduced by C or H_2 to get Co metal.

(*a*) Roasting of Ore:

$$\text{Co As S (Fe, Co and Ni impurity)} + O_2 \longrightarrow Co_3O_4 + Fe_2O_3 + NiO + As_4O_{10} + SO_2\uparrow\text{(gas)}$$

(*b*) Action of H_2SO_4: $Co_3O_4 + Fe_2O_3 + NiO$ dissolve

(*c*) Action of Lime:

$$(Co_3O_4 + Fe_2O_3 + NiO) + Ca(OH)_2$$
$$\downarrow$$

$$Fe(OH)_3 \text{ is precipitated}$$

***(d)* Action of NaOCl:** Solution containing Co^{2+} and Ni^{2+} ions

$$Co^{2+}/Ni^{2+} + Ca(OH)_2 + NaOCl \downarrow Co(OH)_3\downarrow$$

***(e)* Reduction:**

$$Co_3O_4 + 4C \longrightarrow 4CO + 3Co$$

$$Co_3O_4 + 4H_2 \longrightarrow 4H_2O + 3Co$$

Cobalt is very useful specially to get high temperature alloy with steel and these alloys are used in gas turbines as well as in lathes for cutting tools. In addition, cobalt is also used in glass, ceramic and paint industry and has its importance to counteract yellow colour of Fe and give bright white colour.

Since, cobalt has ferromagnetic property, magnetic alloys are made by it *i.e.* Alnico (Al, Ni and Co) and these magnetic alloys are 20–30 times more powerful than the Fe magnets. In biological world, Co is an essential constituent of several enzymes and vitamin B_{12} (cyanocobalamine), also an important component of fertile soil.

Cobalt also has the property of radioactivity and an artificial isotope ^{60}Co is a radioisotope which undergoes β-decay having half life of 5.2 years. In addition to β-radiation, it also gives intense high energy γ-radiation which is very much effective for curing of cancerous tumours through radiotherapy.

$$^{60}_{27}Co \longrightarrow {}^{60}_{28}Ni + {}^{2}_{-1}e + \nu + \gamma$$

^{60}Co is prepared by irradiation of naturally occurring ^{59}Co in a nuclear reactor and by a nuclear explosion it produces ^{60}Co which is dispersed as nuclear fallout and also called as cobalt bomb.

Electronic Configuration and Oxidation States

				Oxidation states
Cobalt	Co	(Ar)	$3d^7 4s^2$	(–1), 0, (1), **II, III**, (IV)
Rhodium	Rh	(Kr)	$4d^8 5s^1$	(–1), 0,(1), II, **III,** IV, (VI)
Iridium	Ir	(Xe)	$5f^{14} 5d^9$	(–1), 0,(1), II, **III, IV,** V(VI)

Note: *Important stable oxidation states are bold, less important oxidation state are shown in normal type and unstable oxidation states are given in brackets.*

General Oxidation States

It has been observed that the elements in the second half of the transition elements do not use all the outer electrons for bonding in the maximum oxidation state. Co(+V) presence has been disproved and Co(+IV) is unstable.

Regarding rhodium and iridium, the maximum oxidation state is found to be (+VI).

Common oxidation states of cobalt Co(+II) and Co(+IIl), are by far the most important oxidation states. While observing the trend for oxidation state in the later elements of 3*d*-block elements, (+11) state is more stable than (+III) state. In several simple cobalt compouds, cobalt exists as Co^{2+} ions as well as hydrated ion $[Co(H_2O)_6]^{2+}$ whereas, in many simple compounds of Co^{3+} compounds are oxidising and relatively less stable. But, the complexes having Co^{3+} are very stable *e.g.,* $[CoF_6]3^-$ and $[Co(NH_3)_6]^{3+}$. Rh(III) and Ir(III) and lr(IV) are the most stable states.

Their oxides and halides are shown in the table.

Table: Oxides and halides

Oxidation states					
(+II)	**(+III)**	**(+IV)**	**(+V)**	**(+VI)**	**Others**
CoO	$(Co_2O_3)^h$	$(CoO_2)^h$	–	–	Co_3O_4
RhO	**Rh_2O_3**	RhO_2	–	–	
–	Ir_2O_3	**IrO_2**	–	(IrO_3)	
CoF_2	CoF_3	–	–	–	
$CoCl_2$	–	–	–	–	
$CoBr_2$	–	–	–	–	
CoI_2	–	–	–	–	
–	**RhF_3**	RhF_4	(RhF_5)	RhF_6	
–	**$RhCl_3$**	–	–	–	
–	**$RhBr_3$**	–	–	–	
–	**RhI_3**	–	–	–	
–	IrF_3	**IrF4**	(IrF_5)	IrF_6	
$(IrCl_2)$	**$IrCl_3$**	$(IrCl_4)$	–	–	
–	**$IrBr_3$**	–	–	–	
	IrI_3	–	–	–	

The most stable oxidation states are shown in bold, unstable ones in brackets; h = hydrated oxide.

General Characteristics

The cobalt is found to be harder and tougher with high tensile strength in comparison to steel. In appearance, cobalt is bluish white, lustrous metal. It is ferromagnetic, but loses it ferromagnetic property when heated above 1000°C. Cobalt does not react with H_2O, H_2 and N_2. In general, cobalt is relatively unreactive, but reacts with steam forming CoO and it can be oxidised when heated in air and burns at white heat to give Co_3O_4.

Though cobalt is unaffacted by water, but it is slowly attacked by dilute acids and get dissolved, and rendered passive by concentrated HNO_3. It combines with halogens readily and four dihalides of cobalt(ll) are very common *i.e.*, CoF_2, $CoCI_2$, $CoBr_2$ and Col_2,. At high temperature, it reacts with S, C, P, As, Sb, and Sn.

The other two metals rhodium and iridium are also found to be hard metals. Like platinum, they are also noble and unreactive. Among all the elements, Iridium is densest element (22.61 g cm^{-3}) known slightly lower (by about 0.12%) than that of Osmium. At high temperature, Rh and Ir react with O_2 and halogens, but are found to be resistant to acids They also form a large number of complexes.

Physical Characteristics

Table: Some physical properties

	Covalent radius (A)	Ionic radius (A)		Melting point (°C)	Boiling point (°C)	Density (g cm-1)	Pauling's electro-negativity
		M2+	M3+				
Co	1.16	0.745^h	0.61^h	1495	3100	8.90	1.8
		0.65^i	0.545^i				
Rh	1.25	–	0.665	1960	3760	12.39	2.2
Ir	1.26	–	0.68	2443	(4550)	22.61	2.2

h = high spin value, i = low spin radius.

Compounds with *n* bonding ligands, *i.e.* CO, PF_3, NO and CN^-, have lower oxidation states (–1) and (0). The compounds with oxidation state (–1) are tetrahedral complexes, *i.e.,* $Na[Co(CO)_4]$, $[Rh(CO)_4]^-$, $[Co(CO)_3NO]$ and $K[Ir(PF_3)_4]$ and the complex $[Co_2(CO)_8]$ is found to be with zero oxidation state, but the existence of such corresponding complex of Ru is doubtful. Other zero valent compounds are:

$$K_4[Co(CN)_4] \qquad \text{and} \qquad [Co(PMe_3)_4]$$

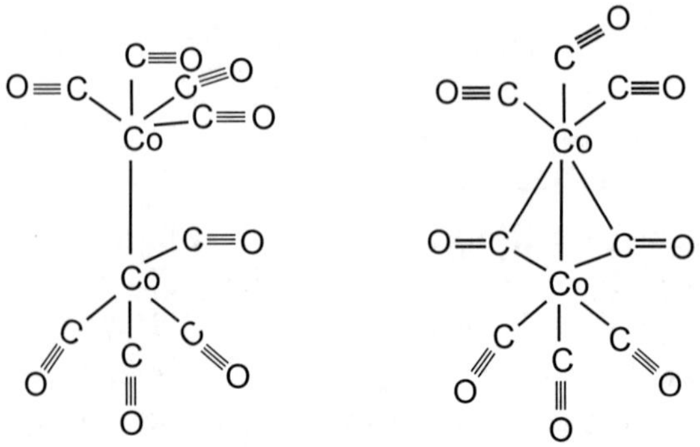

Structure of $[Co_2(CO)_8]$

There is a cluster of four metal atoms having carbonyl, *i.e.*, $[Co_4(CO)_{12}]$, $[Rh_4(CO)_{12}]$ and $[Ir_4(CO)_{12}]$ with M-M bond. In this, there are two possibilities of the CO group which may be apical or bridging. The compounds of Co and Rh has three bridging CO groups, but Ir has none.

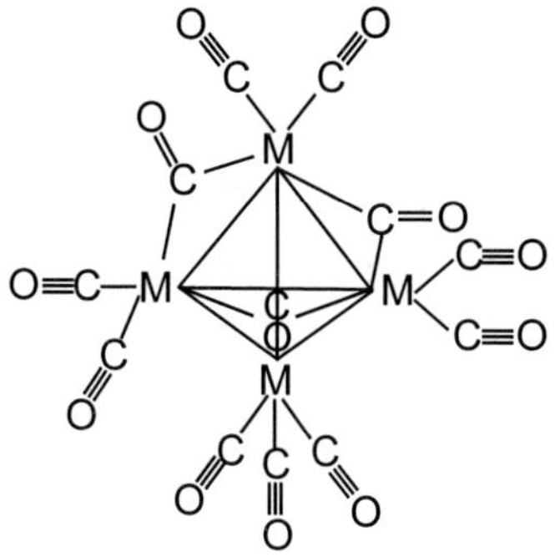

Structures of (a) $[Co_4(CO)_{12}]$ and $[Rh_4(CO)_{12}]$ (b) $[Ir_4(CO)_{12}]$

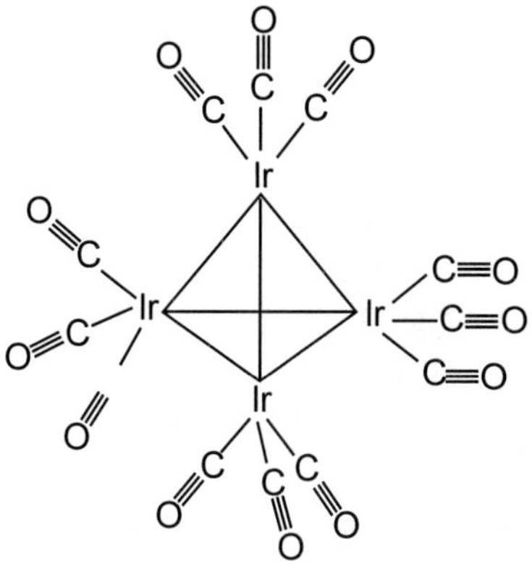

5.1.1 Compounds of (+I) State

The complexes of cobalt with (+1) oxidation state are common. In the first row of transition elements, except for Cu, this state (+1) for Co is well-known. Compounds are made in the presence of ligand and by reducing $CoCl_2$ with Zn or N_2H_4. They offer trigonal bipyramidal or tetrahedral structures. The trigonal bipyramidal structure can be observed in ions of $[Co^ICNR)_5]^+$ and it is formed when ions of $[Co^I(CO)_5]^-$ react with isonitiriles R—NC. There is also formation of dinitrogen complex obtained by uptake of N_2 at atmospheric pressure as shown in the following reaction:

$$[Co(acac)_3] + N_2 + PPh_3 \longrightarrow [Co^1(H)(N_2)(PPh_3)_3]$$

acac stands for acetylacetonate. The structure of the complex having dinitrogen is trigonal bipyramidal.

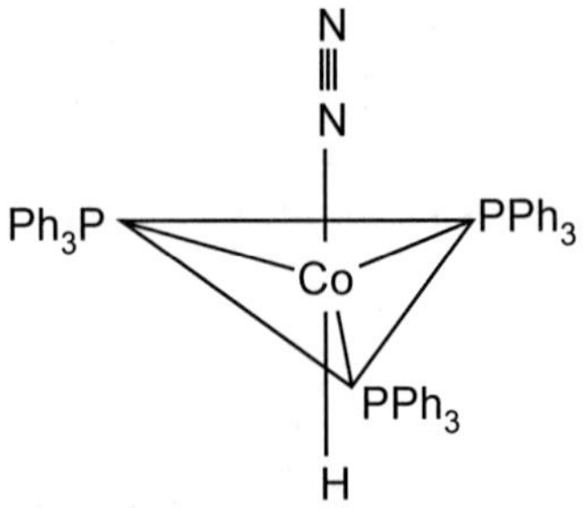

It is also noticed that the reduced form of vitamin B_{12} contains Co(+I).

The complexes of Rh(+I) an Ir(+I) with *pi* bond ligands, *i.e.* CO, phosphines PR_3 and alkenes have usually a square planar structure *e.g.*, Trans-$[IrCl(CO(PPh_3)_2]$ (Vaska compounds) and $[Rh(Cl)(PPh_3)]$ (Wikinson compound) or trigonal bripyramidal $(Rh(H)(CO)(PPh_3)_2]$.

5.1.2 Compounds with (+II) State

Cobalt forms simple compounds in the (+II) state, whereas the complexes are more common in the (+III) state. The (+II) state compounds of Rh and Ir are of little importance. There are several compounds of Co (+II) like CoS, CoO, $Co(OH)_2$, $CoCl_2$, $CoSO_4$, $Co(NO_3)_2$ and $CoCO_3$. Regarding the solubility of cobalt (+II) compounds in water, except $CoCO_3$ all are soluble. The hydrated salts of cobalt consist of $[Co(H_2O)_6]^{2+}$ ions and they are pink or red in colour. When cobalt (+II) solution is reacted with NaOH it gives $Co(OH)_2$ of blue precipitate which on standing turn to pale pink in colour, but it dissolves in very strong NaOH to give blue solution containing $[Co(OH)_4]^{2-}$. It oxidises in air to form brown $Co^{III}O.(OH)$.

On heating $Co(OH)_2$ or $CoCO_3$ in the absence of air, olive green coloured CoO is obtained. When CoO is heated with SiO_2 and K_2CO_3, it results in deep blue glass potassium cobalt (II) silicate. The potassium cobalt silicate is grounded to powder form and called as smalt. The smalt is useful to give blue colour to glass, enamels and glazes. A flame test for distinguishing the potassium from sodium is also done by cobalt glass because yellow colour is absorbed by blue glass and as a result, only the colour of potassium is seen. Hydrated $CoCl_2.6H_2O$ is another (II) state compound of cobalt which is pink in colour and contains $[Co(H_2O)_6]^{2+}$ ion having octahederal structure. When it is partially dehydrated by heating, gives $[Co(H_2O)_4]^{2+}$ ion, which is blue in colour and have tetrahedral structure. But addition of water produces a reverse change in which blue colour changes back to pink colour.

$$\underset{\text{Pink}}{[Co(H_2O)_6]^{2+}} \rightleftharpoons \underset{\text{Blue}}{[Co(H_2O)_4]^{2+}} + 2H_2O$$

Similarly, there is a reaction between octahedral aqua ion and excess of Cl^- which gives blue coloured $[CoCl_4]^{2-}$ tetrahedral ion.

$$[Co(H_2O)_6]^{2+} + 4Cl^- \rightleftharpoons [CoCl_4]^{2-} + 6H_2O$$

Pink Blue

Generally, the complexes of Co(+II) are low spin octahedral. In tetrahedral complexes, the intensity of colour is more than octahedral complexes. The magnetic moment (μ) for both type of complexes, *i.e.* tetrahedral and octahedral is higher than predicted value. Another salt, cobalt (II) acetate $Co(CH_3COO)_2.4H_2O$ having red crystals, is obtained by dissolving $CoCO_3$ in acetic acid which is useful as drying agent for varnish. It is interesting to note that the anhydrous cobalt salts cannot be prepared by heating the hydrated salts otherwise they decompose to form oxide. Therefore, dry methods are adopted. The cobalt fluoride (CoF_2) is anhydrous and pink, obtained by reacting $CoCl_2$ and HF. Another halide $CoCI_2$ is blue, $CoBr_2$, is green and are obtained by heating Co and Cl_2/Br_2. Another halogen compound CoI_2 is blue black and is prepared by heating metal with HI.

There are several complexes of Co(+II), but they less stable than Co(+III) complexes. Among transition metals ion, Co(+II) forms maximum tetrahedral complexes. There is a exceptional type of blue coloured complex $Hg[Co(NCS_4]$ in which Co^{2+} is tetrahedrally coordinated to N atoms. Generally, Co(+II) complexes are high spin, but CN^- produces low spin complexes. When excess of CN^- is treated with a salt solution of Co^{2+} then a green coloured complex $[Co(CN)_5]^{3-}$ is formed which can be isolated as barium salt. This complex has a square pyramidal shape and is paramagnetic with one unpaired electron. Also, it has a property to dimerise to purple $[Co_2(CN)_{10}]^{6-}$ which is diamagnetic in nature and has a structure like $[(CN)_5Co—Co(CN)_5]$. There is further formation of a complex wherein $[Co^{II}(CN)_5]^{3-}$ by oxidising in air gives brown peroxo complex $K_6[(NC)_5—Co^{III}—O—O—Co^{III}(CN)_5]$. It has been found that Co(II) square planar complexes are less common. They do so with bidentate and tetradentate ligands. Tetrahedral complexes have three unpaired electrons and only one in square planar tetradentate. It has been reported that Co^{2+} ions are stable and do not oxidise easily in comparison to Co^{3+} ions which are less stable and can be reduced by water. In contrast, Co(+II) complexes are readily oxidised to Co(+III) complexes which are highly stable

$$[Co^{II}(NH_3)_6]^{2+} \xrightarrow{\text{air}} [Co^{III}(NH_3)_6]^{3+}$$

This is because Co(+III) complexes have d^6 configuration and high CFSE value than Co(+II) with d^7 arrangement with low CFSE value.

Porphyrin complexes of Co(+II) (vitamin B_{12}) are structurally similar to haemoglobin.

Similar to ferrocene, cobalt and rhodium form the organometallic compounds like cobatocene.

$$CoCl_2 + 2NaC_5H_5 \longrightarrow [Co^{II}(\eta^5\text{–}C_5H_5)_2]$$

Cobaltocene
purple dark

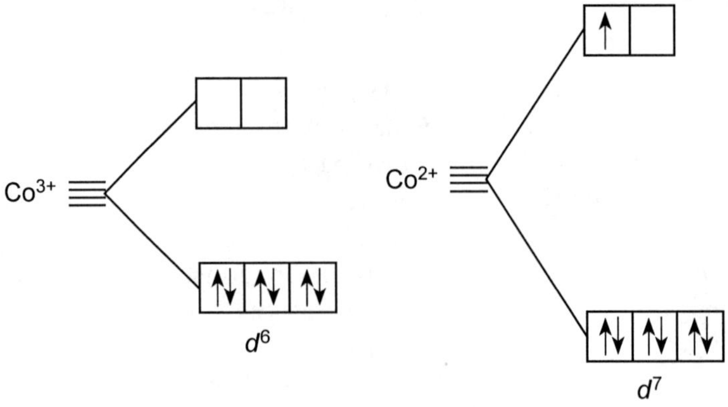

Fig. 5.1: d^6 and d^7 ion in strong octahedral field

Cobaltocene is easily oxidised by air to stable yellow coloured $[Co^{II}(\eta^5\text{-}C_5H_5)_2]^+$. Whereas, rhodocene is less stable and tends to dimerize.Rh and Ir(+II) compounds are very few.

5 1.3 Compound s of (+III) Oxidation State

The oxidation state (+III) is frequently found in all the three elements *i.e.* Co, Rh and Ir of this group. Some simple compounds of cobalt are known for Co(+III) *e.g.,* $Co_2(SO_4)_3.18H_2O$, $KCo(SO_4)_2.12H_2O$ and $NH_4Co(SO_4)_2.12H_2O$. All these are blue coloured with hexa-aqua ion $[Co(H_2O)_6]^{3+}$ and are strong oxidising agents. A compound dicobalt trioxide Co_2O_3 of (+III) is known as hydrated oxide which oxidises water. Another compound CoF_3 is a light brown solid which can be easily hydrolysed by water and used as a fluorinating agent. About anhydrous compound in Co(III) state, at low temperature CoF_3 and non-aqueous N_2O_4 or N_2O_5 are reacted to form $Co^{III}(NO_3)_3$. Also, when Co metal heated in air at 500°C forms black Co_3O_4 which is better represented as $Co^{II}Co^{III}{}_2O_4$.

The preparations of Co(+III) complexes are easier as compared to simple Co(+III) compounds. The number of complexes formed by cobalt is more than any other element.

Some well-known complexes are listed below with their colours.

$[Co(NH_3)_6]^{3+}$	yellow
$[Co(NH_3)_5H_2O]^{3+}$	pink
$[Co(NH_3)_4Cl]^{2+}$	purple
$[Co(NH_3)_4CO_3]^+$	purple
$[Co(NH_3)_3(NO_2)_3]$	yellow
$[Co(CN)_6]^{3-}$	violet
$[Co(NO_2)_6]^{3-}$	orange

All Co(+III) complexes are octahedral where metal has d^6 configuration. Nearly all the ligands cause spin pairing to gain CFSE *e.g.* $t_{2g}^6 e_g^0$. Complexes with N atom as donor are preferred. These complexes are diamagnetic. Generally, these complexes are generated by oxidation of Co^{2+} with H_2O_2 or air in the presence of the desired ligand and these ligands can be easily substituted. These complexes are highly stable.

It has been reported that Co^{3+} has affinity for N donors *i.e.* NH_3, ethylenediamine, EDTA and NO_2^- ligands. In order to conduct qualitative and quantitative analysis for estimation of K^+ in the form of $K_3[Co(NO_2)_6]$, an orange coloured salt, sodium cobaltinitrite $Na_3[Co(NO_2)_6]$, is used. Another complex $[Co(CN)_6]^{3-}$ is extremely stable even in strong alkalis. With bidentate ligands like (en), (acac) or $C_2O_4^{2-}$ ion it form:

$$4Co^{2+} + 12en + 4H^+ + O_2 \longrightarrow 4[Co^{III}(en)_3]^{3+} + 2H_2O$$

The complex ions of potassium *tris* (ethylenediamine) cobalt $[Co^{III}(en)_3]^{3+}$ are optically active and exist in *d* and *l* forms. Similarly, with HC1 it gives a salt of dark green colour *trans* $[Co(en)_2Cl_2]^{1+}$ and also, on evaporation, it forms another purple coloured *cis* isomer. These isomers on heating experience substitution reaction with H_2O to produce $[Co(en)_2Cl(H_2O)]^{2+}$ and later forms $[Co(en)_2(H_2O)_2]^{3+}$. Also, in a similar fashion, the substitution reactions take place with other ligands like NCS^- to give $[Co(en)_2(NCS)_2]^+$. It has seen that complexes with O donors of Co(III) like $[Co(ac\ ac)_3]$ and $[Co(C_2O)_3]^{3-}$ are less stable.

The halogen complexes are not common and $[CoF_6]^{3-}$ is the only familiar hexahalide known. This complex is blue in colour as $[CoF_3(H_2O)_3]$ and unusual with high spin and paramagnetic activity.

The halides of Rh(+Ill) and Ir(-HII) are common. The common RhF_3 is formed by $RhCl_6$ fluorination, whereas IrF_3 is generated by reducing IrF_6 with Ir and rest by direct combination of elements. The rhodium sesquioxide Rh_2O_3 is formed by burning the metal in air and similar oxide of iridium like Ir_2O_3 is not obtained easily because it oxidises to $Ir^{IV}O_2$. $[Rh^{III}(H_2O)_6]^{3+}$ is stable yellow coloured ion. A good number of Rh(+III) and Ir(+III) complexes are known and are prepared by heating Rh or Ir with NaCl and Cl_2

$$2Rh + 6NaCl + Cl_2 \longrightarrow 2Na_3[RhCl_6]$$

Rh on reacting with sodium chloride and chlorine give red coloured $Na_3[RhCl_6]$. $12H_2O$ complex. While on boiling with H_2O, Rh gives $[Rh(H_2O)_6]^{3+}$ and when reacting with NaOH it forms $Rh_2O_3.H_2O$ yellow hydrated oxide which with limited produces $RhCl_3.3H_2O$ and with excess acid it gives $[Rh\ Cl_5]^{3-}$ ion. There are some complexes which are not octahedral, *i.e.* $[Rh\ Br_5]^{2-}$ and $[RhBr_7]^{4-}$ and few complexes show metal-metal bond. It has been noted that reduction of Rh(III) and Ir(III) gives metals while reduction of Co(III) complexes result in favour of oxidation state Co(II).

5.1.4 Compounds with (+IV) Oxidation State

Cobalt has the maximum oxidation state as (+IV). The oxidation of alkaline solution of Co^{2+} results in hydrated CoO_2 and a complex $Ba_2Co^{IV}O_4$. In absence of activated charcoal catalyst for the preparation of $[Co(NH_3)_6]^{3+}$ through air oxidation of Co^{2+}, a brown complex of the type $[(H_3N)_5Co—O—O—Co^{III}(NH_3)_5]^{4+}$ is isolated. The formed complex is stable in concentrated NH_4OH or as solid. However, it can be oxidised by strong oxidising agents like persulphate $(S_2O_8)^{2-}$, as a result green peroxo complex with Co(+IV) oxidation state is formed.

$$[_5(H_3N)\,Co—O—O—Co(NH_3)_5]^{4+} \xrightarrow{\text{Oxidation}} [(H_3N)_5Co^{III}—O—O—Co^{IV}(NH_3)_5]^{5+}$$

It gives d^5 low spin with one unpaired electron.

There are many other common binuclear complexes which use the bridging groups like OH^-, NH_2^- or NH^{2-}.

$$(H_3N)_5Co—NH_2—Co(NH_3)_5]^{5+}$$

blue

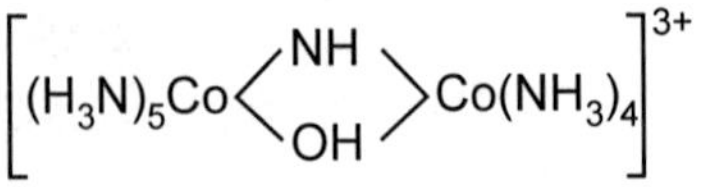

brown

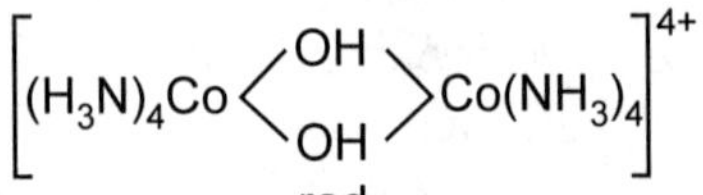

red

Regarding iridium and rhodium, the oxidation state of Ir(+IV) is very stable, but Rh(+IV) is unstable and both form their respective tetrafluorides. $IrCl_4$ is found to be unstable. Regarding their oxides, IrO_2 is produced by burning metal in air, whereas to get RhO_2, strong oxiding agent like sodium bismuthate $NaBiO_3$ is required in an alkaline solution.

There are very few rhodium complexes, *i.e.* $K_2[RhF_6]$ and $K_2[RhCl_6]$, which can react with water to liberate O_2 and finally RhO_2 is formed. About iridium, it forms several halide complexes which are $[IrCl_6]^{2-}$, $[IrCl_5.(H_2O)_3]^+$, $[IrCl_4(H_2O)_2]$ and $[IrCl_5.H_2O]$.

Regarding (+V) and (+VI) oxidation states, cobalt does not form compounds in (+V) state, but Rh and Ir form pentafluorides as $(RhF_5)_4$ and $(IrF_5)_4$, respectively. The complexes represented are $Cs[RF_6]$ and $Cs[IrF_6]$. IrF_6 is stabler than RhF_6. Rh and Ir also exist in (+VI) oxidation state in the form of RhF_6 and IrF_6, but none of them is stable. However, IrF_6 is comparatively more stable than RhF_6.

5.2 THE NICKEL GROUP

On earth, nickel metal occurs in quite good quantity. It forms ferrous and non-ferrous alloys. The most common oxidation state of Nickel is (+II) both in simple ionic compounds and complexes. The geometry of its compounds is mostly square planar or octahedral. The other group members are palladium (Pd) and Platinum (Pt), both have very limited availability and are expensive. In addition, they are noble, usually unreactive and used as catalyst. Generally the oxidation states in which these three elements are found are Ni(+II), Pd(+II), and Pt (+IV).

5.2.1 Occurrence and Extraction

Nickel ores often exist in it sulphide form mixed with Fe and Cu. The most common ore is Pentalandite $(Fe, Ni)_9S_8$. In addition, many other sulphide ores are there, such as millerite (NiS) and arsenide ores such as, NiAs (nickeline) and nickel glance NiAsS. Due to presence of other metal elements as impurity, the extraction process is complex.

The sulphide deposits are processed through froth floatation process and by magnetic separation. The ore is then heated with SiO_2. Due to roasting, the present FeS is converted to FeO which form slag $FeSiO_3$ by reacting with SiO_2 and slag can be eliminated easily. After cooling, the remaining sulphide matter includes Cu_2S as an upper layer and can be removed mechanically. The lower layer is of Ni_2S_3. The separated lower layer of Ni_2S_3 is roasted in air and converted to NiO which can be reduced by carbon in a smelter.

$$Ni_2S_3 + 4O_2 \longrightarrow 3SO_2 + 2NiO$$

$$NiO + C \longrightarrow CO + Ni \text{ (impure)}$$

Pure Ni can be obtained by electrolysis in an aqueous solution of $NiSO_4$ or alternatively by Mond's Process.

Mond's Process

The Mond's process is also called as carbonyl process to extract and purify nickel metal which is mainly by changing nickel oxide into pure nickel. In the process, nickel oxide (NiO) is reduced by water gas (H_2 + CO) and forms Ni. In the process, carbon monoxide forms complex with nickel and finally gives volatile nickel tetracarbonyl $Ni(CO)_4$. The nickel tetracarbonyl get decomposed by heating at 230°C, results in pure nickel and CO. The released CO is recycled.

The impurities are left behind as solid.

$$Ni + 4CO \xrightarrow{50^0C} Ni(CO)_4 \xrightarrow{230^0C} \underset{\text{pure}}{Ni} + 4CO$$

Regarding nickel silicate or garnierite ores are smelted with coke after mixing with gypsum ($CaSO_4$). The obtained CaSiO, slag alongwith sulphide matte are treated as above.

Nickel has its importance in improving the quality of steel and makes it resistant to chemical attack. The percentage of nickel in stainless steel is 12-15 per cent, whereas the steel cutlery also has Ni and Cr in the ratio of 10 per cent and 20 per cent, respectively. Nickel is also an important constituent of Alnico alloys which make permanent strong magnets. Another important alloy of nickel is Monel which primarily composed of 68 per cent Ni and 32 percent Cu with traces of Fe and Mn. Nickel is also used as a catalyst in several industrial reduction processes known as Raney nickel in the form of fine grains.

Regarding palladium and platinum, they are known as rare elements, however Pd is slightly more abundant than Pt, but Pt production is higher than Pd. The platinum occurs as sulphide ores of copper and nickel. Platinum is obtained as a by-product from nickel and copper mining processing. The separation of Pt is quite complicated. The Pt and Pd are obtained from last stages of the process when $(NH_4)_2[PtCl_6]$ and $[Pd(NH_3)_2CI_2]$ are ignited to collect respective metals.

Electronic Configuration and oxidation states

Element		Electronic structure	Oxidation states*
Nickel	Ni	$[Ar]\ 3d^8 4s^2$	–1, 0, (+1), **+2**, (+3) (+4)
Palladium	Pd	$[Kr]\ 4d^{10}$	0, +1, **+2**,
Platinum	Pt	$[Xe]\ 4f^{14} 5d^9 6s^1$	0, +1, +2, (+3), **+4,** (+5), +6)

The most stable and abundant oxidation states are bold, less important in normal type and unstable and doubtful in brackets ( ).

5.2.2 Common Oxidation States

Nickel shows (–I) to (+IV) oxidation states, but (+II) stable is most common. Ni (+II) forms many complexes which are square planar or octahedral. The complexes of Ni are green in solution and simple compounds, such as the ions $[Ni(H_2O)_6]^{2+}$. The most significant oxidation state in platinum is Pt(+II) which exists in the form of hydrated ion as $[Pt(H_2O)_4]^{2+}$ and in complexes. The oxidation states of Pt(+II) and Pt(+IV) are equally important and are not ionic. When studying the geometry of Pt(+II) and Pt(+IV) are square planar and octahedral, respectively.

In all the three metals, zero-valent state exists with CO ligand having π-bonding. In platinum, PtF_6 is an example of maximum oxidation state as (+VI) whereas, $[PtF_6]^-$ is a case of (+V) oxidation state. The Ni and Pd have the maximum oxidation state as (+IV) *e.g.*, NiF_4 and PdF_4, respectively.

5.3 GENERAL PROPERTIES

Ni is silvery white, while Pd and Pt are grey-white. These elements are unreactive. Ni is electroplated on other metal to provide a protective coating.

In Ni electroplating, Ni is deposited on metal and provide the protective coating, but the Ni layer can be lost when heated in air. Finely divided form nickel called as Raney Ni is useful as a catalyst. It is quickly oxidised by air The red hot nickel reacts with steam.

The dilute acids are good solvents to dissolve Ni easily to produce hydrated $[Ni(H_2,O)_6]^{2+}$ ions and H_2. Nickel instantly becomes passive if immersed in concentrated HNO_3 and aquaregia. While comparing the reactivity, Pd and Pt are found to be less reactive compared to Ni. Though slowly and in the presence of O_2 or $CI_{2,}$ Pd is found to be soluble in concentrated HC1, but readily in concentrated HNO_3 forming $[Pd^{IV}(NO_3)_2(OH)_2]$. Pt is non-reactive to acids, but dissolves in aquaregia (3HCl:1 HNO_3) producing $H_2[PtCl_6]$ chloroplatinic acid. Nickel being unaffected by NaOH. It is used to build the apparatus for manufacturing NaOH whereas, Pd and Pt are not resistant to alkali *e.g.,* fused Na_2O and Na_2O_2.

When nickel is heated, though slowly, it reacts with halogens and give NiX_2(X=F, Cl, Br or 1). Monel, an alloy of nickel and copper, is used in the construction of apparatus to handle fluorine and corrosive fluorides. Also, on heating Ni reacts with S, P, Si and B. All the three metals of this group absorb a proportionate quantity of and Pd is found to absorb maximum hydrogen. The absorbed hydrogen can be evoloved on heating, therefore, Ni/H_2, Pd/H_2 and Pt/H_2 form good reducing agents. Pt(+II) and Pt(+IV) are able to form several complexes, specially like Co.

5.3.1 Compounds with Oxidation States (–I), O, (+I)

Though all the three metals form zero valent compounds but nickel with oxidation state of (-I) is available as carbonyl anion.

$[Ni_2(CO)_6]^{2-}$. Th zero valent compounds are:

$$Ni + 4CO \longrightarrow [Ni^{\circ}(CO)_4]$$

$$Ni(CO)_4] + 2PF_3 \longrightarrow [Ni(CO)_2(PF_3)_2]$$

The other common complexes are phosphine derivative as $[Ni^{o}(PF_3)_4]$ and mixed compound as $[Ni(CO)_2(PF_3)_2]$. A complex $K_4[Ni^{o}(CN)_4]$ is formed by reduction of $[Ni^{II}(CN)_4]^{2-}$ through ptassium and liquid ammonia, whereas complex of different oxidation states, nickel $K_4[Ni_2{}^{II}(CN)_6]$ is obtained by reducing with hydrazine sulphate in aqueous media. Also, carbonyl hudride $[Ni(CO)_3H]_2.(NH_3)_4$ is obtained when $[Ni(CO)_4]$ is reduced by Na in liquid NH_3.

Pd and Pt are not similar to Ni as they do not form simple carbonyls of the formula $M(CO)_4$, but form phosphine complexes, such as $[Pt^{o}(PPh_3)_4]$ and $[Pt^{o}(PPh_3)_3]$.

$$2K_2[Pt^{III}Cl^4] + N_2H_4 + 8PPh_3 \longrightarrow 2[Pt^{II}(PPh_3)_3] + 4KCl + 4HCl\ N_2$$

$$[Pd^0(CO)(PPh_3)_3] \quad \text{and} \quad [Pd^0(CO)_2(PPh_3)_2]$$

The absence of carbonyls in Pd and Pt can be attributed that they form π bonds less readily than Ni.

5.3.2 Compounds with (+II) State

All the three elements form compounds in (+II) state and Ni forms a good number of simple compounds of oxidation state (+ll). These compounds contain X_2, O_2, S, Se and Te as well as salts $Ni(ClO_4)_2$ of common acids are known. Many green coloured hydrated nickel salts are obtained from hydrated ion $[Ni(H_2O)_6]^{2+}$ and the anhydrous Ni salts are yellow in colour. Nickel also form double salts *e.g.*, $NiS0_4(NH_4)_2SO_4.6H_2O$, with alkali metals and ammonium ions.

Generally, Ni(+II) complexes are quite complicated which have square planar and octahedral geometry. However, few tetrahedral complexes, trigonal bipyramidal and square pryamidal structures are also known. Pd(+ll) andPt(+II) complexes are all square planar.

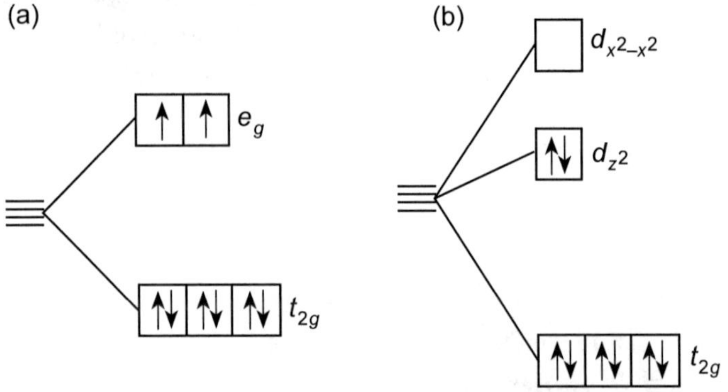

Fig. d^8 arrangement in (a) weak and (b) strong octahedral fields

The octahedral complexes with Ni(II) are $[Ni(NH_3)_6]^{2+}$, $[Ni(H,O)_4(NH_3)_2]^{2+}$ and $[Ni(e\eta)\ )_3]^{2+}$ which are blue in colour and paramagnetic due to d^8 ion having unparied electrons (shown in figure). In these complexes, strong field ligand CN^- are present, where the electrons pair up to form diamagnetic complexes with square planer geometry, such as $[Ni(CN)_4]^{2-}$.

A red colour with geometry of square planar complex is obtained by addition of dimethylglyoxime and Ni^{2+} in a slightly ammoniacal solution. In this complex, the molecules are stacked on top of each other and results in Ni-Ni interaction.

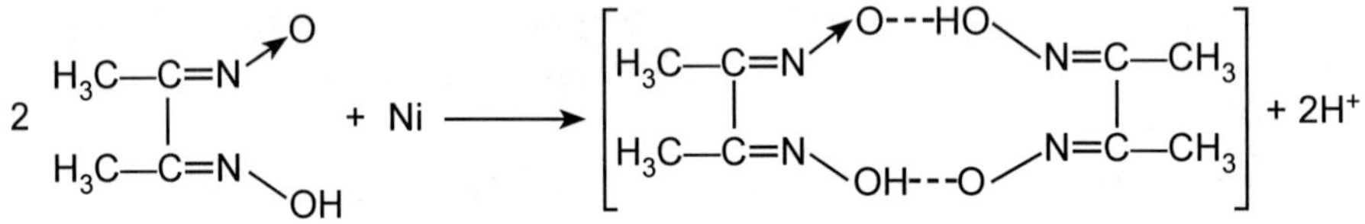

Ni dmg complex

Generally, the sqaure planar complexes are of red, brown or yellow colour. Several tetrahedral complexes containing halides are known *e.g.* $[Ph_4As]_2^+$ $[NiCl_4]^{2-}$; $[(PPh_3)_2\ NiCl_2]$ and $[(Ph_3AsO)_2NiBr_2]$, They are of dark blue colour and paramagnetic.

Regarding (+II) oxidation state of palladium and platinum, both occur in the form of their oxides, halides nitrates and sulphates. The oxide of palladium (PdO) is anhydrous and PtO is hydrated oxide which is unstable. Except PtF_2,both Pd and Pt form their dihalides and PdF_2 is ionic. All Pd(+II) and Pt(+ll) complexes are diamagnetic and their geometry is of square planar. The diamagnetic $[PtCl_4]^{2-}$ ions are formed in HCL The other dihalide compounds are molecular or polymeric as well as diamagnetic.

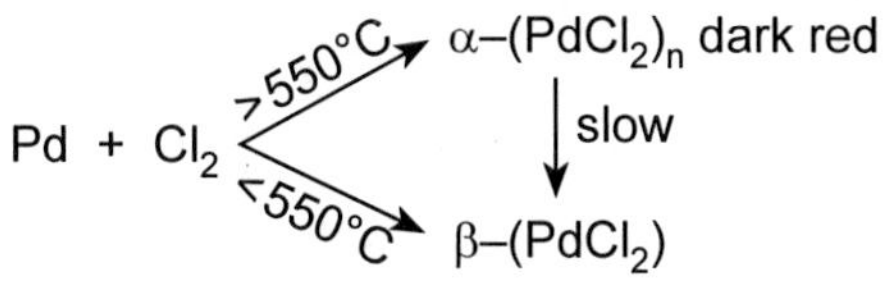

When comparing the chloride of Pd and Pt, it was rioted that α-$PdCl_2$ is dark red with a flat chain polymeric structure. Whereas, α-$PtCl_2$ has olive green colour and its structure is not known, but forms $[PtCl_4]^{2-}$ ions which dissolved in HCl.

While observing the (β-form of $PdCI_2$ and $PtCl_2$, it is based on Pd_6Cl_{12} or Pt_6Cl_{12} unit. Here metal is surrounded by 4 chlorine atoms in a square planar arrangement

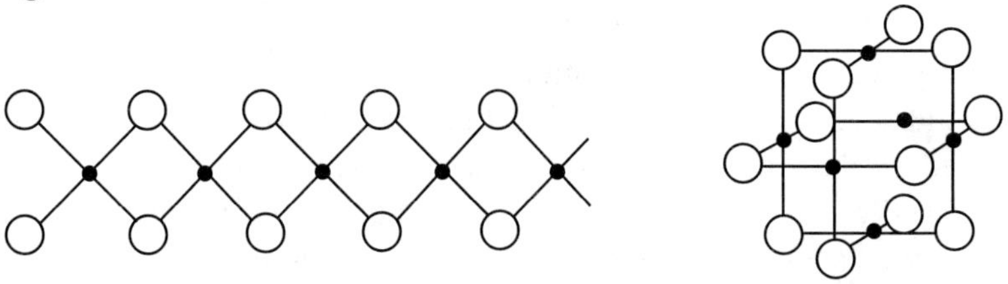

(a) α-$(PdCl_2)_n$ (b) β-$(PdCl_2)_6$

Structure of α and β-$PdCl_2$

There is an important reaction which takes place between $PdCl_2$ and alkenes which form complexes *i.e.,* $[PdCl_3(C_2H_4)]^-$, $[Pd(C_2H_4)Cl_2]$ and $[Pd(C_2H_4)_2Cl_2)]$. Similarly, Pt also reacts to form compounds *e.g.,* Zeise's salt $K[Pt(C_2H_4)Cl_3]$. H_2O yellow coloured crystals.

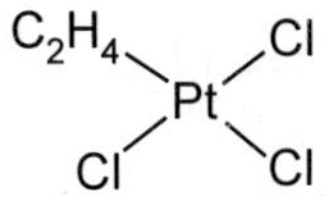

Fig. $[Pt\,(\eta^2\text{–}C_2H_4)Cl_3]^-$

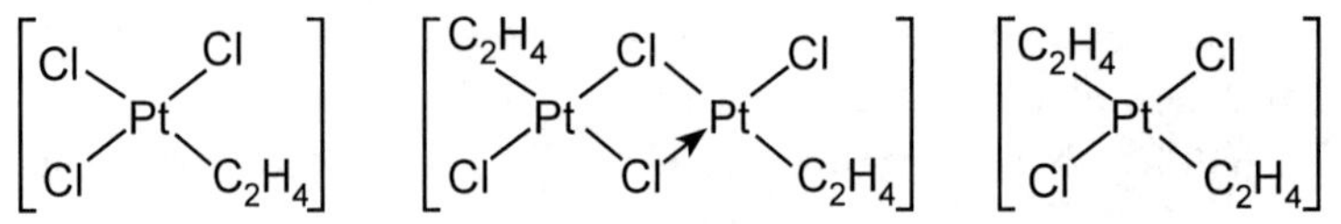

Fig. Structure of $[PtCl_3(C_2H_4)]^-$, $[Pt(C_2H_4)Cl_2]$ and $[Pt(C_2H_4)_2Cl_2)]$

Another colourless well known compound of Pt is $K_2[Pt(CN)_4].3H,O$. This is a stable compound and shows a geometry of square planar $[Pt(CN)_4]^{2-}$ units stacked on top of each other, but its solid form is a poor conductor of electricity. In medicines, Pt(II) plays an important role through its *cis* isomer of $[Pt(NH_3)_2CI_2]$ named *cis*-platin, highly toxic and acts as anticancer drug to treat malignant tumours by injecting into the blood stream whereas, the *trans* isomer of $[Pt(NH_3)_2CI_2]$ is ineffective.

5.3.3 Compounds with (+III) Oxidation State

The (+III) oxidation state does not carry much significance in all the three metals. Though a small number of Ni(+III) compounds are common. In an alkaline solution with Br_2, the oxidation of $Ni(OH)_2$ forms hydrated black solid $Ni_2O_3.2H_2O$ and on dehydration it decomposes to NiO, Ni with (+III) state of oxidation can have stablised complexes *e.g.,* $K_3[NiF_6]$. Regarding palladium, in (+III) state of oxidation it rarely forms compounds and Pt(+III) do not exist. The complexes of palladium with fluorine and sodium also exist, *i.e.* $Na[PdF_4]$ and $NaK_2[PtF_6]$.

5.3.4 Compounds of (+IV) Oxidation States

Nickel rarely occurs in (+IV) oxidation state though it is available in the form of oxide and fluoride. The red coloured complex $K_2[Ni^{IV}F_6]$ occurs which is produced by fluorination of $NiCI_2$ and KCI. The only hydrated oxide of palladium (PdO_2) is known, but is less stable than PtO_2 which occurs in hydrated as well as in anhydrous form. Regarding compounds with halogens and with palladium of (+IV) oxidation state, only one halide as PdF_4 exists, whereas Pt halides PtX_4 do occur with all four halogens. Formation of $PtCI_4$ is carried out by direct combination or by heating $H_2[PtCI_6]$. Also, palladium exists in the form of octahedral complexes *i.e.,* $[PdX_6]^{2-}$ where X = F, Cl or Br.

$$Pt \xrightarrow[\text{regia}]{\text{Aqua}} H_2[PtCl_6] \xrightarrow{\text{Heat}} PtCl_4 + 2HCl$$

Platinum forms a large umber of stable octahedral complexes in (+IV) state of oxidation *i.e.,* $[Pt(NH_3)_6]^{4+}$, $[Pt(NH_3)_5Cl]^{3+}$ and $[Pt(NH_3)_4Cl_2]^{2-}$ and finally $[PtCI_6]^{2-}$. A compound of platinum named Chloroplatinic acid $H_2[PtCl_6]$, $2H_2O$ (red crystals) has commercial requirement. The platinised asbestos used as a catalyst and is obtained by soaking asbestos in solution of $H_2[PtCl_6]$. Platinum is also used to measure conductivity through black electrodes. Platinum forms alkyl derivatives through grignard reagent.

$$4PtCl_4 + 12CH_3MgI \longrightarrow [(CH_3)_3PtI]_4 + 8MgCl_2 + 4MgI_2$$

It is also reported that the organo derivatives of platinum are soluble in organic solvents.

5.4 THE COPPER GROUP OF COINAGE METALS

The elements included in the group 11 are copper, silver and gold. These elements have one *s* electron outside the penultimate complete *d*-subshell. They all have cubic close-packed crystal structure. They are good conductors of heat and electricity, but chemically remain unreactive. In these three elements the most stable oxidation states varies and are Cu(+II), Ag(+I) and Au(+III). Copper is most abundant and forms good number of alloys. The two oxides of copper are common *i.e.* Cu_2O (red) and CuO(black) and the mixed oxides of copper act as super conductors.

Electronic configuration and important oxidation states of copper group metals are shown in the table:

Element		Electronic structure	Oxidation states*
Copper	Cu	[Ar] $3d^{10}4s^1$	+1, **+2**, (+3)
Silver	Ag	[Kr] $4d^{10}5s^1$	**+1**, +2, (+3)
Gold	Au	[Xe] $4f^{14}5d^{10}6s^1$	**+1**, +3, (+5)

5 4.1 Abundance and Extraction

Copper is available in plenty compare to silver and gold, and occurs to the extent of 68 ppm by weight. Silver and gold are rare. Chalcopyrite ($CuFeS_2$) is an important commercial ore of copper and to a lesser extent chalcocite (Cu_2S) is also a common ore of copper. Other ores are basic copper carbonate $CuCO_3$. $Cu(OH)_2$, called as malachite which is green in colour, copper oxide Cu_2O or cuprite a red oxide and Cu_5FeS_4 bornite or peacock. A popular gemstone of copper is Turquoise $CuAl_6(PO_4)(OH)_8.4H_2O$.

The sulphide ore $CuFeS_2$ is crushed and concentrated to 15 percent of copper by froth floatation process and then roasted with air:

$$2CuFeS_2 + O_2 \xrightarrow[1450^oC]{} Cu_2S + Fe_2O_3 + 3SO_2$$

In order to remove iron, the sand is added so that iron is removed in the form of iron silicate $Fe_2(SiO_3)_3$ floating on the surface. Partial oxidation is performed by blowing air through liquid matte of Cu_2S with little FeS and silica.

$$2FeS + 3O_2 \longrightarrow 2FeO + 2SO_2$$

$$2FeO + 3SiO_2 \longrightarrow Fe_2(SiO_3)_3$$

$$Cu_2S + O_2 \longrightarrow Cu_2O + SO_2$$

Air is cut off and after sometime the process of self reduction takes place to form impure blister copper (98% pure). In order to purify further, blister copper is cast into blocks and followed the process of electrolysis. The electrolysis process is done by using Cu electrodes with electrolyte of dilute H_2SO_4.

$$Cu_2S + 2CuO \longrightarrow 4Cu + SO_2\uparrow$$

The lean ores are exposed to air and CuS oxidises to $CuSO_4$ which is leached with water or dilute H_2SO_4. Then scrap iron helps to displace Cu from $CuSO_4$.

$$CuSO_4 + Fe \longrightarrow FeSO_4 + Cu\downarrow$$

Since copper is having high conductivity and grip, it is very useful in the electrical industry. Copper is an inert metal, therefore is used in water pipes. There are several important alloys of copper available which include brass (Cu/Zn 20-50%), nickel silver (Cu 55-60% and Zn 17-27% and Ni 10-18%), Phosphor bronze (Cu 1.25—10% and P 0.35— 1%) and other coinage alloys.

The silver is extracted from the ores of sulphide *i.e.,* Ag_2S (argentite) and AgCl (horn silver). There are three main extraction processes of silver:

(*i*) As a byproduct from the extraction of Cu, P or Zn

(*ii*) By solvent extraction method from molten Pb in Parke's Process.

(*iii*) By making soluble cyanide complexes.

The gold element is available in the ground as lumps of metals named as nuggets. It is extracted by cyanide process where crushed rock is treated with NaCN and air is passed through it.

$$4Au + 8NaCN + 2H_2O + O_2 \longrightarrow 4Na[Au(CN)_4] + 4NaOH$$

Similarly, argentocyanide complex of Ag is soluble. Au is precipitated by adding Zn powder.

$$2Na[Au(CN)_4] + Zn \longrightarrow Na_2[ZN(CN)_4] + 2Au\downarrow$$

All the three elements of this group Cu, Ag and Au show oxidation states of (+I), (+II) and (+III). The hydrated ions found in solution are of Cu^{2+} and Ag^{+}. The Cu^{+} and Au^{+} disproportionate in water and these exist as insoluble compounds or complexes. Cu(+III) and Ag(+II) are strong oxidising agents and can reduce H_2O, therefore, they occur in stabilized complexes or insoluble compounds. All the three elements form their respective oxides and halides shown in the following table:

Oxidation states					
(+I)	(+II)	(+III)	(+IV)	(+V)	Others
CU_2O	**CuO**	–	–	–	
Ag_2O	AgO	(Ag_2O_3?)	–	–	
Au_2O	–	AU_2O_3	–	–	
–	CuF_2	–	–	–	
CuCl	$CuCl_2$	–	–	–	
CuBr	$CuBr_2$	–	–	–	
CuI	–	–	–	–	
AgF	AgF_2	–	–	–	Ag_2F
AgCl	–	–	–	–	
AgBr	–	–	–	–	
AgI	–	–	–	–	
–	–	**AuF_3**	–	(AuF_5)	
AuCl	–	$AuCl_3$	–	–	
–	–	$AuBr_3$	–	–	
AuI	–	–	–	–	

5.5 GENERAL PROPERTIES

All the three metals Cu, Ag and Au have the highest conductivity values of thermal and electricity. Due to their cubic close-packed structure, they are malleable and ductile. Their electronic configuration shows one ns^1 electron which is similar to group I metals. There are similarities between these two groups of elements. Both the group metals have high electrical conductivity.

Due to poor screening of inner *d* electrons, atoms of copper group are smaller in size. Therefore, they have high densities and are harder. They show higher ionization energy and their compounds are more covalent. The *d* electrons are involved in metallic bonding. Thus, their melting points and heat of sublimation are higher than group I metals.

The reason for non-reactivity or noble character of Cu, Ag and Au is the higher enthalpy of sublimation and ionisation energy. The element of group I have high negative standard potentials *i.e.* E^0 values, are most reactive metals, whereas these metals of group II have positive standard potential and are lower than hydrogen in electrochemical series. Therefore, they do not perform any reaction with water or dilute acids to liberate H_2. The noble behaviour increases from Cu to Ag to Au. The inert behaviour of Au is similar to that of Pt. However, Cu reacts with oxidising acids, such as HNO_3 and H_2SO_4.

$$3Cu + \underset{\text{dilute}}{8HNO_3} \longrightarrow 3CU(NO_3)_2 + 2NO + 4H_2O$$

$$\underset{\text{concentrated}}{Cu + 4HNO_3} \longrightarrow Cu(NO_3)_2 + 2NO_2 + 2H_2O$$

$$\underset{\text{concentrated}}{Cu + 3H_2SO_4} \longrightarrow CuSO_4 + 2SO_2 + 2H_2O$$

Some physical properties have been listed in table

	Covalent radius (Å)	Ionic radius (Å)			Melting point (°C)	Boiling point (°C)	Density (g cm^{-3})	Pauling's electro negativity
		M+	M^{2}+	M^{3}+				
Cu	1.17	0.77	0.73	0.54[1]	1083	2570	8.95	1.9
Ag	1.34	1.15	0 94	0.75	961	2155	10.49	1.9
Au	1.34	1.37	–	0.85	1064	2808	19.32	2.4

I = low spin radius

Copper shows the character of reactivity in slow rate with moist air to form basic copper carbonate $CuCO_3.Cu(OH)_2$. Regarding silver, it dissolves in concentrated H_2SO_4, but gold does not dissolve in any acid except in aquaregia ($HC1/HNO_3$ 3:1). Here HNO_3 behaves as an oxidising agent and ions of chloride as a complexing agent. Cu reacts with oxygen, while Ag and Au are inert.

$$2Cu + O_2 \xrightarrow{\Delta} 2CuO \xrightarrow[\text{temperature}]{\text{Still higher}} Cu_2O + \frac{1}{2}O_2$$

Cu and Ag react with H_2S to form their respective sulphide, while Au does not show any reaction.

$$2Ag + H_2S \xrightarrow{\text{air}} Ag_2S + H_2$$

Therefore, silver articles turn black due to traces of H_2S in air to black coloured Ag,S. Ag compounds are reduced easily compared to Cu compounds.

The oxides and hydroxides of alkali metals are strongly basic, while copper gives insoluble oxides and hydroxides that are less basic.

Generally, these three elements occur in oxidation state as of Cu(+Il), Ag(+I) and Au(+III). All these elements form coloured compounds and show their strong behaviour to form complexes.

Copper is found to be useful to act as a catalyst in the manufacture of silicones. As a catalyst $CuCI_2$ is used in Deacon's process of making CI_2.

5.5.1 Compounds in (+I) Oxidation State

Generally, the compounds of (+1) state and complexes are colourless, and diamagnetic due to their d^{10} configuration. Though, there are few coloured compounds *e.g.* Cu_2O is red, Cu_2CO_3 is yellow and Cu_2I_2 is brown. The colour in these compounds is due to charge transfer spectra as *d-d* transitions are not possible. Due to completely filled *d* shell, the (+I) state compounds should have extra stability, but it is not so. Only Ag^+ is stable in solid or in solution form. Cu^+ is stable in solid or solution form. Cu^+ and Au^+ disproportionate water.

$$2Cu^+ \rightleftharpoons 2Cu^{2+} + Cu^0 \quad K = \frac{[Cu^{2+}]}{[Cu^+]} = 1.6 \times 10^6$$

$$3Au^+ \rightleftharpoons Au^{3+} + 2Au^0 \quad K = \frac{[Au^{3+}]}{[Au^{3+}]} = 1.6 \times 10^6$$

It has been observed that the value of equilibrium constant for Cu^+ is high, and concentration for Cu^+ ion in solution is low.

The compounds of Cu(I) and Au(I) which are stable towards water are either insoluble in water or form their complexes *e.g.,* CuCI, CuCN, CuSCN.

$$2Cu^{2+} + SO_3^{2-} + 2SCN^- + H_2O \longrightarrow 2Cu^ISCN + H_2SO_4$$

The gravimetrically estimation of copper can be done by CuSCN and is formed in reducing atmosphere of Na_2SO_3. In the Fehling's solution test, Fehling solution A (Cu(II) tartarate) and Fehling solution B(NaOH) are mixed with some aldehyde solution then yellow or red Cu_2O precipitates are obtained, Cu_2O reacts with HX to form insoluble CuCl, CuBr and Cul.

When KI is added to the Cu^{2+} solution, it results in the I^- ions reducing Cu^{2+} to Cu^{1+} iodide, at the same time oxidising of I^- take place to I_2. The estimation of I_2 can be done by titrating with $Na_2S_2O_3$(hypo) solution:

$$2Cu^{2+} + 41^- \longrightarrow 2CuI + I_2$$

$$I_2 + 2Na_2S_2O_3 \longrightarrow 2NaI + \underset{\text{Sodium tetrathionate}}{Na_2S_4O_6}$$

The halides of copper (I) are partly covalant having Zn blende structures. The vapour phase of CuCl and CuBr are polymeric. Cu(I) halides are insoluble in H_2O and become soluble in presence of excess halide ions forming complexes, such as $[CuCI_2]^-$, $[CuCl_3]^{2-}$ and $[CuCl_4)^{3-}$. The geometry of these complexes as well as of other Cu(I) are tetrahedral in solid form. The HCl, HNO_3 and aqueous NH, are very good solvents for Copper(I) halides. The dissolution of CuCl in concentrated HCI and NH_4OH is important as they absorb CO. Therefore, ammoniacal CuCI is used to estimate CO in gas samples.

For extraction Ag and Au, cyanide complexes play their role as they are soluble complexes and the metals are obtained from the complex by reduction with Zn.

$$4Au + 8CN^- + 2H_2O + O_2 \longrightarrow 4[Au^I(CN)_2]^- + 40H^-$$

On adding CN^- to $CuSO_4$ solution, precipitate of CuCN is formed which dissolves in excess CN^- forming $[Cu(CN)_4]^{3-}$ soluble complex which has the geometry of tetrahedral structure.

$$Cu^ICN + \underset{\text{Excess}}{3CN^-} \longrightarrow [Cu(CN)_4]^{3-}$$

There are several polymeric complexes of Cu(+I) where a cluster of four Cu atoms are present at the tetrahedron corners, but do not involve in metal-metal bonding *e.g.,* phosphinie $[Cu_4I_4(PR_3)_4]$ and arsine $[Cu_2I_4(AsR_3)_4]$ complexes.

The maximum stable state of oxidation of silver is Ag(+I). All Ag(+I) salts do not dissolve in water, though exceptions are there *i.e.,* $AgNO_3$, AgF and $AgClO_4$. In silver, generally salts are anhydrous except AgF. $4H_2O$.

Silver nitrate ($AgNO_3$) is a common natural substance and an important salt. Though, the oxide of silver Ag_2O can dissolve in acids, but, is basic in nature. However, it has been found that Ag_2O also dissolves in NaOH, therefore, it also possess slightly acidic properties. The moist Ag_2O forms Ag_2CO_3 absorbing CO_2.

Ag_2,O absorbs CO_2 to form Ag_2CO_3.

Silver halides are used in photographic film and photographic paper. Regarding the property of solubility, silver halides are insoluble in water except AgF. In qualitative salt analysis, silver nitrate with dil. nitric acid are used to test for the halide ions *i.e.,* Cl^-, Br^- and I^- with formation of white precipitates of AgCl which are soluble in ammonia. AgBr with pale yellow precipitates are partially soluble and yellow Agl precipitates are insoluble. The dissolution of AgCl and AgBr is due to the formation of a linear ammine complex *i.e.* $[H_3N \rightarrow Ag \leftarrow NH_3]^+$.

In silver, it has been noted that there are some colourless compounds, but their anions are coloured *e.g.,* Agl, Ag_2CO_3 and Ag_3PO_4 are yellow, while Ag_2S is black. Thuere are linear complexes also which inlcude $[Ag(NH_3)_2]^+$, $[Ag(CN)^-]$ and $[Ag(S_2O_3)_2]^{3-}$. The polynuclear complexes are formed due'to bidentate ligands.

Au(+I) is less stable in its compounds *i.e.,* Au_2O, AuCI and AuBr. They are obtained by slow heating of AuX_3 halides. The linear complexes of Au also exist *e.g.,* $[NC \rightarrow Au \leftarrow CN]^-$ $[Cl \rightarrow Au \leftarrow Cl]^-$. By reacting $Au_2Cl_6 + R_3P$ in ether solution forms phosphine complexes $[R_3PAuCl]$. The Cl group may be replaced by other groups SCN or Me.

There is an important role of Au(+I) in manufacturing the drugs to treat rheumatoid arthritis where linear complexes are formed *i.e.,* $[RS \rightarrow Au \leftarrow SR]$ or $[R_3P \rightarrow Au \leftarrow PR_3]$.

5.5.2 Use of Ag Compounds in Photography

The process of photography is the chemical reaction where photographic film and photographic paper is treated after the photographic exposure in order to generate a negative or positive image. Silver halides are quite useful in photography. Light sensitive film contains fine particles of silver salts in gelatine spread on a celluloid strip. The particle size or grain affects the quality of the picture. AgBr is the main light sensitive material. Some Agl is also used when light intensity is low for moving objects and exposure is very short. Therefor, some AgCl may aLso be present in the emulsion.

On exposure, the light from the object may travel, enter the camera focussed by the lens to give image on the film. The incident light starts a photochemical reaction. It excites X^- ion to lose electron which travels to the grain where Ag^+ is reduced to metallic Ag.

In the present day photography, exposure of l/l00th of second is used. Thus only a few atoms of silver are produced in each grain exposed to light. Some parts of the film are exposed to bright spots of the object contain lot of grains, some are exposed to the paler parts of the object and produce few grain, while parts not exposed produce none. Therefore, a latent image of the object is contained in the film. However, the number of atoms of silver product are so small that the image is not visible to the naked eye.

The film is then placed in a developer solution which is a reducing agent usually quinol. It reduces silver halides to metallic Ag. Ag is deposited where there are already some Ag atoms.

Thus, developer solution intensifies the latent image to become visible to the eye. The quality of impression depends upon the concentration of the developer solution, the pH, the temperature and the amount of time the film was kept in the developer solution.

When the film is brought out into the day light the unexposed parts are turned black and destroy the picture. To prevent this unexposed silver halides are removed by putting in fixer solution. In the earlier times, ammonia was used to dissolve unchanged AgBr, but now sodium thiosulphate is used as fixer.

$$AgBr + 2Na_2S_2O_3 \longrightarrow \underset{\text{Soluble complex}}{Na_3[Ag(S_2O_3)_2]} + NaBr$$

$Na_2\ S_2O_3$ forms a soluable complex with silver halides. After fixing, the film is brought out in the daylight. Parts blackened by silver show the light parts effect, it is called the 'Negative'.

To get image where light and dark are in the original form, light is passed through the negative on the piece of paper coated with AgBr emulsion. Therefore, a negative of the negative is formed and photograph in the original form is fixed.

5.5.3 Compounds with (+II) Oxidation State

The stability for copper compounds in the (+II) state of oxidation is very high. The Cu^{2+} has d^9 configuration with unpaired electron. It is paramagnetic *d-d* transitions are possible. The hydrated copper (II) salts are blue *e.g.* $CuSO_3.5H_2O$.

Due to strong heating, $Cu(NO_3)_2$ decomposes into black CuO and when it is very strongly heated (>800°C) gives red Cu_2O.

$$2CuO \xrightarrow{>800^0C} Cu_2O + \frac{1}{2}O_2$$

If NaOH is added to Cu^{2+} solution, a blue precipitate of the hydroxide is obtained. When $CuCO_3$ or $Cu(OH)_2$ dissolve in acids or $CuSO_4$ and $Cu(NO_3)_2$ in water results in hydrated ions formed $[Cu(H_2O)_6]^{2+}$. These ions show blue colour and has a distorted octahedral structure (tetragonal distortion). Such geometry of octahedral structure causes crystal field splitting. The d^9 configuration splits into $(t_{2g})^6$ and $(e_g)^3$. Since e_g is not symmetrical therefore, Jahn-Teller distortion occurs. Similar distortion is common in Cu^{2+}compounds, same arguments hold for Cu^{2+} octahedral complexes *e.g.,* $[Cu(H_2O)_5NH_3]^{2+}$, $[Cu(H_2O)_4(NH_3)_2]^{2+}$), $[Cu(H_2O)_3(NH_3)_3]^{2+}$ and $[Cu(H_2O)_2(NH_3)_4]^{2+}$. The tetragonal distortion does make 5th and 6th bond long and weak. With bidentate ligand ethylenediamine $[Cu(en)_2(H_2O)_2]^{2+}$ and $[Cu(en)(H_2O)_4^{2+}$ are made easily, but not $[Cu(en)_3]^{2+}$ so easily.

The halide complexes exhibit two dissimilar stereochemistry as the complex $(NH_4)_2[CuCl_4]$, the complex ion $[CuCl_4]^{2-}$ has square planar geometry, but $Cs_2[CuCl_4]$, $Cs_2[CuBr_4]$, the ion $[CuX_4]^{2-}$ has tetrahedral appearance. Also, the tetrahedral ions *i.e.,* $[CuCl_4]^{2-}$ are orange and square planar ions $[CuCl_4]^{2-}$ are yellow in colour. The octahedral Cu(+II) compounds and complexes are blue or green in colour.

There is similarity for being dimeric and hydrated, copper (II) acetate $Cu_2(CH_3COO)_4.2H_2O$ with the carboxylate complexes of Cr^{2+}, Mo^{2+}, Rh^{2+} and Ru^{2+}, however with an important difference of absence of M-M bonding in Cu. In Cu^{2+}, four acetate groups act as bridging ligands between two Cu atoms and the fifth position round each Cu atom is occupied by O of the water molecule. The bond distance of Cu—Cu is 2.64Å which is more than the bond distance in metallic Cu *i.e.* 2.55 Å. Thus, there are only interactions between the 2Cu atoms and no permanent bond, but in Cr and other metals, there is M—M bond which advocates the pairing of unpaired electrons and the complex being diamagnetic. The magnetic moment value of Cu at 25° is 1.4BM/Cu atom.

In (+II) state of oxidation, silver halide AgF_2 is a brown solid formed by heating Ag and F_2. AgF_2 is a strong oxidising as well as fluorinating agent.

$$AgF_2 \xrightarrow{\Delta} AgF + \frac{1}{2}F_2$$

Ag(+II) oxidation state forms relatively more stable complexes *e.g.,* $[Ag(py)_4]^{2+}$ and $[Ag(dipy)_2]^{2+}$. These complexes are square planar and paramagnetic. Black AgO is formed by oxidising Ag_2O. It is strange that AgO is diamagnetic. AgO is actually $Ag^IAg^{III}O_2$.

About gold (+II) state of oxidation is observed in dithiolene compounds and in $[Au(B_9C_2H_{II})_2]^{2-}$

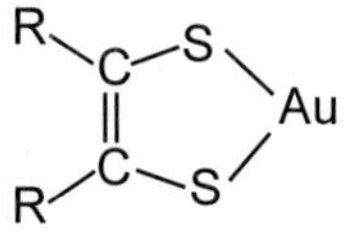

Ditholene Compound of Au(+II)

5.5.4 Compounds in (+III) Oxidation States

The (+III) state is very rare in Cu and Ag, but very common and stable for gold. Au^{3+} had d^8 configuration like Pt^{2+} and forms square complexes. These compounds decompose to the metal on heating

$$[Au^{III}Cl_4]^- + OH^- \longrightarrow Au(OH)_3 \xrightarrow{\text{Dehydrate}} Au_2O_3$$

$$Au_2O_3 \xrightarrow{150°C} Au + Au_2O_3$$

All the halides AuX_3 are known, $AuCl_3$ dissolves in aqua-regia

$$Au + HNO_3 + HCl \longrightarrow H_3\overset{+}{O}\ [AuCl_4]^- \cdot 3H_2O \longrightarrow AuCl_3$$

AuF_3 is obtained by fluorination

$$2Au + 2BrF_3 \longrightarrow 2AuF_3 + Br_2$$

$AuCl_3$ is $AuBr_3$ are dimeric.

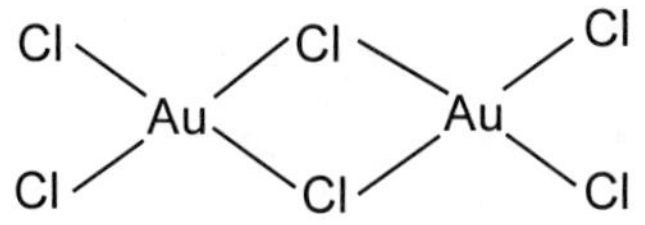

Au_2O_3. H_2O dissolves in alkali to form $NaAuO_2.H_2O$ and in strong acids to give $H[AuCl_4]$, $H[Au(NO_3)_4]$ and $H[Au(SO_4)_2]$ with respect to the acids used. The complexes of Au(+lll) *e.g.*, $[Au(NH_3)_4](NO_3)_3$ are common which have geometry of square planar. It has been found that dialkyls R_2AuX are stable organometallic compounds.

5.5.5 Compounds in (+V) State

There is only one compound AuF_5 known in (+V) oxidation state. The compound is unstable and decompose to AuF_3 and F_2 when heated above 60°C.

$$Au + 3F_2 + O_2 \xrightarrow[\text{High press}]{370°C} O_2AuF_6 \xrightarrow{\Delta} AuF_5 + \tfrac{1}{2}F_2 + O_2$$

5.5.6 Biological Use of Copper

Copper is present in adult humans to an extent of 100 mg which is next to Fe (4g) and Zn(2g). Large amounts of copper are toxic, but small quantity is required by the human body as daily intake of copper is about 4-5 mg. The deficiency of copper results in the inability to store the iron in liver which results in anaemia. The role of copper is also extended as it binds as metalloproteins or to enzymes in the form of oxidases and blue proteins *e.g.* amine oxidase ascorbate oxidase, cytochrome oxidase, galactose oxidase. In Wilson's disease the copper gets accumulated in liver, kidney and brain which is treated by chelating agents like EDTA. Copper forms complex with EDTA and is excrected from the body.

Despite d^9 configuration of Cu^{2+}, copper is diamagnetic due to antiferromagnetic coupling between the Cu^{2+} ions. Haemocyanine, as respiratory protein occur in the pattern of metalloproteins with two Cu atoms binding with aoxygen molecule. The haemocyanine is found in the blood of snails, crabs lobsters and scorpions, the oxygenated haemocyanin is blue in colour. Plastocyanin pigment is present in green plants and contains one copper atom. Here the copper plays an important role in photosynthesis as an electron carrier. Azurin is an enzyme found in bacteria containing one Cu per molecule and is structurally similar to plastocyanin.

■ ■ ■

6

The Zinc Group

6.1 INTRODUCTION

This zinc group is constituted by the elements *viz.,* zinc, cadmium and mercury. These elements have $d^{10}s^2$ type of electronic configuration with the most common oxidation state of (+II). Overall compounds of Hg(+II) are more covalent and forms stable complexes in comparison to Cd and Zn(+II) complexes. These elements and their stable ion have d^{10} configuration therefore, they are not true transition elements and show similarity with group 2 elements though, Zn and Mg show some similarity. Zinc is found to be less reactive because of its smaller atomic radii and higher nucleus charge hence, it has a tendency to form covalent compounds. In broad sense, Zn and Cd show similarity in their compound formation. Mercury is found to be different than these two members specially in its physical state as it is liquid at room temperature, noble and the only element which forms stable (+I) compounds. Hg(+I) salts are stable akin to p-block elements. Zn is a rust proof metal and used for casting and making alloys. Zn is considered as second biologically important metal and ZnO is a commercially important compound.

Electronic configuration and oxidation

Element		Electronic structure	Oxidation states
Zinc	Zn	[Ar] $3d^{10}4s^2$	<u>2</u>
Cadmium	Cd	[Kr] $4d^{10}5s^2$	<u>2</u>
Mercury	Hg	[Xe] $4f^{14}5d^{10}6s^2$	1, <u>2</u>

The most important and stable oxidation state is underlined and less important in normal type

6.1.1 Occurrence

Zinc is available on the earths's crust to an extent of 132 ppm by weight. Cd and Hg are rare in comparison to zinc. The most common ore of zinc is ZnS called Zinc blende. Another ore is sphalerite (ZnFe)S and other where it occurs as galena PbS. ZnCO, is calamine. Another ore is Hemimorphite $Zn_4(OH)_2(Si_2O_7)$. H_2O but is commercially less important. Cd is found as traces in Zn ores and Hg is mined as cinnabar HgS.

6.1.2 Extraction

(a) Zinc: ZnS ore is concentrated by froth floatation process and then roasted in air

$$2\,ZnS + 3O_2 \longrightarrow 2ZnO + 2SO_2$$

Smelting: The smelting process is performed at 1200°C in which ZnO is reduced by reacting CO and ZnO which is reversible reaction. As a result, Zn metal is obtained in gaseous state.

$$ZnO + CO \rightleftharpoons CO_2 + Zn$$

The process of reoxidation is done after removing and cooling the gaseous Zn and CO_2 and finally, large amount of Zn is obtained. In modern smelters, there are two ways to reduce reoxidation:

(*i*) Minimised by excess carbon so that CO_2 formed is converted to CO.

(*ii*) Sudden cooling of gases leads to 99 per cent pure Zn with droplets of Pb impurity. If Cd is present it is separated by distillation.

A second method is also used by heating ZnS at lower temperature to obtain ZnO and $ZnSO_4$. They are dissolved in H_2SO_4 and Zn dust is added to precipitate Cd. Finally, $ZnSO_4$ is electrolysed to obtain pure Zn.

(b) Cadmium: Cd is present in traces in most zinc ores. The ore is treated to get $ZnSO_4$ and $CdSO_4$. Zn powder is then added to the solution containing $ZnSO_4$ and $CdSO_4$, Cd is precipitated while, Zn dissolves. While observing the electrochemical series in this group, Zn is found to be higher than Cd. The elements high in the series displace the elements lower in the series.

$$Zn_{(solid)} + Cd^{2+}{}_{(solution)} \longrightarrow Zn^{2+}{}_{(solution)} + Cd_{(solid)}$$

The $Cd_{(solid)}$ is dissolved in H_2SO_4 and purified by electrolysis.

(c) Mercury: The ore cinnnabar is crushed and concentrated by sedimentation. It is then heated in air, as a result Hg vapours arc formed which are condensed. In the process SO_2 forms H_2SO_4.

$$HgS + O_2 \xrightarrow{(600°C)} Hg + SO_2$$

Also, raw ore HgS is heated with scrap iron or quicklime

$$HgS + Fe \longrightarrow Hg + FeS$$

$$4HgS + CaO \longrightarrow 4Hg + CaSO_4 + 3CaS$$

Hg obtained may contain the traces of impurities of Pb, Zn and Cd. To get pure Hg, blow hot air through the metal at 250°C. Impurities form their respective oxide and float and then are easily removed by:

(*i*) Scraping off the scum

(*ii*) When oxides are dissolved treat with dilute HNO,.

(*iii*) Pure metal is removed from the mixture of other metals and oxides by distillation. The boiling points of the present impurities and Hg are Pb(1751°C), Zn(908°C), Cd(765°C) and Hg(357°C).

6.1.3 Uses of Zinc, Cadmium and Mercury

Zinc is used tor galvanization of iron to prevent rusting. Its coating can be applied electrolytically. The object may be coated with powdered zinc, or it may be sprayed. Important alloys include Cu/Zn called brass (20% Cu, 50% Zn). It is used as a negative pole in sealed batteries (Dry cells). ZnO is used as a white pigment in paints. It absorbs UV light.

Cadmium is used for protecting steel from corrosion. It absorbs neutrons and is used in nuclear reactors. Ni/Cd storage batteries are used in diesel locomotives. CdS is a yellow pigment of paints. Mercury is an important constituent of electrolytic cells for producing NaOH and Cl_2, used in CFL bulbs and tubes. Hg also forms amalgams with Ag and Au. In order to prevent contamination specially in marine environment HgO is quite useful for making antifouling paint for ships. All Hg compounds are toxic and have lasting ecological effects. Mercury is very sensitive towards heat, that is why it is used in thermometers to measure temperature, also in barometers and manometers for measuring pressure. Mercury has its role in several medicines also.

6.1.4 Important and Common Oxidation States

The elements in the group have completely filled *d* orbitals and two ,*s* electrons, removal of two *s* electrons give the most important oxidation state (+II) to these elements. Most stable oxidation state in case of Hg compounds is +I which are dimeric. Hg(I) chloride is written as Hg_2Cl_2 in which two Hg^+ species with $6s^1$ configuration are bonded together. Hg(+I) compounds are diamagnetic, Cd_2^{2+} has been isolated, but disproportionate in water:

$$Cd_2^{2+} \longrightarrow Cd^{2+} + Cd$$

Thus, Hg—Hg bond is stronger than Cd—Cd bond. Oxidation states higher than +2 do not occur due to symmetry of completely filled *d* shell.

6.1.5 Ionic Size and Some Physical Properties

The ionic radii of M^{2+} ions are smaller due to intervening electrons which have poor shielding effect.

$Zn^{2+} = 0.74$ Å $Ca^{2+} = 1.00$ Å

$Cd^{2+} = 0.95$ Å $Sr^{2+} = 1.18$ Å

$Hg^{2+} = 1.02$ Å $Ba^{2+} = 1.35$ Å

The effect of lanthanide contraction is observed in Cd and Hg, but to a lesser extent. It has been noticed that the increase in size between Cd^{2+} to Hg^{2+} is smaller than between Zn^{2+} and Cd^{2+}

	Covalent radius (Å)	Ionic radius M2+ (Å)	Melting point (°C)	Boiling point (°C)	Density (g cm^{-3})	Pauling's electro-negativity
Zn	1.25	0.740	420	907	7.14	1.6
Cd	1.41	0.95	321	765	8.65	1.7
Hg	1.44	1.02	–39	357	13.534	1.9

Ionization energy of Zn, Cd and Hg group is higher due to small size and poor shielding by d shell. The filled 4*f* shell increases the binding energy of the outer electrons, therefore, first ionisation energy of Hg is higher than any other metal.

	Promotion energy $s^2 \rightarrow s^1p^1$ (kJ mol^{-1})	Ionization energy (kJ mol^{-1})		
		1st	2nd	3rd
Zn	433	906	1733	3831
Cd	408	876	1631	3616
Hg	524	1007	1810	3302

6.2 GENERAL CHARACTERISTICS

Zn, Cd and Hg have some properties different from transition elements because of complete *d* shell which is not available for bonding.

1. Zn, Cd and Hg show no variable valency.
2. They have d^{10} electronic configuration hence, no *d-d* transition and hence, they are white. Some Hg(+III) and Cd(+II) compounds are coloured due to charge transfer from ligands to the metal.
3. The metals are soft in comparison to other transition elements there is no participation of *d* electrons in metallic bonding.
4. The melting and boiling points are very low, for this reason they are more reactive than copper groups. Inertness is due to high heat of sublimation, high ionisation energy and low heat of hydration.

Mercury is the liquid metal at room temperature and the liquid has vapour pressure, that is why mercury metal may vapourise easily and cause poisoning. The gas is monoatomic like noble gases.

In this group, the hydrated sulphates are isomorphous with double salts. $K,SO_4.HgSO_4.6H_2O$ are analogous to K_2SO_4. $MgSO_4.6H_2O$. Zinc group is more noble, more covalent and has a greater ability to form complexes and less basic.

The reactivity decreases from Zn to Cd to Hg. This is shown by their standard reduction potential values given in the table.

Oxidation state

+II		+I		0	+II		+I		0	+II		+I		0
Zn2^{+}		−0.76		Zn	Cd^{2+}		−0.40		Cd	Hg^{2+}		+ 0.91	Hg_2^{2+}	+0.79 Hg
												+0.85		

Standard Reduction Potential Values

Zinc and cadmium are electropositive metals. Hg is noble owing to its highest first ionisation energy. Hg has a positive potential.

$$Zn^{2+} + 2e^- \longrightarrow Zn \quad E° = -0.76\ V$$

$$Cd^{2+} + 2e^- \longrightarrow Cd \quad E° = -0.40\ V$$

$$Hg^{2+} + 2e^- \longrightarrow Hg \quad E° = +0.85\ V$$

Both elements Zn and Cd are similar by having silvery solid state which is damaged in moist air, whereas Hg silvery shine does not damage readily. Zn and Cd dissolve in non-oxidising acids liberating H_2, but Hg does not. All the three metals give their respective salts and oxides of nitrogen and sulphur when reacted with oxidising concentrated HNO_3 and concentrated H_2SO_4, however mercury (I) reacts with dilute HNO_3 to form $Hg_2(NO_3)_2$.

Zinc is amphoteric like aluminium. It is soluble in alkalis forming zincates similar to aluminates *e.g.* $Na_2[Zn(OH)_2]$, $[NaZn(OH)_3]$ or $[Na\ Zn\ (OH)_3].H_2O$.

On heating, these three metals form their oxides MO, sulphides MS and halides MX_2. On further heating HgO decomposes and this is used for the preparation of O_2:

$$Hg + O_2 \xrightarrow[350°C]{} 2HgO + \xrightarrow[400°C]{} 2Hg + O_2$$

In qualitative analysis, by passing hydrogen sulphide gas in acidified solution of these three metals, they form different coloured precipitates *i.e.,* ZnS (white), CdS (yellow). HgS(black). The halides $ZnCl_2$ and $CdCl_2$ are of ionic linkage whereas, $HgCl_2$ is covalent. Zn and Cd react with P to give their respective phosphides, however Hg does not react with phosphorus. All the three metals play an important role in the formation of alloys with other metals *e.g.,* Cu and Zn form brass. Alloys formed with Hg are called amalgams and Na(Hg) is a strong reducing agent. In this group Zn and Cd are very similar but with respect to Hg they are comparatively different.

6.2.1 Oxides

ZnO occupy very important role in shortening the time taken for vulcanisation in production of rubber. In addition, ZnO is used as white pigment in paint, but less common than TiO_2 because it has better covering power. Zinc soaps are used to stabilise plastics. ZnO is amphoteric, forms zincates with alkali, *i.e.* $Na_2[Zn(OH)_4]$ and $Na[Zn(OH)_3].H_2O$ whereas CdO is basic, forms $Na_2[Cd(OH)_4]$ with strong alkali and HgO decomposes on heating above 400°C.

$$2HgO \xrightarrow{>400°C} 2Hg + O_2$$

When we move from Zn to Hg, thermal stability of oxides decreases. The colour of oxides of these metals are ZnO is white, CdO is yellow, green or brown due to crystal defects and HgO is yellow. The red HgO has the same structure and is formed by heating $Hg(NO_3)_2$. The colour difference is due to particle size.

Dihalides: There are 12 dihalides (MX_2) that exist. Among these, the fluorides are ionic and have high melting points. Other halides are partly covalent with low melting points.

Table for melting points of dihalides

ZnF_2	872°C	CdF_2	1049°C	HgF_2	645°C (decomp.)
$ZnCl_2$	283°C	$CdCl_2$	568°C	$HgCl_2$	276°C
$ZnBr_2$	394°C	$CdBr_2$	567°C	$HgBr_2$	236°C
ZnI_2	446°C	CdI_2	387°C	HgI_2	259°C

Fluorides are not water soluble due to high lattice energy. The chlorides, bromides and iodides are hygroscopic and soluble in H_2O because the crystal lattice is not very strong. Another reason for solutbility is that metal ions from variety of aqua complexes such as $[Zn(H_2O)_6]^{2+}$, $[ZnCl_4(H_2O)_2]^{2-}$ and solution of Zn^{2+} and Cd^{2+} are acidic due to hydrolysis.

$$H_2O + [Zn(H_2O)_6]^{2+} \longrightarrow [Zn(H_2O)_5{-}OH]^{+} + H_3^{+}O$$

Cd salts are less hydrated and halides may undergo self complexing. CdI_2 gives $[CdI_3]^-$ and $[CdI_4]^{2-}$. Hg(II) salts are anhydrous and do not ionise in water and are called corrosive sublimate. The role of Hg_2Cl_2 calomel is in making medicines for laxative purposes.

6.2.2 Complexes

All the three metal elements form complexes and the two metals Zn^{2+} and Cd^{2+} form complexes with O, N and S donor ligands as well as with halide ions whereas, Hg(+II) does not form complex with O donor ligand, but forms with N P and S donor ligand. In case of complexes formed by Hg, greater stability is due to smaller ions. No complex has been reported for *pi* bonding ligands *i.e.* CO, NO or alkenes. But Zn and Cd do form complexes with CN^- *e.g.*, $[Zn(CN)_4]^{2-}$ and $[Cd(CN)_4]^{2-}$. $[Cd(CN)_4]^{2-}$ ion, in very dilute solution gives CdS precipitate only on passing H_2S gas even in the presence of Zn^{2+} solution. Cd and Hg form coloured complexes due to charge transfer. There is no CFSE (crystal field stabilization energy) value because these elements have d^{10} configuration.

The tetrahedral complexes for Zn^{2+} and Cd^{2+} are reported like $[MCl_4]^{2-}$, $[M(H_2O)_4]^{2+}$, $[M(NH_3)_4]^{2+}$, $[M(NH_3)_2Cl_2]$ $[Zn(CN)_4]^{2-}$. In case of zinc, when the ligand is bonded with N atoms then the complex is $[Zn(NCS)_4]^{2-}$ but in Cd metal, the ligand is bonded with S atoms and have the molecular structure as $[Cd(SCN)_4]^{2-}$.

Zinc does not have very stable octahedral complexes, but the octahedral complexes of Cd are more stable because it is larger in size.

In case of mercury, the octahedral complexes of Hg(+II) do occur, but they are distorted with four long bonds and two short bonds. Some tetrahedral type of complexes in Hg(+II) halide complexes exist *e.g.,* $[Hg(NH_3)_2CI_2]$ and Nessler's reagent $K_2[HgI_4]$, Hg is detected by Nessler's reagent forming a brown precipitate in ammonia.

Two coordinate complexes are rare in Zn and Cd, but well known in case of Hg *e.g.,* $[Hg(SCN)_2]$ and $Hg(CN)_2$.

6.2.3 Mercury (+1) Compounds

Hg(+I) compounds, are also called mercurous compounds. They have $[Hg—Hg]^{2+}$ and not Hg^+. The two Hg atoms join because of pairing of $6s^1$ electrons in the outermost shell.

They are prepared by reducing Hg(+II) compounds with Hg metal. Hg(I) nitrate can be prepared by dissolving Hg in dilute HNO_3. Other mercury salts are produced by adding $NaHCO_3$ and by precipitating Hg_2CO_3. Hg_2CO_3 is treated with HF, HCl, H_2SO_4 etc.

Hg_2F_2 is decomposed when hydrolysed

$$[Hg_2]F_2 + H_2O \longrightarrow 2HF + \underset{\text{Unstable}}{[Hg_2](OH)_2}$$

$$[Hg_2(OH)_2 \xrightarrow{\text{Disproportionates}} Hg^{II}O + Hg^0 + H_2O$$

Hg_2Cl_2, Hg_2Br, and Hg_2I_2 are insoluble in H_2O. $Hg_2(NO_3)_2$ is water soluble and contains linear structure $[H_2O—Hg—Hg—H_2O]^{2+}$ Oxidising agent like HNO_3 can convert Hg° to Hg(+II) rather than Hg(I).

$$Hg^{2+} \longrightarrow Hg^0 \quad E° = +0.85 \text{ V}$$

$$[Hg_2]^{2+} \longrightarrow 2Hg \quad E° = +0.79\text{V}$$

Reduction potential shows

$$Hg^{2+} + Hg° \longrightarrow Hg_2^{2+} \quad E° = +0.13\text{V}$$

Mercury (I) compounds can be prepared by heating Hg(+II) compounds with 50% excess of Hg.

$$HgCl_2 + Hg \longrightarrow [Hg_2]^{2+}Cl_2$$

The presence of Hg_2Cl_2 is confirmed when a black precipitate is obtained with NH_4OH which may be $[Hg(NH_3)_2Cl_2].HgNH_2Cl$ or $Hg_2NCl.H_2O$ or their mixture alongwith Hg.

6.2.4 Organometallic Compounds

In 1849, Edward Frankland was the first who synthesised useful organometallic compounds *e.g.,* (I) ZnR, (II) RZnX

$$2C_5H_5I + Zn \xrightarrow[N_2]{\text{Inert}} 2C_5H_5ZnI \xrightarrow{\text{Heat}} (C_5H_5)_2Zn + ZnI_2$$

The reaction was very promising with alkyl or aryl iodides, however RBr is found to be a cheaper proposition. R_2Zn are liquids or solids with low melting points. Alkyl zinc halide has the following structure:

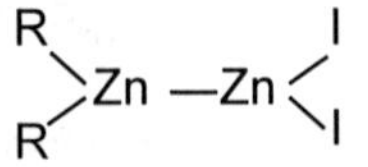

It has been found that on exposure to air and water, the organometallic compounds of Zn and Cd decompose rapidly. Presently, alkyl of Zn and Cd are easily synthesised through grignard reagent or alkyl lithium reagent or alkyl mercury compounds.

$$CdCl_2 + 2RMgCl \longrightarrow R_2Cd + 2MgCl$$

$$ZnCl_2 + 2RLi \longrightarrow ZnR_2 + 2LiCl$$

$$Zn + HgR_2 \longrightarrow R_2Zn + Hg$$

According to requirement, Grignard reagent type compounds can also be prepared as follows:

$$CdR_2 + CdI_2 \longrightarrow 2RCdI.$$

R_2Zn and R_2Cd are unaffected by CO_2. Arylmercury compounds can be produced by

$$C_6H_6 + (CH_3COO)_2Hg \longrightarrow \underset{\text{Phenyl mercury (II) acetate}}{C_6H_5.Hg\ CH_3COO}$$

6.2.5 Biological Role of Zn

Zinc has a very important role in biological world. In humans, the amount of iron is found to be 4.0g and Zn 2.0g. There are around twenty enzymes which contain Zn. Some are given as follows:

1. Carbonic anhydrase, found in red blood cells, is involved in respiration. It enhances the absorption of CO_2 by red blood cells in muscles and other tissues. Also, it reverses reaction by release of CO_2 in the lungs. In addition it also regulates pH.

$$CO_2 + OH^- \rightleftharpoons HCO_3^-$$

2. Carboxypeptidase has its role in the digestion of proteins and found in pancreatic juice. It also works in plants for protein metabolism. The

enzyme catalyses the hydrolysis of terminal peptide (—$CONH_2$) and carboxyl end of the chain.

3. Alkaline phosphatase is responsible for removing phosphate groups from many types of molecules, including nucleotides, proteins, and alkaloids.
4. Dehydrogenase and aldoses is mainly involved in sugar metabolism.
5. Alcohol dehydrogenase (metabolism of alcohol) are a group of dehydrogenase enzymes which facilitate the interconversion of alcohols to aldehydes and ketones.

6.2.6 Toxic Nature of Cd and Hg

Cd and Hg are extremely toxic, but Zn is an essential element. Cd may escape as dust with flue gases, it may be present in smelters, Nicad batteries and cigarette smoke. Cd affects kidney by accumulating there and causes malfunctioning of the kidneys. It also replaces zinc from enzymes and thus, stops the enyzme to function.

Mercury vapour is toxic and in case of inhalation or ingestion, it causes giddiness, tremors, lung and brain damage. The spillage maybe treated with flowers of sulphur to form HgS. The compounds of mercury like $HgCl_2$, Hg_2Cl_2 and HgO are also toxic. They cause cumulative poison with Cd, Pb, Sn and Sb. The enzymes having thiol group are inhibited by Hg. The role of inorganic mercury compounds is appreciated in controlling the growth of slime mould in wood pulp and in paper industry. They are also used as antifouling agents in paints and as fungicides for treating seeds and plants. $Hg(Me)_2$ and MeHgX are common fungicides. Aryl mercury compounds are found to be troublesome and risky as they can damage brain resulting in numbness, loss of vision, deafness, madness and death.

EXERCISE

1. Write the ground state electronic configurations of: Fe°, Ti^{4+}, V^{2+}, Cr^{5+}, Co^{+}, Ni^{2+}, Cu^{3+} and Mn^{6+}.

2. Which of the above ions will give tetrahedral, octahedral and square pyramid complexes?

3. Explain Jahn-Teller distortion in the octahedral complexes.

4. Differentiate between the structures of $TiOSO_4.H_2O$ and $VOSO_4.5H_2O$.

5. Give two examples each of MCl_4 and MCl_3 behaving as Lewis acids.

6. Give reactions when acidified $K_2Cr_2O_7$ is added to solutions containing

(*a*) F^- (*b*) Cl^- (*c*) Br^- (*d*) I^- (*e*) OH^- (*f*) NO_2 (*g*) SO_4^{2-} (*h*) H_2O_2.

7. The M^{3+} solution of transition elements are acidic. Discuss.

8. Draw the structure of $Cr_2(CH_3COO)_4.2H_2O$ red crystals.

9. Draw the structure of $[Ni(ac\ ac)_2]_3$, $CrCI_3(THF)_3$ and $CrO_5(py)$.

10. $La2O_3$ is not reduced by Al. Explain.

11. How do we obtain pure Ti?

12. Zr and Hf show similar physical and chemical properties than Ti and Zr.

13. Explain the formation of condensates of $[VO_4^{3-}]$ with reducing pH values.

14. How do NaOH and $H{,}SO_4$ react with CrO, Cr_2O_3 and CrO_3?

15. Describe extraction of iron and conversion into steel.

16. Why does Fe corrode? How can it be prevented.

17. $Mo(OH)_6$ is pink, MnO_2 is black and MnO_4^- is dark purple explain.

18. You are given a green solution of Mn(+VI) salt. On passing CO_2 turns brown precipitate. Illustrate by chemical equations.

19. Explain the difference between

(*i*) pig iron

(*ii*) wrought iron and

(*iii*) stainless steel.

20. From the point of view of crystal field splitting write the electronic arrangement of:

(*a*) $[Fe(H_2O)_6]^{2+}$

(*b*) $[Fe(CN)_6]^{4-}$

(*c*) $[Fe(CN)_6]^{3-}$

Indicate the number of unpaired electrons against each complex and their expected magnetic moments.

21. NMR spectra shows only one peak of $[Fe(CO)_5]$ at low temperature. Two peaks are expected. Explain.

22. Compare the haem, vitamin B_{12} and chlorophyll structures.

23. Draw the possible isomers of:

(*a*) $[Co(en)_3]^{3+}$

(*b*) $[Co(en)_2(SCN)_2]^+$.

24. Draw square planar, tetrahedral complexes of Ni(*II*) and explain the shapes.

25. Why noble metals are relatively unreactive? Explain.

26. Give structures of compounds showing metal-metal bonds. Write examples.

27. What are electronic configurations of Zn, Cd and Fig in the (+2) state? Do you expect them to behave like transition elements?

28. Explain how Zn is sacrified in the extraction of Cd.

29. $ZnSO_4 + NH_3 + H_2O \longrightarrow$ White precipitate.

$ZnSO_4 + NH_3 + H_2O + NH_4Cl \longrightarrow$ No white precipitate.

30. Freshly prepared hydroxide of Mn^{2+} is white and on warming it turns dark brown; Co^{2+} is blue and turns pink on warming; Cu^{2+} is blue but turns black on heating. Explain with the help of chemical equations.

31. Write the number of unpaired electrons in.

(*i*) diamagnetic Mn^{2+}
(*ii*) tetrahedral Cr^{4+}
(*iii*) tetrahedral Co^{2+}
(*iv*) octahedral V^{3+} and Co^{3+}
(*v*) diamagnetic Fe^{2+} and
(*vi*) paramagnetic Mn^{2+} .
Give examples.

32. How is oxygen bonded in

(*a*) $[Cs_3(TiO_2F_5)]$
(*b*) $K_3[CrO_g]$ and
(*c*) $[Co_2O,(NH_3)_{10}](SO_4)_2$.

33. Draw the possible isomers of $[Co(en)_2(SCN)_2]^+$ and write the names of each.

34. M—O—H bonds are angular but in $[(NH_3)_5$—Cr—O—$Cr(NH_3)_5]^{4+}$ they are linear explain.

35. $[Mn(acac)_3]$ has M—O axial bond with bond length (~ 1.94 Å) than equatorial (~ 2.00 Å). Explain.

36 Complete and balance the following equations:

(*i*) $[Co(H_2O)_6]^{2+} + Na_2edta \longrightarrow$
(*ii*) $[Fe(H_20)_6]^{2+} + NaHCO. \longrightarrow$
(*iii*) $Mn^3 + H_2O \xrightarrow{\text{Reduction}}$
(*iv*) Oxidation of $[Fe(H_2O)_6]^{2+} \longrightarrow$
(*v*) $TiCl_4 + 4H_2O \longrightarrow$
(*vi*) $Ti^{3+} + H_2O_2 \xrightarrow{\text{Oxidatlon}}$
(*vii*) $V_2O_5 + NaOH \longrightarrow$
(*viii*) $VO(OH)_2 + HNO_3 + H_2O \longrightarrow$
(*ix*) Heating Cr in air $\longrightarrow$
(*x*) Preparation of CrO_3.
(*xi*) $CrO_2CI_2 + H_2O \longrightarrow$
(*xii*) $Mn^{2+} + H_2O + PbO_2 \longrightarrow$
(*xiii*) $Cu^{2+} + I^- \longrightarrow$
(*xiv*) Aqueous Cu^{2+} with CN $\longrightarrow$

37. Bring out the number of unpaired electrons in $[Fe(H_2O)_6]^{2+}$ and $[Fe(CN)_6]^{4-}$. Give reasons.

38. Draw the structures of *cis*-$[VO_2Cl_4]^{3-}$, *cis*-$[VO_2(C_2O_4)_3]^{3-}$, $[TiCl_4(OPCl_3)]$ and $[TiCl_4(POCl_3)_2]$,

39. Predict the geometries for $[Ni(CN)_5]^{3-}$.

40. Describe the hybridisation involved in Ni(*ii*) complexes. Draw crystal field diagrams.

41. Write balanced equation, for the preparation of the complexes obtained by the action of:

$CoCl_2 + [NO(CN)_4]^{2-}$ *(aq)* + KCN + en

42. The complex $[NiCl_4]^{2-}$ is paramagnetic while $[Ni(CN)_4]^{2-}$ is diamagnetic. Deduce the structures of these two complexes and apply ligand field theory.

43. What geometries do you expect for four coordinate Zn(II) complexes?

44. Why are tetrahedral complexes not low spin?

45. Discuss the tendency of Ti^{4+} compounds to be octahedrally coordinated.

46. Describe the geometry of $[CuCl_5]^{3-}$. Why Cu—Cl axial bond is shorter than the equatorial bond.

47. Compare the geometries of $[CuCl_5]^{3-}$, $[Ni(CN)_5]^{3-}$ and $[MnCl_5]^{2-}$, draw the crystal field diagrams.

48. For the complexes $[NiX_2(PRPh_2)_2]$, state the isomer preference when X = C^-, Br^-, or I^-. In which cases square planar or tetrahedral geometries are adopted?

49. Which of the above geometries lead to paramagnetic or diamagnetic complexes. Explain with the help of crystal field diagrams.

50. In the oxovanadium (IV) dichloride complex, why is the coordination geometry around the oxovanadium (IV) compound a trigonal bipyramid instead of the usual square pyramid.

51. Prepare crystal field diagrams with suitable electronic configuration in case of:

$[Ru(NH_3)_5py]^3V$, $[Co(NH_3)_5py]^{3+}$, $[Fe(CN)_6]^{4-}$, $[Fe(CN)_6]^{3-}$ and $[Ru(NH_3)_6]^{3+}$.

52. How is cadmium extracted? What are the uses of Cd and how does cadmium impurity in water acts as a health problem?

53. What are the different ways in which mercury is discharged into the environment? List the poisonous inorganic and organomercury compounds.

54. Why is the mercury (I) ion written as Hg_2^{2+} whereas copper (I) ion is written as Cu^+? Give reason.

55. When a limited amount of oxidising agent is used (for excess of Hg) then Hg(I) compounds are formed, while an excess oxidising agent forms Hg(II) compounds. Explain.

56. Compare and contrast the differences between the physical properties namely:

(*a*) atomic and ionic radii

(*b*) oxidation states

(*c*) metal-metal bonds

(*d*) stereochemistry and

(*e*) magnetic properties of first transition series with 4*d* and 5*d* series.

57. Why are physical and chemical properties of Zr and Hf compounds so similar?

58. Give three examples of elements which form cluster compounds in lower oxidation states.

59. Draw structures of $[Re_2Cl_8]^{2-}$, $[TaCl_5]$, $[NbF_5]$ and $[Rh_2Cl_2(CO)_4]$,

60. Describe preparation, chemical and physical properties of RuO_4 and OsO_4.

61. Discuss the position and oxidation states of the 'platinum metals' in the periodic table.

62. Describe the true nature of dihalides of Mo and W.

63. Explain why technetium is available in trace amount on the earth.

64. Show evidence of metal-to-metal interaction in square planar complexes of Ni^{2+}, Pd^2 and Pt^{2+}.

65. Draw the structure of $Pd^{(III)}$ acetate.

66. Compare and contrast the chemistry of Cu with that of Ag and Au.

67. Prepare $[Pt(PPh_3)_4]$ and show its reaction with CH_3I.

68. Compare the properties of Ag^{1+} and Au^{1+}.

69. Write balance of chemical equation in the following:

(*i*) leaching of Au with KCN in the presence of O_2

(*ii*) Agl and a solution of $Na_2S_2O_3$

(*iii*) Aqueous $AgNO_3$ with $S_2O_3^{2-}$ in the presence of excess pyridine.

70. Name the silver salts that are:

(*i*) soluble in H_2O

(*ii*) insoluble in H_2O

71. Ni-Cu sulphide contains lot of Pt. How will you isolate Pt.

72. How would you:

(*i*) dissolve Ta metal

(*ii*) precepitate Zn from aqueous solution containing A1

(*iii*) prepare Mo(V) chloride from MoO_3

(*iv*) make $[K_2MoOCl_5]$ from MoO_3

(*v*) dissolve MoO_3

(*vi*) prepare Re(III) chloride.

73. How would you separate Ag and Au from its alloy?

74. Give two bridge clearing reaction of Rh or Pt.

75. Write equations when $RuCl_3.3H_2O$ is:

(*i*) dissolved in concentrated HC1

(*ii*) warmed with excess PPh_3 in ethanol

(*iii*) heated with NH_3 in ethanol

(*iv*) boiled with sodium acetate in ethanol.

■ ■ ■

7

Inner Transition or *f*-block Elements: The Lanthanides

7.1 INTRODUCTION

It has two sets of elements, each set comprising of 14 elements. The 4*f*-block elements are termed as lanthanons or lanthanides. The elements are differentiated from others by filling of antipenultimate 4*f* orbitals which has a capacity to fill 14 electrons. The element Lanthanum belonging to 5*d* block element is the parent member of lanthanides and hence, the name is Lanthanides. In earlier days, they were called rare earth elements, but many of these elements are not rare. For 5*f*-block elements, the parent actinium, Ac belongs to 5*f* series and itself Ac belongs to 6*d* block element series. The 14 elements belonging to 5*f* block series are called Actinides and named after the element actinium.

7.1.1 The Electronic Configuration

The electronic structures of lanthanides are given in the table. Lanthanum of 5*d* series has the structure (Xe) $5f^1 6s^2$. The 14 electrons, 1 to 14 are added to 4*f* orbital. The 4*f* is inner to 5*d* therefore, they are also called inner transition elements. Due to increased stability of half filled and fully filled 4*f* orbitals, the electronic configuration shows a definite bias to retain these configurations in Gadolinium $4f^7 5d^1 6s^2$ and Lutetium $4f^{14} 5d^1 6s^2$. The oxidation state (+III) is ionic for Ln^{3+} (Lanthanons). Ln^{2+} and $Ln4^+$ ion are less stable than Ln^{3+}. The oxidation states of (+II) and (+IV) are shown when they lead to f°, f^7 and f^{14} configurations.

Element		Electronic structure of atoms			Electronic structure of M^{3+}	Oxidation states*		
Lanthanum	La	[Xe]	$5d^1$	$6s^2$	[Xe]		+III	
Cerium	Ce	[Xe] $4f^1$	$5d^1$	$6s^2$	[Xe] $4f^1$		+III	+IV
Praseodymium	Pr	[Xe] $4f^3$		$6s^2$	[Xe]$4f^2$		+III	(+IV)
Neodymium	Nd	[Xe]$4f^4$		$6s^2$	[Xe]$4f^3$	(+II)	+III	
Promethium	Pm	[Xe]$4f^5$		$6s^2$	[Xe]$4f^4$	(+II)	+III	
Samarium	Sm	[Xe]$4f^6$		$6s^2$	[Xe]$4f^5$	(+II)	+III	
Europium	Eu	[Xe]$4f^7$		$6s^2$	[Xe]$4f^6$	+II	+III	
Gadolinium	Gd	[Xe]$4f^7$	$5d^1$	$6s^2$	[Xe]$4f^7$		+III	
Terbium	Tb	[Xe]$4f^9$		$6s^2$	[Xe]$4f^8$		+III	(+IV)
Dysprosium	Dy	[Xe]$4f^{10}$		$6s^2$	[Xe]$4f^9$		+III	(+IV)
Holmium	Ho	[Xe]$4f^{11}$		$6s^2$	[Xe]$4f^{10}$		+III	
Erbium	Er	[Xe]$4f^{12}$		$6s^2$	[Xe]$4f^{11}$		+III	
Thulium	Tm	[Xe]$4f^{13}$		$6s^2$	[Xe]$4f^{12}$	(+II)	+III	
Ytterbium	Yb	[Xe]$4f^{14}$		$6s^2$	[Xe]$4f^{13}$	+II	+III	
Lutetium	Lu	[Xe]$4f^{14}5d^1$		$6s^2$	[Xe]$4f^{14}$		+III	

The most important oxidation states (generally the most abundant and stable) are shown in bold. Other well-characterized but less important states are shown in normal type. Oxidation states that are unstable, or in doubt, are given in parentheses.

The resemblance occur in lanthanide elements more closely in their row than the row of transition elements. This is explained with the reason that they have only one stable oxidation state of (+III). Hence, it is possible to observe changes due to changes in size and nuclear charge.

7.1.2 Abundance and Isotopes

Lanthanides are not considered to be rare, ceruim is as abundant as copper. The element promethium (Pm) does not exist in nature and all other elements in the series are more abundant than iodine. It has been noticed that abundance of elements as well as naturally existing isotopes vary regularly, described in the Table. It is noticed that the elements with even number of protons in the nucleus are more abundant than those with odd numbers. Also, two or more than two stable isotopes do exist in those elements which have even atomic number. If two elements, which have consecutive atomic numbers, have an isotope of the same weight then only one of the isotopes will be stable. There are seven isotopes for the elements 60 and 62 each. Pm (atomic number 61) do not have many stable mass numbers and the isotopes of Pm made so far are radioactive.

Abundance of the elements in the earth's crust, by weight and number of natural isotopes

Atomic number	Element	Abundance (ppm) in earth's crust	Relative abundance	Naturally occurring isotopes
58	Ce	66	26	4
59	Pr	9.1	37	1
60	Nd	40	27	7
61	Pm	0		0
62	Sm	7	40	7
63	Eu	2.1	49	2
64	Gd	6.1	41	7
65	Tb	1.2	56	1
66	Dy	4.5	42	7
67	Ho	1.3	55	1
68	Er	3.5	43	6
69	Tm	0.5	61	1
70	Yb	3.1	44	7
71	Lu	0.8	59	2

Numbers of stable nuclei with odd and even numbers of neutrons and odd and even atomic number

Atomic number	Number of neutrons	Stable nuclei
Even	Even	164
Even	Odd	55
Odd	Even	50
Odd	Odd	4

Element 60	142,	143,	144,	145,	146,		148,		150		
Element 62			144,			147,	148,	149,	150,	152,	154

7.1.3 Extraction

Monazite sand is the most important mineral for the isolation of rare earths (78%). It is a mixture ot Lanthanum phosphate and triphosphates of lighter lanthanides, *i.e.,* Ce, Pr and Nd, with little quantity of Y and heavier lanthanides and thorium phosphate. This a weak radioactive element with traceable amount daughter product as Ra which is more radioactive.

Treatment of Monazite Sand

The sulphuric acid digestion has been used in order to treat monazite with hot concentrated H_2SO_4. The sulphates of Th, La and lanthanides are dissolved and then separated from the insoluble material. By partial neutralisation with NH_4OH, ThO_2 gets precipitated. Na_2SO_4 solution is then added to salt out La and light lanthanides as sulphates leaving heavier lanthanides in solution. The bleaching powder is used to oxidise light lanthanides. Ce^{3+} is oxidised to Ce^{4+} which is precipitated as $Ce(IO_3)_4$ and removed. Tri-*n*-butylphosphate is used to remove La^{3+} by solvent extraction. The individual element is separated by ion exchange method.

Whether the different-lanthanide elements are completely or partially separated, the element may be further acquired by the following methods:

(i) Electrolysis of fused $LnCl_3$ with addition of NaCl and $CaCl_2$ to lower the melting point.

(ii) La and elements from Ce to Eu are acquired by reducing $LnCl_3$ with Ca at 1000—1100°C in vessel filled with Ar. The heavier elements melt above 1400°C. At this temperature, we use LnF_3 and Li is used instead of Ca. The unseparated mixture of La and lanthanides is called Misch metal.

Separation of Lanthanides

Lanthanides are all trivalent and identical in ionic size therefore, they have almost identical chemical properties. The separation of lanthanide elements is a tough job. For separation, slight difference in basic properties, solubility or stability is exploited. Modem methods used recently are ion exchange and valency.

Precipitation Method

Hydroxyl ions are added to a solution of $(LaNO_3)_3$. The weakest base $Lu(OH)_3$ is precipitated first and strongest base $La(OH)_3$ is precipitated last. The common precipitate for all elements will contain more of the elements at the right of the series. By filtration, process of partial separation is performed *i.e.* the solution contains more elements at the left and precipitate elements at the right of the series. The precipitate is redissolved in HNO_3 and the process repeated many times to obtain purity.

Heating

Mixture of $Ln(NO_3)_3$ is fused till the least basic nitrates change to oxides and then mixture is dissolved in water. The dissolved nitrates are separated by filtration leaving behind the insoluble oxides. The oxides are redissolved in HNO_3 and the process is repeated.

Fractional Crystallisation

Solubility of salts decrease when we move from La to Lu, salts of Lu side of the series crystallize out first. Salts like nitrates, sulphates, bromates. perchlorates and oxalates are used as well as the double salts such as, $Ln(NO_3)_3.3Mg(NO_3)_2$ $24H_2O$ as they form good crystals. The process is repeated many times to achieve good separation. Sometimes, non-aqueous solvents such as, diethyl ether is used to separate $Nd(NO_3)_3$ from $Pr(NO_3)_3$.

Complex Formation

A complexing agent EDTA is used to treat the mixture of lanthanide ions. As a result complexes are formed by all the Ln^{3+} ions. Ions on the right of the lanthanide series form strong complexes as they have smaller ions. If some acid is added to the above solution, the least stable EDTA complexes are dissociated. The released ions at the left of the series such as, Ce^{3+}, Pr^{3+} and Nd^{3+} are precipitated as oxalates. These are filtered off and the process repeated many times.

Solvent Extraction

It has been noted that in tri-*n*-butylphosphate, the heavier Ln^{3+} ions are more soluble in comparison to the lighter ions. The solubility in water or ionic solvents are reversed. The ratio of partition coefficients of $La(NO_3)_3$ and $Gd(NO_3)_3$ in concentrated HNO_3 and tri-n-butylphosphate is 1:1.06. The difference is too small, but by counter current apparatus a large number of partitions can be performed. In this way, Gd upto 95% purity can be obtained.

Valency Change

Some lanthanons exist in (+2) and (+4) oxidation states and their properties are quite different from Ln^{3+} and this difference makes their separation quite easy. Ce^{4+} is the only lanthanide which is stable in aqueous solution. Oxidising a mixture of Ln^{3+} ions with NaOCl in alkaline conditions give Ce^{4+}. Ce^{4+} has high charge/radius ratio hence, it is less basic than the Ln^{3+} ions. It can be separated by precipitation of CeO_2 or $Ce(IO_3)_4$ leaving Ce^{3+} in solution. Similarly, properties of Eu^{3+} are quite contrasting from Ln^{3+} solution. $EuSO_4$ is insoluble in water while Ln^{3+} sulphates are soluble. Eu^{2+} can be obtained electrolytically by using Zn/Hg or mercury cathode. In case H_2SO_4 is present, $EuSO_4$ will be precipitated. Sm^{2+} and Yb^{2+} may be separated in a similar way.

Ion Exchange

The procedure of ion exchange is the most important as it is quite efficient and fast among the general methods for separation and purification of lanthanides. A column of ion-exchange resin *i.e.* Dowex-50 is used where a solution of lanthanide ions is run down. It is a sulphonated polystyrene and contains $—SO_3H$ group. The absorption of Ln^{3+} ions takes place on the resin and replaces H atom in the $—SO_3H$ group.

$$Ln^{3+}_{(aq)} + 3H(resin)_{(s)} \longrightarrow Ln(resin)_{3(s)} + 3H^+_{aq}$$

The H^+ ions produced are washed through the column and Ln^{3+} ions are washed in selective way with the help of a complexing agent. The complexation takes place in a buffered solution: Citric acid/ammonium citrate or a solution of $(NH_4)_3(H.\ edta)$ maintaining pH 8.

$$Ln^{3+}(resin)_3 + 3H^+ + Citrate^{3-} \rightleftharpoons 3H(resin) + Ln(Citrate)$$

When citrate solution flows down the column, Ln^{3+} are removed from the resin forming citrate complex. The metal travels down and comes out from the bottom as citrate complex. Smaller Ln^{3+} form stronger complexes with citrate ions than the larger La^{3+} ions. Thus, smaller and heavier ions spend more time in solution and less time on the column. The different metal ions present separate into bands and pass down the column. This may be spectroscopically monitored. The solution passing the column is accumulated in automatic fractions in another flask. The metals from these fractions may be precipitated as oxalates and heated to give oxides. The process of long ion exchange column is an efficient method to have the purity of elements upto 99.9%.

7.1.4 Lanthanide Contraction

When we move in a period from left to right, the value of covalent and ionic radii declines as extra orbital electrons shield the nuclear charge less effectively. As a result, the electrons are pulled in closer and the order of shielding effect of orbitals is $s > p > d > f$. The contraction in size for one element to another is very small. The overall decrease in the size from Ce to Lu is about 0.2Å which is known as lanthanide contraction. The value of hardness, melting and boiling points show an increase because the atom-atom attraction increases due to decrease in size.

The size and charge of an ion contribute towards the properties of that particular ion. In Ln^{3+}, the changes from ion to ion are at a very low level. Therefore, their chemical properties are very similar. Though lanthanides do not form complexes very frequently but due to small size, Lu^{3+} forms strongest complexes. The La^{3+} and Ce^{3+} possess largest ions and they also form strong bases as $La(OH)_3$ and $Ce(OH)_3$, respectively.

It is noted that lathanide contraction lowers the size of the last four elements below Y_b in the preceding transition series. Also, Dy^{3+} and Ho^{3+} are heavier ions and show similarity with Y_b^{3+}. Hence, they show similar chemical properties and consequently, these elements are separated with great difficulty.

Sc														
0.745														
Y														
0.900														
La	Ce	Pr	Nd	Pm	Sm	Eu	Gd	Tb	Dy	Ho	Er	Tm	Yb	Lu
1.032	1.02	0.99	0.983	0.97	0.958	0.947	0.938	0.923	0.912	0.901	0.890	0.880	0.868	0.861

In covalent radii, similar change in size is observed across a period.

Sc 1.44		Ti 1.32	V 1.22	Cr 1.17	Mn 1.17	Fe 1.17	Co 1.16	Ni 1.15
Y 1.62		Zr 1.45	Nb 1.34	Mo 1.29	Tc –	Ru 1.24	Rh 1.25	Pd 1.28
La 1.69	14 Lanthanide elements	Hf 1.44	Ta 1.34	W 1.30	Re 1.28	Os 1.26	Ir 1.26	Pt 1.29

Since the lanthanide contraction in size is along the lanthanide series, it is expected that the elements which follow the third transition series are smaller, but it does not. The normal increase from Sc → Y → La is not observed after the lanthanides *e.g.,* Zr/Hf; Nb/Ta and Mo/W have almost identical size and identical properties. Therefore, the separation of these elements is very tough. In third row of transition elements, the size of the elements is very similar to the second row transition elements. Thus, 4*d* and 5*d* elements resemble closely with each other, whereas 3*d* and 4*d* properties are not so.

7.1.5 Properties of (+III) Compounds

Lanthanides exist in soft and silvery white metals and are electropositive, hence are very reactive. It has been found that due to an oxide layer on the surface, the heavier metals are less reactive. Trivalent ions are most stable and considered to be most important. There is variation in the sum of first three ionisation energies as the minima at La^{3+}, Gd^{3+} and Lu^{3+} have three stages of *f* shell *i.e.,* empty, half filled and fully filled *f* shell, whereas maxima occurs at Eu^{3+} and Yb^{3+} with a half filled or fully filled shell. For all elements, the standard reduction potential E° are high ranging between –2.48 to –2.26 volts, depending upon the size of the ions. Lanthanides are more reactive than Al (E^0 = –1.66 volts) and slightly more than Mg *(E°* = –2.37 volts) Therefore, the rate of reaction with cold water is slower than with hot water.

$$2Ln + 6H_2O \longrightarrow 2Ln(OH)_3 + 3H_2$$

When NH_4OH is added in the aqueous solution then hydroxides ($Ln(OH)_3$) are resulted which is in the form of gelatinous precipitates.

These precipitates of $Ln(OH)_3$ are less basic in comparison with $Ca(OH)_2$ but more basic compared to $Al(OH)_3$, which is amphoteric. These metal oxides and hydroxides are soluble in dilute acids to give their respective salts. Also, $Ln(OH)_3$ are found to be basic as they can absorb CO_2 from air to give carbonates. The degree of basicity decreases as the ionic radius decreases from Ce to Lu. $Ce(OH)_3$ is most basic and $Lu(OH)_3$ is least basic, but falls in between Sc and Yb in basic strength. The later elements dissolve in concentrated NaOH forming complexes.

$$Yb(OH)_3 + 3NaOH \longrightarrow 3Na[Yb(OH)_6]^{3-}$$

$$Lu(OH)_3 + 3NaOH \longrightarrow 3Na^* + [Lu(OH)_6]^{3-}$$

The metallic lustre gets damaged readily in air or heating in O_2 as it forms metal oxides Ln_2O_3. Yb and Lu form protective oxide layer. The metal Ce is an exception as it forms $Ce^{IV}O_2$ instead of Ce_2O_3. The oxides are ionic and basic, their basic strength decreases with decrease in size of ions. The metals react with H_2, at 300-400°C and the compound formed is LnH_2. Eu and Yb form their hydride salts like EuH_2 and YbH_2 containing M^{2+} and two H^-. The other metal hydrides are represented by LnH_2. There is formation of non-stoichiometric hydride compound with Yb as $YbH_{2.5}$. They are stable upto 900°C and can be decomposed by water.

$$CeH_2 + 2H_2O \longrightarrow CeO_2 + 2H_2$$

The hydrides LnH_2 are all ionic solids except EuH_2.

The anhydrous halides are prepared by heating metal and halogen or metal oxide with ammonium halide:

$$Ln_2O_3 + 6NH_4Cl \xrightarrow{300^0C} 2LnCl_3 + 6NH_3 + 3H_2O$$

The fluorides can be precipitated by Ln^{3+} and NaF or HF and in qualitative analysis, it is a test for the presence of lanthanides. With smaller metals ions, Ln^{3+} ion can form complexes $[LnF(H_2O)_n]^{2+}$. The chlorides are found to be deliquescent ($LnCl_3.6H_2O$) and on heating form oxohalide:

$$LnCl_3.6H_2O \xrightarrow{Heat} LnOCl + 5H_2O + 2HCl$$

However, heating $CeF_3.(H_1O)_n$ results in CeO_2. The bromides and iodides resemble chlorides.

At high temperature, Ln reacts with B giving LnB_4 and LnB_6. Reaction of Ln with C in electric arc and in inert atmosphere produces carbides of formula LnC_2 and $Ln_4(C_2)_3$. Ln_2O_3 and C in electric furnace can also give carbides. LnC_2 hydrolyse in H_2O to form C_2H_2, C_2H_4 and C_2H_6.

$$LnC_2 + 6H_2O \longrightarrow 2Ln(OH)_3 + 2C_2H_2 + H_2$$

$$C_2H_2 + H_2 \xrightarrow{Ni} C_2H_4 \xrightarrow{H_2} .\ C_2H_6$$

At high temperature, the nitrides undergo hydrolysis very easily and the metal reacts with N, P, As, Sb and Bi.

The salts such as, oxalates, nitrates, carbonates, sulphates and phosphates are well known. $Ln(ClO_4)_3$ salt is also reported.

7.1.6 Compounds with (+IV) Oxidation State

Among lanthanides, only Ce^{4+} ion has a good amount of aqueous chemistry. Due to high charge it is heavily hydrated. Hydrated Ce^{4+} is easily hydrolysed to give polymeric species. Ce(+IV) is used in volumetric analysis as an oxidising agent like $KMnO_4$ and $K_2Cr_2O_7$ Aqueous Ce(IV) can be prepared by oxidising Ce^{3+} solution with $(NH_4)_2S_2O_g$. Ce(+IV) is used in oxidising aldehydes and ketones. CeO_2 can be acquired by heating $Ce(OH)_3$ or $Ce_2^{IH}(oxalate)_3$ in air. It is insoluble in acids or alkalis.

$$Ce + O_2 \longrightarrow CeO_2$$

$$Ce(OH)_3 + \frac{1}{2}O_2 \longrightarrow 2CeO_2 + 3H_2O$$

The pure CeO_2 is white and $CeO_2.(H_2O)_n$ is yellow, gelatinous precipitate. Ceric sulphate $Ce(SO_4)_2$ is yellow, whereas CeF_4 obtained from CeF_3 and

F_2 is white and can be rapidly hydrolysed. The complex ammonium cerium (IV) nitrate is well known $(NH_4)_2[Ce(NO_3)_6]$. The other (+IV) compounds of lanthanides are unstable in H_2O and are known as oxides, fluorides and fluorocomplexes PrO_2, PrF_4, $Na_2[PrF_6]$, TbO_2, TbF_4, PyF_4. Pr, Nd, Tb and Dy also form (+IV) compounds, but these are unstable.

7.1.7 Compounds of (+II) Oxidation State

The Eu^{2+}, Sm^{2+} and Yb^{2+} are the only elements in lanthanides which exist in (+II) oxidation state in solution. Eu^{2+} being the most stable of all is stable in water, but its solution can be strongly reduced. By electrolysis of $Eu^{III}SO_4$ solution we can have $Eu_2(SO_4)_3$ precipitate.

$Eu^{II}Cl_2$ can be precipitated by passing H_2 through $Eu^{III}Cl_3$.

$$2EuCl_3 + H_2 \longrightarrow 2EuCl_2 + 2HCl$$

The Mg, Zn and Zn/Hg amalgam have the capability to reduce the aqueous solutions of Eu^{3+} or through electrolysis form Eu^{2+} which resembles calcium due to the following reasons:

(*i*) Their sulphates and carbonates are insoluble in water.

(*ii*) Their dichloride is insoluble in strong HCl.

(*iii*) The metals are soluble in liquid NH_3

The two metals Eu and Ca mainly differ in their magnetic moment values. EuX_2 has a magnetic moment value of 7.9 BM having seven unpaired electrons, while Ca compounds are diamagnetic.

Standard reduction potential value of $Eu^{3+}/Eu^{2+} = -0.41$ volts which is the same as for Cr^{3+}/Cr^{2+} and both are strong reducing agents.

Yb^{2+} and Sm^{2+} can be obtained by reducing their trivalent ions electrolytically in aqueous solution. Ln^{2+} ions are readily oxidised by air. These two elements form their respective hydroxides, carbonates, halides, sulphates and phosphates.

Nd(+II), Pm(+II), Sm(+II) and Gd(+II) are also found in solid dihalides *e.g.,* $LnCI_2$ and LnI_0. These can be obtained from their trihalides by reducing with H_2 or Na(Hg). They show metallic conduction *i.e.,* $Lu^{2+} + 2I^- + e^-$. They are non-stoichiometric.

7.1.8 Solubility

There is water of crystallization in lanthanide salts and their solubility depends upon the difference in lattice energy and solution energy. The solubility of their salts have similar pattern as of group II elements. Hence, their chlorides and nitrates are soluble in water, while oxalates, carbonates and fluorides are insoluble. But their sulphates are soluble which is contrary to group 2 elements. Property of forming double salts in many lanthanides is similar to the group I elements or NH_4^+ salts *e.g.,* $Na_2SO_4.Ln_2(SO_4)_3.8H_1O$. These double salts crystallize out well which help in separating lanthanides from each other.

7.1.9 Colour and Spectra

The trivalent lanthanide ions show colour both in the solid state and in the aqueous solution. The property of colour depends upon presence of unpaired electrons in *f* sub-shell and similar colour is there with the elements having *nf* electrons and those with (14-*n*) *f* electrons. Elements in other oxidation states do not show colours similar to (+III) state.

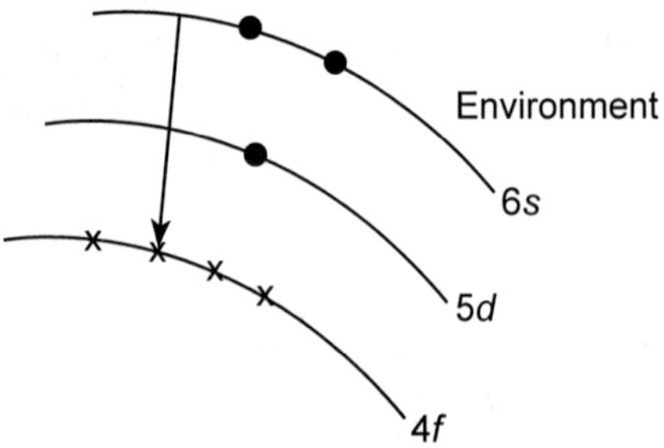

Colour appears because of absorption of a particular wave length from the visible region. The wave length absorbed 'λ' promotes electron to a higher energy level. In transition elements, crystal field splitting is most important whereas, in lanthanide spin orbits coupling is important. Lanthanide ions do not show absorption in the visible or near UV region. The colours arises out of *f-f* transitions. These transition are Laporte forbidden (change in subsidiary quantum number is zero). Thus, the colours are pale. The *f* orbitals are quite deep inside an atom and are covered by environmental factors such as, nature and number of ligands which form complexes and vibration of the ligands. Thus, because of no disturbance due to external factors the peaks of the absorption band are sharp. If *f* is four then $l = 3$ or m_l will have values –3, –2, –1, 0, +1, +2, +3 therefore, large number of transition are possible. In comparison, in *d-d* transitions broad absorption bands are observed instead of sharp peaks in *f-f* transition owing to the fact that *d* orbitals are close to the environment and there arc disturbances due to environmental factor and nature of ligands. Absorption spectra is used for the estimation of lanthanides. Lanthanide elements are useful as biological tracers as they can be easily tracked or identified by spectroscopy.

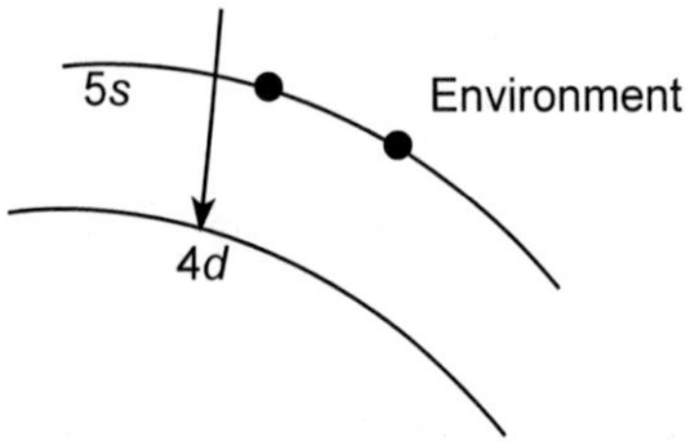

In elements Ce^{3+} and Yb^{3+}, do not show colour as there is no *d-d* transition (visible region). If excitation of electron is from 4*f* to 5*d* then absorption is from UV region, so, there is no colour. In case of Ce^{3+}, configuration is f^1 and Yb^{3+} is f^8 (one electron loss can give f^0 or $f^7$with extrastability).

In case an electron is transferred from ligand to metal then there is possibility of charge transfer spectra. Charge transfer is easy if metal has high charge and ligand has reducing property and charge transfer gives strong colours. Ce^{4+} is yellow due to charge transfer as there arc no *f*-*f* transitions. Sm^{2+} is blood red due to charge transfer.

7.1.10 Magnetic Moment

While observing Cc^{4+} and Lu^{3+}, Ce^{4+} has an f^0 configuration, while Lu^{3+} has f^{14} configuration. These metal ions are diamagnetic with paired electrons, therefore $\mu = 0$. In all other oxidation states, the metal ions contain unpaired *f* electrons in *f* subshell and behave as paramagnetic.

Paramagnetism is measured in Bohr Magnetons (BM) unit.

For transition elements

Magnetic moment $\mu = \mu_{s+L}$

Magnetic moment is the sum total of spin quantum number *S* and orbital quantum number *L*. In transition elements, the orbital motion is quenched by electric field of the ligands in environment. Therefore, magnetic moment μ is given by spin only motion:

$$\mu = \sqrt{S(S+1)}$$

for number of unpaired electrons '*n*'

$$\mu_S = \sqrt{n(n+2)}$$

La^{3+} and Lu^{3+} have no unpaired electrons or $n = 0$ and $\mu_S = \sqrt{0(0+2)} = 0$. For Gd^{3+}, $n = 7$.

Therefore $\mu_S = \sqrt{7(7+2)} = \sqrt{63} = 7.9\,\text{BM}$.

For lanthanides *f* electrons are buried deep into an atom and not affected by external factors. Therefore, orbital motion is not quenched out. The magnetic moment μ has both spin and orbital contributions but different from transition elements. For lanthanides, spin contribution and orbital contribution couple together to give new quantum number J.

J = L – S (Shell is less than half)

J = L + S (Shell is more than half)

Therefore $$\mu = g\sqrt{J(J+1)}$$

where $$g = \frac{S(S+1) - L(L+1)}{2J(J+1)}$$

The μ valiue for lanthanides has the coupled spin plus orbital contribution at 300°K.

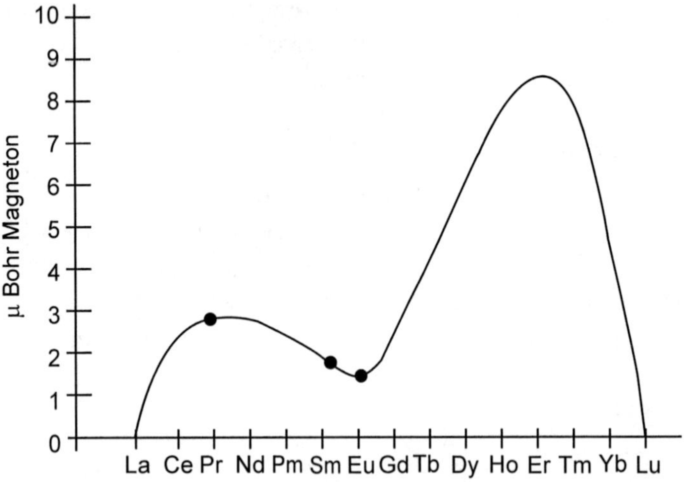

Magnetic moment spin plus orbital motion at 300°K μ value for Ln^{3+} ions

Element		Electronic structure of M3+	Magnetic moment Calculated	Observed (BM) (BM)
Lanthanum	La	[Xe]4f0	0	0
Cerium	Ce	[Xe]4f1	2.54	2.3-2.5
Praseodymium	Pr	[Xe]4f2	3.58	34-3.6
Neodymium	Nd	[Xe]4f3	3.62	3.5-36
Promethium	Pm	[Xe]4f4	2.68	2.7
Samarium	Sm	[Xe]4f5	0.84	1.5-1.6
Europium	Eu	[Xe]4f6	0	34-3.6
Gadolinium	Gd	[Xe]4f7	7.94	7.8-8.0
Terbium	Tb	[Xe]4f8	9.72	94-9.6
Dysprosium	Dy	[Xe]4f9	10.63	10.4-10.5
Holmium	Ho	[Xe]4f10	10.60	10.3-10.5
Erbium	Er	[Xe]4f11	9.57	9.4-96
Thulium	Tm	[Xe]4f12	7.63	7.1-7.4
Ytterbium	Yb	[Xe]4f13	4.50	4.4-4.9
Lutetium	Lu	[Xe]4f14	0	0

The experimental values of p measured at 300°K are in close agreement with the calculated values (S + L). In the graph, Eu^{3+} is poor and Sm^{3+} is not good, thus difference in energy between ground state and the next state is very small. The thermal motion is enough to promote electrons and put it in higher levels. At low temperature, the higher levels are less populated by electrons as g is determined by ground state configuration. The unusual curve values are attributed to:

(*i*) less than half *f* shell J = L – S

(*ii*) more than half *f* shell J = L + S

In (*i*) case, the spin and orbital work in opposition and in (*ii*) case, spin and orbital work together.

7.1.11 Coordination Complexes

Complex formation is guided by charge/radius ratio. If the ratio is higher, complexes are readily formed and if small, complex formation is slow. In lanthanide ions, Ln^{3+} charge is high, but the size is also large (1.03 – 0.86A). Due to low charge to radius ratio, complexes are not formed readily. They do not tend to form ammine complexes in aqueous medium as water is a stronger ligand than NH_3. Ammine complexes, can be obtained in non-aqueous conditions. Very few stable complexes are formed with CO, CN^-, alkyl and aryl groups. This happens because the metal is entrenched deep inside an atom and is reluctant to form π back bonding.

The most stable complexes are formed with chelating oxygen ligands such as, oxalic acid, citric acid, edta and acetyl acetone. Coordination number below 6 are possible with bulky ligands *e.g.,* $[N(SiMe_3)_2]$ and (2, 6 dimethyl phenyl) coordination number 6 is not common in contrast to transition elements. The most common coordination numbers are 7, 8 and 9. High coordination number 10 and 12 are obtained from larger lanthanides and smaller ligands NO_3^- and SO_4^{2-}.

Chelated complexes are more stable than non-chelates. Fluoride complexes are obtained in water and not Cl^- complexes. The complex $[Ce(NO_3)_6]^{2-}$ is formed in non-aqueous N_2O_4. Higher coordination number 10 and 12 involve in participation of *f* orbitals.

The common alkyls and aryls of lanthanides are prepared in ether solvent.

$$LnCl_3 + 3LiR \longrightarrow 3LiCl + LnR_3$$

$$LnR_3 + LiR \longrightarrow Li[LnR_4]$$

Cyclopentadienyl compounds $[Ln(C_5H_5)_3]$, $[Ln(C_5H_5)_2Cl]$ and $[Ln(C_5H_5)Cl_2]$ do exist, but are susceptible to air and water.

EXERCISE

1. Explain the position of lanthanides in the periodic table. Write the electronic configuration of Eu, Gd, Yb, Ce and Lu.
2. What is lanthanide contraction? Explain its effect on the properties of lanthanides.
3. Write the electronic configuration of ions Ce(+4), Gd(+3), Sm(+2), Tb(+4).
4. Why +3 is the most common oxidation state of lanthanides?
5. Which elements show other than +3 oxidation state in lanthanides and why?
6. Describe the precipitation reactions of +2, +3 and +4 lanthanide ions.

7. The electronic absorption spectra shows sharp bands in lanthanide ions while *3d* elements spectra show broad bands. Explain.

8. How do we get anhydrous lanthanide chlorides?

9. Write equations:

(*i*) Reduction of CeO_2 in aqueous HCl solution by Sn^{2+}

(*ii*) $CeO_2.nH_2O$ in aqueous HCl

(*iii*) Oxidation of Ce^{3+} in aqueous HNO_3 by $K_2S_2O_g$.

(*iv*) Reduction of aqueous Eu^{3+} by Zn.

10. Hydrolysis increases from La to Lu with the decrease in size due to lanthanide contraction? Explain.

11. On the basis of magnetic moment 'μ' values indicate the number of unpaired electrons in Nd atom and the formal charge on the atom.

12. Indicate the number of unpaired electrons in the ground state of following ions:

La^{3+}, Ce^{4+}, Lu^{3+}, Gd^{2+}, Eu^{2+} and Tb^{4+}

13. Bring out the stereochemistry and coordination numbers of common lanthanide complexes. Compare with transition metal complexes.

■ ■ ■

8

The Actinides

8.1 INTRODUCTION

These elements are also called second inner transition elements as the 5*f* orbital which is inner to 6*d* is getting filled with electrons. The 5*f* series starts from the element thorium and ends with lawrencium with a total of 14 elements. The series starts after the transition element actinuim (Ac) $6d^1 7s^2$ which resembles in properties with 5*f* series therefore, it is taken as a parent element and hence, the name actinides was coined.

Lanthanide and Actinide Element

Transition elements	Lanthanides													
La	Ce	Pr	Nd	Pm	Sm	Eu	Gd	Tb	Dy	Ho	Er	Tm	Yb	Lu
Ac	Th	Pa	U	Np	Pu	Am	Cm	Bk	Cf	Es	Fm	Md	No	Lr

The electronic structures of actinides do not follow the simple trend as in lanthanides, 5*f* become lower in energy to 6*d* orbitals. In the first four elements Th, Pa, U and Np, the difference in energy between 5*f* and 6*d* is very small. Therefore, electrons keep shunting between 5*f* and 6*d* or may occupy both. From Pa onwards, the 5*f* shell fills regularly and the elements show striking similarity. Due to the increase in the number of oxidation states, inverted pyramid as shown in the table is adopted.

Oxidation States of Actinides (+)

Ac	Th	Pa	U	Np	Pu	Am	Cm	Bk	Cf	Es	Fm	Md	No	Lr
Ac	Th	Pa	U	Np	Pu	Am	Cm	Bk	Cf	Es	Fm	Md	No	Lr
						2			2	2	2	2	2	
3	3	3	3	3	3	3	3	3	3	3	3	3	3	3
	4	4	4	4	4	4	4	4						
		5	5	5	5	5								
			6	6	6	6								
				7	7									

Oxidation states of actinides are in contrast to stable +3 oxidation state in lanthanides.

Uranium is found to be the heaviest naturally occurring element. The trans-uranuint elements are man-made elements with atomic number higher than uranium (92).

Because of the following characteristics they were clubbed with *f*-block elements:

(*i*) Sharp visible and UV spectra
(*ii*) Magnetic studies
(*iii*) Importance of (+III) oxidation state

Actinide series starts with Th and ends with Lr. Actinides resemble lanthanides in many ways as:

(*a*) There is a close resemblance between Cm and Gd in their configuration $f^7d^1s^2$.
(*b*) Sharp lines in absorption spectra due to *f*-*f* transition are observed.
(*c*) The most important oxidation state is +III for both lanthanides and actinides.
(*d*) The 5*f* orbitals extend farther in space and therefore participate in bonding compared to 4*f* orbitals.
(*e*) Participation of 5*f* orbitals in bonding leads to higher oxidation states.

8.1.1 Electronic Configuration

Actinides do not show any consistency in electronic configurations owing to very small difference in energy between 5*f* and 6*d* orbitals. Only the electron is able to detect which orbital becomes relatively lower in energy and occupies the same. Therefore, it is quite natural that electron will keep shunting between 5*f* and 6*d* orbitals as indicated in the Table.

Atomic Number	Name	Symbol	Electronic Structure	Radii A°	
				M3+	M4+
89	Actinium	Ac	$6d^17s^2$	1.11	
90	Thorium	Th	$6d^27s^2$		0.90
91	Protactinium	Pa	$5f^26d^17s^2$ or $5f^16d^27s^2$		0.96
92	Uranium	U	$5f^3d^17s^2$	1.03	0.93
93	Neptunium	Np	$5f^57s^2$	1.01	0 92
94	Plutonium	Pu	$5f^67s^2$	1.00	0.90
95	Americium	Am	$5f^17s^2$	0.99	0 89
96	Curium	Cm	$5f^76d^17s^2$	0.985	0 88
97	Berkelium	Bk	$5f^86d^17s^2$ or $5f^97s^2$	0.98	

98	Californium	Cf	$5f^{10}7s^2$	0.977
99	Einsteinium	Es	$5f^{11}7s^2$	
100	Fermium	Fm	$5f^{12}7s^2$	
101	Mendelevium	Md	$5f^{13}7s^2$	
102	Nobelium	No	$5f^{14}7s^2$	
103	Lawrencium	Lr	$5f^{14}6d^17s^2$	

From the electronic configurations it is quite obvious that regular filling of orbitals is not maintainable rather it is erratic because of very little energy difference between $5f$ and $6d$ orbitals. To maintain the prejudice for half filled and fully filled $5f$ and $6d$ orbital is involved in Bk and Lr respectively.

8.1.2 Important Oxidation States

The various oxidation states are given in the inverted pyramid table.

Existence of (+II) oxidation state is quite occasional. Only the element Am^{2+} is having half filled f^7 configuration and Eu^{2+} element is its analogue in the lanthanides. Am^{2+} is available in solid state in the form of AmF_2, whereas other elements of actinides, *i.e.* Cf^{2+}, Es^{2+}, Fm^{2+}, Md^{2+} and No^{2+} occur in the form of solution. Due to the similarity in the size of atoms they resemble alkaline earth metals particularly Ba^{2+}. The element No is most stable as +II because of f^{14} configuration. The oxidation state (+III) is shown by all actinides, but it is not the most stable state. U^{3+} is oxidised in air and solution though, in the later elements, +III oxidation state is most stable. In the first four elements, the most stable oxidation states are Th(+IV), Pa(+V) and U(+V1). Np(+VII), but (+V) is the most stable oxidation state. Pa shows (+III) to (+VII), but (+IV) is most stable. Am shows (+II) to (+VI), but (+III) is most stable as well as for other elements.

The higher oxidation state is shown in the form of oxocations UO_2^+ between pH 2 to 4. These oxocations disproportionate in solution.

$$2UO_2^+ + 4H^+ \longrightarrow U^{4+} + UO_2^{2+} + 2H_2O$$

They are however, quite stable in the solid compounds. The lower ones are ionic and higher covalent.

Hydrolysis of M^{2+}, M^{3+} and M^{4+} ions is quite quick. Hydrolysis of (+V) and (+VI) give MO_2^+ and MO_2^{2+} ions.

8.1.3 Occurrence

In p block elements, $_{83}Bi$ onwards the elements have unstable nuclei and undergo radioactive decay. Upto $_{92}U$, the elements occur in nature as they have long half lives; Th and U are no longer rare. Uranium is found frequently on earth with long half life, $^{238}_{92}U$ $t_{1/2}$ is 4.5 × 10^9 years and $^{235}_{92}U$ is 7.04 × 10^8

years. After uranium, the elements have shorter half lives and are no longer present on the earth.

Thorium is a natural radioactive element and available quite abundant in the earth crust. Upto 10%, Th is found in Monzite sand form which is mixed with lanthanides (Th Ln)PO_4. Thorium also exists as ore in the form called Thorite (Th SiO_4). Uranium is obtained from ore Pitchblende UO_2. Very limited quantity of other actinides are present in these ores. Plutonium (Pu) is a transuranic radioactive element, it is created from uranium in nuclear reactors from uranium-238 to give plutonium-239, hence considered as man made element, but also available in trace amount in nature.

The elements having higher atomic number than uranium are termed transuranium elements and can be produced artificially by irradiating the suitable element with neutrons in a nuclear reactor. They can also be produced by bombarding a sample with α-particles and light atoms like C, B, N, O or Ne.

8.1.4 Preparation of Transuranium Elements

The nuclei of the heavier elements are unstable due to high n/p ratio. The decay of the nuclei may take place by bombarding the nuclei with neutrons which may increase the number of neutrons in the nuclei leading to the decay:

$$n \rightarrow p + \beta\text{-particle}$$

This increases the atomic number by one giving new element towards the right of the periodic table. The $_{92}U$ fuel rod captures the slow neutrons increasing n/p ratio, thus increasing the mass number with the release of β-particle or γ-radiation associated with energy.

(*i*) $$^{235}_{92}U + ^{1}_{0}n \longrightarrow ^{238}_{92}U + ^{1}_{0}n \longrightarrow ^{237}_{92}U$$

$$^{237}_{92}U \xrightarrow[\frac{1}{2}6.7 \text{ days}]{} ^{237}_{93}Np + \beta$$

(Neptunuim element is formed)

(*ii*) $$^{238}_{92}U + ^{1}_{0}n \longrightarrow ^{239}_{92}U \text{ (Unstable)}$$

$$^{239}_{92}U \xrightarrow[+\frac{1}{2}23.3 \text{ min}]{\beta} ^{239}_{93}Np \xrightarrow[+\frac{1}{2}23.3 \text{ min}]{\beta} ^{239}_{94}Pu$$

The yield of the product element depends upon:

(*i*) Half life of the isotopes

(*ii*) Capacity of the nuclei to absorb neutrons

Elements after $_{94}Pu$ can be prepared by number of (*n*, γ) reactions:

$$^{235}_{92}PU + ^{1}_{0}n \xrightarrow{\gamma} ^{240}_{94}Pu + ^{1}_{0}n \xrightarrow{\gamma} ^{241}_{94}Pu$$

$$^{241}_{94}Pu \xrightarrow[t\frac{1}{2}\ 13.2 \text{ years}]{\beta} ^{241}_{95}Am$$

The rapid isolation of new transuranium element can be obtained by taking high density of fast neutrons. The elements einsteinium and fermium were prepared by this method. In reactor, fuel ^{238}U add a fast neutron and then lose two neutrons.

$$^{239}_{92}U \xrightarrow[(n,\ 2^1_0n)]{(n,\ 2n)} {}^{237}_{92}U \qquad ^{241}_{94}U \xrightarrow[t_{\frac{1}{2}}\ 6.7\ \text{days}]{\beta} {}^{241}_{94}Np$$

Presently bombarding the sample with small high speed ions is gaining preference. The ions may be accelerated with linear accelerator or cyclotron. The simplest ion in use is α particle.

$$^{239}_{92}Pu + {}^4_2He \longrightarrow {}^{248}_{96}Cm$$

Addition of ^4_2He upsets n/p ratio and neutrons are emitted *e.g.*

$$^{239}_{92}Pu + {}^4_2He \longrightarrow {}^{241}_{96}Cm + 2^1_0n$$

Other nuclear reactions in which particles added or lost are given in brackets:

$$^{243}_{95}Am \xrightarrow{(a,\ n)} {}^{246}_{97}Bk$$

$$^{242}_{96}Cm \xrightarrow{(a,\ 2n)} {}^{244}_{98}Cf$$

$$^{249}_{98}Cf \xrightarrow{(a,\ 2n)} {}^{251}_{100}Fm$$

$$^{253}_{99}Es \xrightarrow{(a,\ n)} {}^{256}_{101}Fm$$

The heavier unstable nuclei can also be obtained by bombarding the sample element with accelerated ions, *e.g.*, N^{7+}, B^{5+}, C^{6+} or O^{s+}

$$^{243}_{95}U + {}^{14}_{7}N \longrightarrow {}^{249}_{99}Es + 3^1_0n$$

$$^{238}_{92}U + {}^{16}_{8}N \longrightarrow {}^{250}_{100}Fm + 4^1_0n$$

$$^{246}_{96}Cm + {}^{12}_{6}C \longrightarrow {}^{254}_{102}No + 4^1_0n$$

Elements after $_{100}Fm$ have short half life and only a small quantity is prepared.

8.2 GENERAL CHARACTERISTICS

All elements are silvery white metals with slightly high melting points, but lower than the transition elements. Due to poor shielding of electrons, size of the M^{3+} ion shows a decrease 'Actinide contraction' similar to Lanthanides. Therefore, their chemical properties are similar. Actinides have higher densities and greater tendency of formation of complexes and are reactive metals. Actinides when reacted with hot water in air produce oxide coating. Though, in thorium, it is a protective coating, however it is not observed with others. These metals react quickly with HCl, but with other acids react very slowly. Concentrated HNO_3 when reacted with Th, U and Pu make them inactive due to the formation of non-reactive oxide layer and do not react with NaOH. They form non-stoichiometric hydrides MH_2 or MH_3.

8.2.1 Thorium, Protactinium and Uranuim

Thoruim is no longer rare and is quite abundant, obtained from monazite sand as phosphate mixed with Uranium.

Monazite is treated with NaOH. At pH 6, hydroxides of Th(IV), U(IV) and Ce(IV) are precipitated. Hence they are separated from +III lanthanides by dissolving $Th(OH)_4$, $U(OH)_4$ and $Ce(OH)_4$ in HCl. They are further extracted with tributyl phosphate and kerosene. They can be obtained by reducing $ThCl_4$ with Ca or Mg.

$$ThCl_4 + 2Ca \longrightarrow 2CaCl_2 + Th\downarrow$$

Th(IV) salts and Th^{4+} ions occur in solids as well as in solutions. $Th(NO_3)_4.5H,O$ crystals are well known and in its dilute solution hydrated ions exist $[Th(H_2O)_n]^{4+}$. By heating thorium nitrate we get ThO_2. The ThO_2 is white having highest melting point 3220°C. It dissolves in a mixture of HNO_3 and HF. Its halides have high melting point and are white in colour.

$$ThI_4 \xrightarrow{\Delta} Th + 2I_2$$

The above reaction is used to purify metal. The halides hydrolyse in moisture to $ThOX_2$. It has no electrons in *d* and *f* sub-shells. Due to high charge, they favour complex formation *e.g.*, $(NH_4)_4[ThF_g]$, $K_4[Th(C_2O_4)_4]$. $4H_2O$ and $Mg[Th(NO_3)_6]$,

ThI_3 and ThI_2 can be obtained by heating.

$$2Th\,I_4 + Th \xrightarrow{\Delta} 2ThI_3 + ThI_2.$$

ThI_2 is a good electrical conductor.

Thorium is used in gas mantles as ThO_2 + 1% Ce, and burns with a brilliant flame. Naturally occurring Th is not fissionable.

In a nuclear reactor, it can be irradiated to get ^{233}U.

$$^{232}_{90}Th \xrightarrow{(n,\gamma)} {}^{233}_{90}Th \xrightarrow[t_{\frac{1}{2}}\,22\text{ min}]{\beta} {}^{233}_{91}Pa \xrightarrow[27\text{ days}]{} {}^{233}_{92}U$$

$^{233}_{92}U$ has a half life of 1.6×10^s years.

Protactinuim is considered to be one of the rarest and expensive naturally occurring element. It is also found in uranium ore UO_2. Pa is generated in the form of decay products of ^{235}U. The process to separate Pa from minerals is quite difficult. Generally, Pa is obtained from thorium as given above. In addition, it can also be prepared from spent uranium fuel from the nuclear reactors. The compounds of Pa has ability to hydrolyse rapidly. Their ions can polymerise and result in colloidal form of precipitates, but these colloidal precipitates do not exist in fluoride solution as Pa forms fluoride complexes.

The metal looks silvery which is malleable and gets tarnished in air. Pa can be obtained from solutions in HCl or HNO_3 through tributyl phosphate.

$$PaF_4 + 2Ba \xrightarrow{1400^0C} 2BaF_2 + Pa$$

The oxidation state of (+V) is the most stable state. When Pa compound is ignited in air we get white solid Pa_2O_5 which is a weakly acidic and can be dissolved by fused alkali. High heating or reduction with hydrogen form black PaO_2 which is non-stoichiometric.

$$2PaF_4 + F_2 \longrightarrow PaF_5$$

$PaCl_5$ and its oxohalides of the type $PaOX_3$ and PaO_2X can be obtained. There are a several complexes of Pa(V) with oxalates, citrate, tartarate, SO_4^{2-} and PO_4^{3-} ions. Also it shows some uncommon fluoride complexes *e.g.*, $Rb[PaF_6]$, $K_2[PaF_7]$ and $Na_3[PaF_8]$.

Uranium is nautally occurring element and available in earth crust. Also, uranium ore is a source of Ra required in the treatment of cancer in conducting radiotherapy. UO_2 is used for pale yellow or green coloured glass. In the present scenario it is used as a nuclear fuel.

Carnotite $K_2(UO_2)_2(VO_4).3H_2O$ is the chief ore for uranium and is mined as oxide ores namely uraninite and pitchblende (UO_2). The ore is concentrated by froth floatation process and then roasted and dissolved in $H_2SO_4(MnO_2)$ to oxidise U to (+VI) state. This is further precipitated as $Na_2U_20_7$ a bright yellow solid. This dissolves in HNO_3 giving uranyl nitrate $UO_2(NO_3)_3(H_2O)_n$. It is purified by NH_3 or by tributyl phosphate and finally is converted to U by reduction of UF_4 with Ca or Mg.

$$UF_4 + 2Ca \longrightarrow 2CaF_2 + U$$

Nuclear fission was first observed in 1938 by Otto Hahn.

$$^{235}_{92}U + ^{0}_{1}n\,(slow) \longrightarrow ^{138}_{53}I + ^{95}_{39}Y + 3\,^{1}_{0}n + energy$$

The neutrons evolved disintegrate other U nuclei to start a chain reaction. Three neutrons evolved per fission will split three more U nuclei and due to chain reaction lot of energy is evolved. In order to run the reaction according to required limits, the checks are there to ensure that reaction does not go out of control. The rods of B, Cd and Hf may be used in the reaction.

$$^{11}_{5}B + ^{0}_{1}n \longrightarrow ^{12}_{5}B + \gamma$$

$$^{113}_{48}Cd + ^{0}_{1}n \longrightarrow ^{113}_{48}Cd + \gamma$$

The chances of neutron fission increases if the amount of fissile isotope is increased. This is obtained by enrichment of Uranium by 2 to 4%. When ^{238}U absorbs neutron, Pu is produced which is fissile and can be used as a nuclear fuel.

8.2.2 Chemical Characteristics

U, Np, Pu and Am show oxidation states from (+III) to (+VI) and hence show similar properties. The most stable oxidation state for U is (+VI); Np is (+V), Pu(+IV) and Am(+III). The ions in (+III), (+IV), (+V) and (+VI) states are M^{3+}, M^{4+}, MO_2^{+}, MO_2^{2+}, respectively. U is a reactive metal and forms UH_3 and UO_2 with boiling water. It reacts with hydrogen, oxygen, halogens and dissolves in acids.

Uranium reacts with hydrogen at 250°C to produce UH3 black powder. The hydride is used to make other compounds.

$$2UH_3 + 4H_2O \longrightarrow 2UO_2 + 7H_2$$

$$2UH_3 + 4Cl_2 \longrightarrow 2UCl_4 + 3H_2$$

$$2UH_3 + 8HF \longrightarrow 2UF_4 + 7H_2$$

Uranium oxides show very stable oxidation state, but are often non-stoichiometric. UCU is brown, U_3O_g is greenish black and UO_3 is orange-yellow.

$$UO_2\ (NO_3)_2\, 2H_2O \xrightarrow{350^0C} UO_3 \xrightarrow[350^0C]{CO} UO_2$$

$$3UO_3 \xrightarrow{700^0C} U_3O_8 + \tfrac{1}{2}O_2$$

$$3U + 4O_2 \xrightarrow{heat} U_3O_8$$

The above mentioned three oxides dissolve in acids *e.g.*,

$$UO_3 + HNO_3 \longrightarrow \underset{\text{Uranyl nitrate (yellow)}}{UO_2(NO_3)_2.2H_2O}$$

The number of molecules of water may be 2, 3 or 6 depending upon the use of dilute, concentrated or fuming HNO_3.

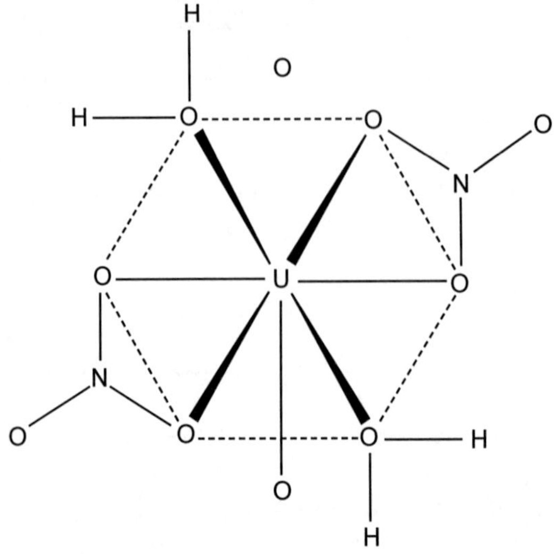

Structure of uranyl nitrate dihydrate

Uranium halides and their colours are given in the table. The fluorides are prepared from UO_2 as shown:

Oxidation state	Fluorides	Chlorides	Bromides	Iodides
+111	UF_3 green	UCl_3 red	UBr_3 red	UI_3 black
+IV	UF_4 green	UCl_4 green	UBr_4 brown	UI_4 black
+V	UF_5 white-blue	U_2Cl_{10} red-brown		
+VI	UF_6 white	UCl_6 black		
	U_4F_9 black			
	U_5F_{22} black			

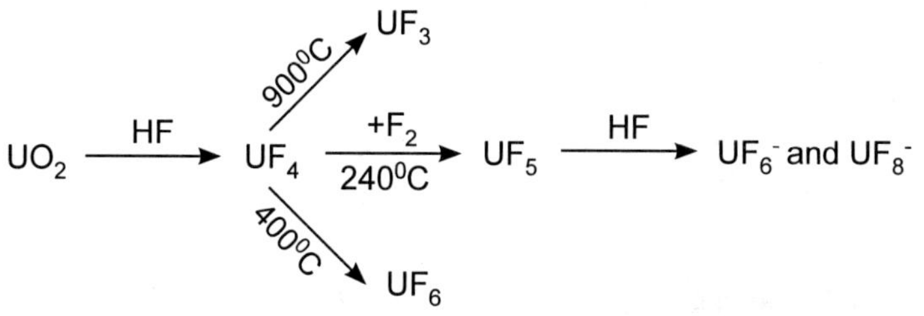

The halides are hydrolysed by water to give UO_2^{2+} ion. UF_6 and UCl_6 are octahedral halides of uranium.

At higher pH they hydrolyse through (-OH) bridge.

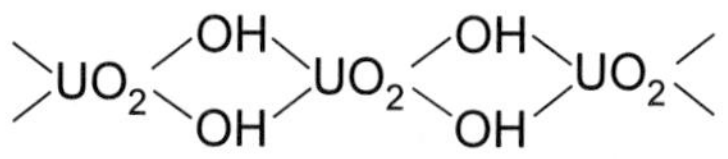

UF_6 is colourless crystal line solid and melts at 64°C. It is a fluorinating agent. It hydrolyses briskly. UF_5 disproportionates to UF_4 and UF_6. MX_4 is the most stable form.

8.2.3 Neptunium, Plutonuim and Americium

$^{229}_{94}$Pu is an important nuclear fuel and is also used for bombs. The other trans-uranuim elements have no uses apart from research. Very limited amount can be obtained for $^{237}_{97}$Np and $^{239}_{94}$ Pu *i.e.,* separated in kilogram quantities, while $^{241}_{95}$ Am and^{242} Am are collected in 100 g quantities from spent uranium fuel rods. The process of separation of these elements require a good care as the separation of these elements is from nuclear waste which is hazardous. Pu is poisonous, carcinogenic even in smaller doses. Am is extracted from spent-fuel rods by reprocessing in breeder reactors and thermal reactors. The fission products may absorb neutrons in the chain reaction of ^{235}U and ^{239}Pu. In this process about 30 different elements are produced including Sr, I, Xe, Cs, Ba and La.

Purex process or the process dealing for separation of uranium and plutonium by the reprocessing of spent nuclear fuel. Fuel rods are dissolved in HNO_3 and separated with tributyl phosphate. UO_2^{2+} and Pu(+IV) are obtained while other elements NP, Am and Cm are present with fission products in aqueous solution. On reduction with SO_2, NFROH or iron sulphamate Pu(+III) is obtained which is separated from UO_2^{2+} and U^{4+} through solvent extraction. The precipitates of U are obtained as uranyl nitrate and oxalate or fluoride of Pu are acquired. Similar to uranium, Np, Pu and Am are also reactive, dissolve in acids and react with H_2,O_2 and halogens and other elements. The +II to +VI are the oxidation states that exist as M^{3+}, M^{4+}, MO_2^{+} and MO_2^{2+}. The +VI state AmO_2^{2+} works as strong oxidising agent in the same way as $KMnO_4$. Am(+III), Pu(+IV) and Np(+V) are the most stable states.

NpO_2^{2+} and PuO_2^{2+} are oxidised in alkaline solutions with HIO_4 or O_3 to (+VII) state. The oxides are in (+IV) state MO_2, but are non-stoichiometric. The halides are quite similar to that of U. ^{241}Am is a source of α-particles. ^{237}Np and Pu are a power source in satellites.

Table Halides of Np, Pu and Am, Remaining actinides

+III	NpF_3 purple-black	PuF_3 purple	AmF_3 pink
	$NpCl_3$ white	$PuCl_3$ emerald green	$AmCl_3$ pink
	$NpBr_3$ green	$PuBr_3$ green	$AmBr_3$ white
	NpI_3 brown	PuI_3 brown	AmI_3 yellow
+IV	NpF_4 green $NpCl_4$ red-brown $NpBr_4$ red-brown	PuF_4 brown	AmF_4 tan
+VI	NpF_6 brown	PuF_6 red-brown	

Regarding elements like curium, einsteinium, fermium, mendelevium, nobelium and lawrencium, we have limited descriptive knowledge about them because of their poor availability and instability. Only Cm is available in good amount. The later actinides show similarity to lanthanides. The point of difference from lanthanides is that actinides are quick to form complexes *e.g.,* even with concentrated HCl they form complexes. Similar to lanthanides, citrate solutions are frequently used for the separation of actinides. The (+III) state is common to all lanthanides and actinides except actinides from curium (96) to lawrencium (103). The element No has most stable (+II) oxidation state because of f^{14} configuration (completely filled *f* shell). The elements from Cf to No, also show (+II) state. Curium (+IV) is shown in CmF_4 and $Rb[CmF_6]$ solids, but not in solution form. Bk(+IV) exists both in solid and solution form *e.g.,* BkO_2 and $Cs_2[BkCl_6]$. Lr has stable (+III) oxidation state, resists oxidation and reduction because of f^{14} configuration.

The elements Cm, Bk and Cf are available in traces (mg). Only limited chemistry is studied about them. The rest of the demerits are also available in minute amounts therefore, radioactive tracer methods have been used to understand their chemistry. Element $_{100}$Fm undergo decay emitting a and p particles. With higher atomic numbers, the elements show a very short half life and undergo self decay (No has $t^{1/2}$ of 3 seconds). These heavier elements become a source of neutrons.

Electronics structure	Ac	Ti	V	Cr	Mn	Fe	Co	Ni	Cu	Zn
	d^1s^2	d^2s^2	d^3s^2	~~d^4s^2~~ d^5s^1	d^5s^2	d^6s^2	d^7s^2	d^8s^2	~~d^9s^2~~ $d^{10}s^1$	$d^{10}s^2$
Oxidation states	II III	II III IV	II III IV VI	I II III IV VI VII	II III IV VI	II III IV	II III IV	II III IV	I II III IV	II

8.3 COMPLETING AND EXTENDING THE PERIODIC TABLE AND ITS SERIES

After the actinide series, elements with atomic numbet 104-109 belong to 6*d*-series. IUPAC names have been proposed for these elements. Each element name ends with -ium as shown in the Table.

Atomic number	Name	Symbol	Half life
104	un-nil-quadium	$^{259}_{104}$Unq	3s or 255 257s
		$^{261}_{104}$Unq	65s
105	un-nil-pentium	$^{258}_{105}$Unp	4s
		$^{260}_{105}$Unp	1 5s
		$^{262}_{105}$Unp	34s
106	un-nil-hexium	$^{260}_{\text{log}}$Unh	180s
		$^{263}_{\text{log}}$Unh	0.9s or 259 264s
107	un-nil-septium	$^{262}_{107}$Uns	0.12s
108	un-nil-octium	$^{265}_{108}$UnO	2×10^{-6} s or 263 264s
109	un-nil-ennium	$^{266}_{109}$Une	5×10^{-6} s

Hyphens have been put in the names for clarity. They should be omitted.

In order to derive the names and symbols for these elements latin and greek roots have been used to maintain them. In this practice for naming, the hyphens are avoided and hyphen is inserted for clarity. It has been observed that the elements with even atomic numbers are more stable and abundant than those with odd atomic numbers. Those nuclei with atomic numbers 2, 8, 20, 28, 50, 82 and 126 are the most stable and their atomic numbers are called 'Magic Numbers'. The nucleus seems to have shell structure with different energy levels. It prefers even number of protons and neutrons over odd numbers. Therefore, ${}^{208}_{82}Pb$ has neutrons equal to $n = 208 - 82 = 126$ neutrons and hence very stable nuclei. Such situation advocate for the existence of nuclides ${}^{278}_{114}Uuq$, and ${}^{310}_{126}Ubh$. Many attempts to prepare these nuclides have been made, but success is eluding. It is reported that through the process of bombarding of ${}^{209}_{83}Bi$, the element 109 was made by accelerated ${}^{56}_{26}Fe$ to form ${}_{109}Unq$.

EXERCISE

1. Write the electronic configuration of Am, Cm, Th, No and Lw and indicate the bias of some elements for half filled and fully filled configuration.
2. Which elements has +3 ion having half filled 5*f* shell?
3. What are the main ores of (a) Thorium and (*b*) Uranium? How are the two elements isolated?
4. How do we convert uranyl nitrate to uranium metal?
5. Write the structure for uranyl nitrate dihydrate.
6. What is most common oxidation state of actinide elements?
7. Th(+IV) and Pa(+IV) are qiute stable. Explain.
8. Compare the properties and uses of UF_6 and uranium hydride.
9. What are the main principles followed in the separations of Np, Pu and Am from U.
10. Compare and contrast the chemistry of dioxo ions of U, Np, Pu and Am.
11. Describe the fluoride and tributyl phosphate method of separation and extraction of Np, Pu from U.
12. Write the chemical equations for the hydrolytic prepartions of UOF_4 and UO_2F_2.
13. How is water obtained in stoichiometric properties for the limited hydrolysis of UF_6?
14. Write three important complexes and organometallic compounds of actinide elements.
15. The heavier elements do not find a place in the main body of the periodic table and their present position is controversial. Suggest some possibilities.
16. Rutherfordium resembles which element out of atomic numbers 105,106, 107 and 112.

1. T. Moeller, Inorganic Chemistry, Wiley Interscience, New York 1982.

Bibliophile

2. F.A. Cotton and G. Wilkinson, Advanced Inorganic Chemistry, Wiley Interscience, New York 1988.
3. Molecular Orbital Theory for Transition Metal Complexes H.B. Gray J. Chem. Educ. 1964.
4. The Nature of the Chemical Bond by L. Pauling Cornell Ithaca, New York 1960.
5. Theoretical Inorganic Chemistry by M.C. Day and J. Selbin, New York 1962.
6. Valence Bond Theory in Coordination Chemistry. J. Chemical Educ. 1964.
7. 'Crystal Field Explain Inorganic Behaviour, Chem. Eng. News, 1959.
8. Molecular Orbital, Valence Bond and Ligand Field A.d. Leihr J. Chem. Educ. 1962.
9. Experimental Inorganic Preparation by W.G. Palmer, Cambridge, New York, 1954.
10. Synthetic Inorganic Chemistry by W.L. Jolley, Princeton-Hall, N.J. 1960.
11. The Determination of Stability Constants by F.J.C. Rossotti and H. Rossotti, McGraw Hill, New York 1961.
12. Rates and Mechanisms of Substitution Reactions in Complexes in Solution by H. Taube, Chem. Rev. 1952.
13. Crystal Field Theory and Substitution Reaction of Metal Ions J Chem. by R.G. Pearson Educ. 1961.
14. Advanced Inorganic Chemistry by J.E. Huheey 1983.
15. Advanced Inorganic Chemistry by Mackay and Mackay 1983.

16. Advance Inorganic by K.F. Purcell J.C. Kotz 1973.
17. Comprehensive Coordination Chemistry, *Elsevier,* www.elsevier.com.
18. Coord. Chem. by SFA Keattle www.Amazon.com.
19. Coord. Chem. by Basolo www.scilet.com.
20. Synthetic Coord. Chem. by JA Davies www.worldcibooks.com.
21. Russian Journal of Coord. Chem. www.springer.com.
22. Integrated App. to Coord. Chem. Rosmary A Murabak, Kate Doon Scott D. Cummings.
23. 5 Steps to 5 App. Chemistry John T. Moore, Richard Layeley.
24. The Inor. Chem. by A Wiley www.wiley.com.
25. Adv. Inorg. Chem. by Albert Cotton and Wilkinson 6th edition.
26. Adv. Inorganic Chem. by Albert Wilkinson, Aurillo A Carlos Bochman Manfred.
27. Coord. Chem. by Basolo and Johnson.
28. A Textbook Inorganic Chemistry by Newth, George Samuel.
29. A Textbook of Inorganic Chemistry by PCL Thorxe.
30. Mod. Inorg. Chem. by KM Mackay, RA Mackay, Wilkinson.
31. Organometallic Chemistry by Helmet, Fritz.
32. Introduction to Coord. Chem. by Wiley-VCH Verlag comb H and Co KGaA, Weinheium.
33. Introduction to Coord. Chem. by Lawrence, Geoffrey Alan.
34. Journal of Coord. Chem. www.tandf.co.uk.
35. Coord. Chem. Wikipedia free encyclopedia.
36. Concise Inorganic Chem. by JD Lee.
37. Darensbourg, M.Y., Ionpairing Effects on Transition Metal Carbonyl Anions in Progress in Inorganic Chemistry vol 33. Wiley-Interscience New York 1985.
38. Hoffmann, R, Theoretical Organometallic Chemistry, Science 1981.
39. Wilkinson, Gis Stone F.G.A and R. West, Eds, Academic New York 1983.
40. Malatesta, L. and Cenini, S., Zero Valent Complexes of Metals. Academic, New York, 1974.
41. Enemark, J.H. and Feltham, R.D., Nitric oxide complexes', Coordination Chemistry Rev. 1974.
42. Davies, S.G., Organotransition metal chemistry. Applications to Organic Synthesis, Pergamon Press, New York 1982.
43. Elsenbroich, C and Salzer, A., Organometallics, 2nd ed. VCH, New York 1990.

44. Hartley, F.R., and Pati, S., Eds, The chemistry of Metal-Carbon Bond, Vol. 3, Wiley New York 1985.

45. Yamamoto, A., Orangotransition Metal Chemistry, Wiley Interscience, New York, 1986.

46. Pruchrick, F.P Organometallic Chemistry of Transition Elements, Plemum Press, New York. 1990.

47. Lukehart, C.M., Fundamental Transition Metal Organometallic Chemistry, Brooks/cole. Monterey, C.A., 1985.

Index

+11 Oxidation State Compounds 294
+IV Oxidation State Compounds 284
+V State Compounds 286
+VII State Compounds 287
+Vl State Compounds 286
6-Coordinate Octahedral Complexes 62

A

Abundance and Isotopes 346
Additional Compounds 5
Alloy Formation 250
Allyl Complexes 207
Antiferromagnetism 246
Application of Polymer-Metal Complexes as Catalysts 147
Applications of Metal Complexes 143
  Applying EAN Rule to $[Ni(CO)_4]$ Complex 178
  Artificial Silk 165
Asymmetric Synthetic and Optical Resolution 150
  Atomic and Ionic Radii 235
  Atomic Size 251
  Atomic Size 257
  Atomic Size 262
  Atomic Volume and Density 236
  Attack on DNA 167

B

Behavior of Calcium and Magnesium 129
  Behaviour of Ti Group Compounds in Different Oxidation States 258
  Bessemer Process 290
  Binding to Meta Thiamine 169
  Bioinorganic Chemistry of Iron 302
  Biological Enzymes 144
  Biological Use 288
Bismuth Anticancer Drugs 171
Blomstrand-Jorgensen Chain Theory 26
  BO_3^{3-} Ion Test 162
  Bonding and Structure 215
  Bonding in $[Fe(CN)_6]_4$ ion 87
  Bonding in $[PtCl_3(C_2H_4)]^-$ 215
Bonding in Complexes 61
  Boron 127
  Br^- and I^- 161
Bridging Ligands 15

C

Carbene and Alkalidyne Complexes 208
  Carbene Compounds 208
  Carbenes and Alkylidene Complexes 208
  Catalytic Activities of Polymers-Metal Complex 146
  Catalytic Property 248
  Chelates 19
  Chemical Characteristics 209
  Chemical Characteristics 210
Chemical Characteristics 252
Chemical Characteristics 257

Chemical Characteristics 366
Chromates, Molybdates and Tungstates 278
Chromium Group 269
Chromium Trioxide or Chromic Anhydride 274
Classification of Inert and Labile Complexes on the Basis of Electrons in d Orbitals 106
Classification 202
Colour and Spectra 354
Colour of Transition Metal Ions 242
Common Oxidation States 291
Completing and Extending the Periodic Table and Its Series 369
Complex Formation Tendency 247
Complex Stability 82
Complexes 253
Complexes 295
Complexes 298
Complexes 3
Compound 45
Compounds in (+6) State 273
Compounds in (+l) Oxidation State 324
Compounds of V, Nb and Ta 263
Compounds with (+II) Oxidation State 327
Compounds with (+IV) Oxidation State 352
Compounds with Oxidation State (III) 296
Conclusion 172
Configuration Symbols 35
Confirmatory Test for Ag^+ Ions 155
Confirmatory Test for Cl^- Ion 161
Confirmatory Test for Fe^{3+}, Al^{3+}, Cr^{3+} 157
Confirmatory test for mercury (1) ion 155
Confirmatory Test for NH^{4+} ion 160
Confirmatory Test for Pb^{2+} Ions 155
Confirmatory Tests 158
Coordinate Mononuclear Metal Carbonyls 177
Coordination Complex in Analytical Chemistry 153
Coordination Complexes and Biological Systems 121
Coordination Complexes 357
Coordination Number (C.N.) 11
Coordination or Complex Compounds 7
Coordination Position Isomerism 44
CoOrdination Sphere 8
Correct Empirical Formula of Complexes 23
Crystal Field Theory (CFT) 69
Cyanide and Co poisoning 123
Cyclopentadienyl Compounds 301

D

Definition 201
Detecting Anions 160
Detecting cations 154
Detoxification by Metal Ion and Chelation 139
Diamagnetism 246
Dicobalt Octacarbonyl $[Co_2(CO)_8]$ 186
Difference Between Double Salts and Complex Compounds 4
Diironnonacarbonyl $[Fe_2(CO)_9]$ 185
Dimanganese decacarbonyl $[Mn_2(CO)_{10}]$ 184
Distortion 13
Distortional Isomerism 47
DNA Base Recognition by Pt Complexes 168

E

Effective Atomic Number (EAN Rule) 25
Effective Collision 104
Effects of Crystal Field Splitting 74
Eight Coordinate Complexes 14
Electronic Configuration and Oxidation States 262
Electronic Configuration and Oxidation States 270
Electronic Configuration and Oxidation States 279
Electronic Configuration of Transition Metal Ions 234
Electronic Configuration 232
Electronic Configuration 252

Electronic Configuration 255
Electronic Configuration 360
Electronic Structures and Oxidation States 288
Enhancing the Effect of Absorption of Metals (even trace elements) 137
Enzymes Inhibition 139
Essential Metals 121
Extent of Cross-Linking 153
Extraction of Metal from Complex Ores 163
Extraction of Metals 256
Extraction 347

F

Factors Influencing Stability 98
Ferromagnetism 246
Five Coordinate Geometry 12
Food and Trace Elements 135
Formation of a Complex 8
Four Coordination Complexes 12
Fourth Transition Series or 6d-series 234

G

General Characteristics of d-block Elements 229
General Characteristics 281
General Characteristics 292
General Characteristics 334
General Characteristics 363
General Characteristics 271
General properties 316
general properties 323
General Rules 33
Gold Antipathetic Drugs 170
Graded General Properties 262
Group 3 or Group 13 Elements 213
Group I Cations 154
Group II Cations 155
Group III Anions 161
Group III Cations 157
Group IV cations 158
Group of II Anions 160
Group Reagent 158
Group reagent 156
Group Study of Physical and Chemical Characteristics of Transition Elements 250
Group V cations 159
Group VI Cations 159
Halides 253
Halides 259
Halides 264
Halides 275
Halides 295
Halides 297

H

Hemoglobin the Oxygen Carrier 124
High Coordination Complexes 14
High Temperature Chemical Reduction 164
History of Complexes 3
History 3
Hydrate Isomerism 45
Hydroformylation of Olefins 149
Hydrogenation 147
Hydrogenation 217
Hydrolysis 167
Hydrolysis 148

I

Inert and Labile Complexes 105
Important Organometallic Compounds 210
Important Oxidation States 361
Inert and Labile Complexes on the Basis of CFSE 106
Initiation of Radical Polymerisation 149
Inner Orbital Complexes 63
Inner Transition or f-block Elements: The Lanthanides 345
Interstitial Compound 249
Ionisation Energies 240
Ionisation Isomerism 44
Iron Pentacarbonyl [$Fe(CO)_5$] 179
Isolation 94
Isomerism and Stereochemistry 43
Isomerism 43
IUPAC System of Nomenclature 33

K

Kinetics or Inert and Labile

Complexes 103
Lanthanide Contraction 350

L

Lewis Acid-Base Theory 61
Ligand Field Theory (LFT) 81
Ligand Isomerism 46
Ligands 8
Limitations of VBT (Valency Bond Theory) 69
Linkage Isomerism 45
Low Coordination Numbers 11
Low Oxidation State Compounds 292
Magnetic Moment 355
Magnetic Properties 244
Magnitude of A_0 72

M

Mechanism of Substitution 112
Mechanisms of Substitution 114
Melting and Boiling Points 237
Metal Carbonyls and Nitosyls 173
Metal Complexes as Drugs and Chemotherapeutic Agents 172
Metal Complexes in Medicine 165
Metal Complexes in the Cure of Cancer 142
Metal Ions Strong and Transportation in the Human System 125
Metal-Alkene Complexes of π-Bonded Organometallics 214
Metal-Benzenoid Complexes 207
Metallic Behaviour 236
Metallo enzymes as acid catalysts 131
Metallo Thionein and Glutathione 169
Mn(III) Oxidation State Compounds 284
Mn(lll) Complexes 284
Modern Techniques 162
Molar Conductivity Method 24
Molecular Orbital Theory (MOT) 222
Molecular Orbital Theory (MOT) 84
Molybdenum 127
Monofunctional Adducts 167
Naming the Inner Complexes 39

N

Naming the Ligand 34
Naming the Metal Ion 33
Nature of Metal-Ligand Bond 61
Neptunium, Plutonuim and Americium 367
Nomenclature of Coordination Compounds 33
Nomenclature 202
Non-Chelates 19

O

Occurrence and Extraction 288
Occurrence and Uses 280
Occurrence 361
Organometallic Compounds of Beryllium 211
Organometallic Compounds 260
Organometallic Compounds 267
Organometallics 201
Organometalliic Compounds of Mg 211
Other Metalloenzymes 130
Other Uncommon Isomerism 46
Outer Orbital Complexes 63
Oxidation States 252
Oxidation States 256
Oxidation States 270
Oxidation States 280
Oxides 265
Oxides 294
Oxides 297
Oxygen Binding and Oxygen Enzymes 132
Oxygen Carriers 125

P

Phosphate Storage and Muscles 131
Photography 165
Physical Characteristics 179
Physical Characteristics 183
Physical Characteristics 209
physical Characteristics 210
Physical Characteristics 215
Physical Characteristics 217
Physical Characteristics 251
Physical Characteristics 255

Physical Characteristics 261
Physical Characteristics 269
Physical Constants 256
Placement in the Periodic Table 229
Platinum Anticancer Agents 166
PO_4^{3-} Ion Test 162
Polymerisation Isomerism 46
Polymer-Metal Complexes as Catalysts 145
Polymer-metal complexes as Heterogeneous biologically active catalysts 152
Polynuclear Metal Carbonyls 184
Porosity and surface area 152
Precipitation Method 23
Preparation and Isolation of Complexes 91
Preparation of Metal Carbonyls 174
Preparation of Metal Carbonyls 92
Preparation of Metallocenes 216
Preparation of Organometallic (Sandwich Compounds) 93
Preparation of Transition Metal a Bonded Complexes 93
Preparation of Transuranium Elements 362
Preparation of Zeise's Salt 206
Preparation 214
Preparation 91
Properties of (+III) Compounds 351
Properties 211
Quantitative Analysis 162
Quantitative Vs. Qualitative 163
Reactivity 250
Reducing Action 240
Roie of a Cataiysi (Kinetics) 107

R

Role of HCl in Group II Analysis 157
Role of Some Metals in the Human Body 126

S

Sandwich Organometallics or Metallocenes 216
Second Transition Series (4d-series) 233
Selectivity 129
Six Coordinate Complexes 13
SO_4^{2-} Ion Test 161
Sodium Potassium Equilibrium 128
Solubility 353
Spacer Groups 153
Spin Isomerism 47
Square Planar 78
Stability of Complex Compounds 248
Stability of Complex Ion 97
Steel Manufacture 290
Stereoisomerism 48
Structural Isomerism 43
Structure and Bonding in $[M(CO)_5]$ 181
Structure and Bonding in Carbonyls 177
Structure 211
Substitution in Non-Aqueous Solvents 113
Substitution Reactions in Complexes 111
Substitution When No Solvent Is Used 114
Summation Isomerism 47
Synthesis of Alkene Complexes 205
Synthesis of Alkyne Complexes 206
Synthesis of Organometallic Compounds 203
Synthesis of Sandwich Compounds 205

T

Targeting Via Transferring Receptor: Ruthenium and Titanium 169
Test for $C_2O_4^{2-}$ Ion 161
Test for NO^{3-} Ion 161
Test for NO^{3-} or NO^{2-} 296
Tetragonal Distortion or Jahn-Teller Distortion 75
Tetrahedral Complexes 79
Tetranuclear Metal Carbonyls $[M_4(CO)_{12}]$ 190
The Actinides 359
The Cobalt Group 305

The Copper Group of Coinage Metals 321
The decomposition of H_2O_2 150
The Iron Group 288
The Manganese Group 279
The nickel group 315
The Zinc Group 331
Third Transition Series or 5d-series 233
Thorium, Protactinium and Uranuim 364
Titanium Group 255
Toxic Metal Ion 137
Trace Elements Absorption 135
Trans Effect 91
Transport of Oxygen 124
Tungsten Bronzes 279
Use of Ag Compounds in Photography 326

U

Uses and Applications of Coordination Complexes 143

V

Valence Bond Theory 62
Valence Isomerism 47
Vanadates 266
Vanadium Group 261
Vanadium in Biological Systems 130
Variable Oxidation States 238
Vitamis B_{12} or Cyanocobalamine 132

W

Werner's Coordination Theory 26
Zero Group cations 159
Zero, (–I) and (–II) State Compounds 277